Electrical Memory Materials and Devices

RSC Polymer Chemistry Series

Editor-in-Chief:
Professor Ben Zhong Tang, *The Hong Kong University of Science and Technology, Hong Kong, China*

Series Editors:
Professor Alaa S. Abd-El-Aziz, *University of Prince Edward Island, Canada*
Professor Stephen Craig, *Duke University, USA*
Professor Jianhua Dong, *National Natural Science Foundation of China, China*
Professor Toshio Masuda, *Shanghai University, China*
Professor Christoph Weder, *University of Fribourg, Switzerland*

Titles in the Series:
  1: Renewable Resources for Functional Polymers and Biomaterials
  2: Molecular Design and Applications of Photofunctional Polymers and Materials
  3: Functional Polymers for Nanomedicine
  4: Fundamentals of Controlled/Living Radical Polymerization
  5: Healable Polymer Systems
  6: Thiol-X Chemistries in Polymer and Materials Science
  7: Natural Rubber Materials: Volume 1: Blends and IPNs
  8: Natural Rubber Materials: Volume 2: Composites and Nanocomposites
  9: Conjugated Polymers: A Practical Guide to Synthesis
  10: Polymeric Materials with Antimicrobial Activity: From Synthesis to Applications
  11: Phosphorus-Based Polymers: From Synthesis to Applications
  12: Poly(lactic acid) Science and Technology: Processing, Properties, Additives and Applications
  13: Cationic Polymers in Regenerative Medicine
  14: Electrospinning: Principles, Practice and Possibilities
  15: Glycopolymer Code: Synthesis of Glycopolymers and their Applications
  16: Hyperbranched Polymers: Macromolecules in-between of Deterministic Linear Chains and Dendrimer Structures
  17: Polymer Photovoltaics: Materials, Physics, and Device Engineering
  18: Electrical Memory Materials and Devices

How to obtain future titles on publication:
A standing order plan is available for this series. A standing order will bring delivery of each new volume immediately on publication.

For further information please contact:
Book Sales Department, Royal Society of Chemistry, Thomas Graham House, Science Park, Milton Road, Cambridge, CB4 0WF, UK
Telephone: +44 (0)1223 420066, Fax: +44 (0)1223 420247
Email: booksales@rsc.org
Visit our website at www.rsc.org/books

Electrical Memory Materials and Devices

Edited by

Wen-Chang Chen
National Taiwan University, Taipei, Taiwan
Email: chenwc@ntu.edu.tw

THE QUEEN'S AWARDS
FOR ENTERPRISE:
INTERNATIONAL TRADE
2013

RSC Polymer Chemistry Series No. 18

Print ISBN: 978-1-78262-116-4
PDF eISBN: 978-1-78262-250-5
ISSN: 2044-0790

A catalogue record for this book is available from the British Library

© The Royal Society of Chemistry 2016

All rights reserved

Apart from fair dealing for the purposes of research for non-commercial purposes or for private study, criticism or review, as permitted under the Copyright, Designs and Patents Act 1988 and the Copyright and Related Rights Regulations 2003, this publication may not be reproduced, stored or transmitted, in any form or by any means, without the prior permission in writing of The Royal Society of Chemistry, or in the case of reproduction in accordance with the terms of licences issued by the Copyright Licensing Agency in the UK, or in accordance with the terms of the licences issued by the appropriate Reproduction Rights Organization outside the UK. Enquiries concerning reproduction outside the terms stated here should be sent to The Royal Society of Chemistry at the address printed on this page.

The RSC is not responsible for individual opinions expressed in this work.

The authors have sought to locate owners of all reproduced material not in their own possession and trust that no copyrights have been inadvertently infringed.

Published by The Royal Society of Chemistry,
Thomas Graham House, Science Park, Milton Road,
Cambridge CB4 0WF, UK

Registered Charity Number 207890

For further information see our web site at www.rsc.org

Printed in the United Kingdom by CPI Group (UK) Ltd, Croydon, CR0 4YY, UK

Foreword: Introduction to Organic Memory Technologies

SEBASTIAN NAU[a] AND EMIL J. W. LIST-KRATOCHVIL*[a,b]

[a]NanoTecCenter Weiz Forschungsgesellschaft mbH, A-8160 Weiz, Austria; [b]Institute of Solid State Physics, Graz University of Technology, A-8010 Graz, Austria
*E-mail: e.list@tugraz.at

The discovery of how to store information in the form of numbers and letters is generally considered as one of the most important steps of social evolution in human history as it allowed passing of information from generation to generation. The simplest forms appeared at least forty thousand years ago in the form of unary numeral systems (tally marks). Real alphabets developed, however, much later around 4000 BC in Mesopotamia. The invention of a moveable metal type printing press by Johannes Gutenberg (1450) is considered as another big leap since it enabled a relatively cheap, large scale duplication of information. Nowadays, information is stored digitally encoded in a binary language consisting of zeros and ones, *i.e.* binary logical states. The origin of the rapid development in this field dates back to the 1940s, starting from the first computer prototypes with an implemented memory – Zuse3 (1942, Germany) and ENIAC (1946, USA) – *via* the launch of the first personal computers (1970s) to the present state where digital memory is found in nearly every electronic device. Recent calculations showed that approximately 300 exabytes (10^{18} bytes) of information is stored worldwide.[1]

As predicted 50 years ago memory capacity develops according to Moore's law,[2] meaning that roughly every 18 months a doubling of the number of transistors per chip at approximately equal cost is found. Although this observation has been doubtlessly valid for the past decades, a strong convergence to an upper integration density is predicted. It is expected that within the next 10 to 20 years it will not be possible to further downsize the traditional complementary metal-oxide-semiconductor (CMOS) architecture due to physical restrictions at such small dimensions. In order to overcome this fact, two parallel strategies are usually presented. First, *more Moore*, dealing with the continuous miniaturization of classical CMOS building blocks as well as post-CMOS materials and devices. Second, the *more-than-Moore* approach which targets system integration and functional diversification rather than scaling issues: the inclusion of non-digital functionalities like sensors and actuators, bio-chips, photonic applications, *etc.* interfaces the integrated circuit to the outside world. Materials and devices presented in this book actually address both approaches by investigating organic and hybrid material memory technologies based on capacitors, transistors and resistive switching elements and their advantages (*more Moore*) and secondly demonstrating how organic and hybrid memory devices may be integrated into flexible electronic applications (*more-than-Moore*).

However, there are numerous prerequisites that novel materials, device concepts and memory elements based on these materials should fulfil to outperform established technologies. The most important ones are listed in the following:

- Non-volatility: state should be preserved at least for several years even if the supply voltage is turned off
- Fast read, write and erase times in the range of the CPU clock (~GHz): this prevents a bottleneck effect in the communication between the individual components in an electronic application
- Cyclability: 10^{12}–10^{15} write–read–erase–read cycles should be possible
- CMOS compatibility
- Operation temperature not higher than 85 °C
- ON/OFF ratio: clear difference between '0' (OFF) and '1' (ON) state which enables for simple peripheral electronics and a reliable read-out
- High integration density (down to $4F^2$, where F denotes the smallest lithographically obtainable feature size)
- Low power consumption
- Compatibility with existing semiconductor fabrication plants
- Non-destructive read-out
- 3D-integration on multiple layers, multiple bits per cell
- Random Access: each bit cell should be accessible directly and sequential read-out is necessary.

As discussed in Chapter 1 in detail, today's dominant memory technologies do not meet all of these criteria simultaneously and the decision about

Foreword: Introduction to Organic Memory Technologies vii

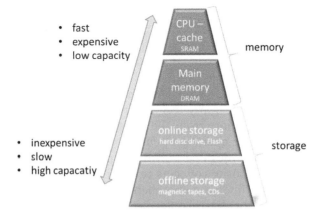

Figure 1 Memory-storage hierarchy. Costs and speed are increasing from bottom to top whereas information capacity is decreasing.

which memory or storage technology is used strongly depends on its actual application case. Figure 1 depicts the memory-storage hierarchy. Usually storage and memory are considered separately, where the first term includes non-volatile, high capacity technologies and the latter term is used for fast, lower capacity and volatile technologies. Along these lines, the main task in the long run is the development of a technology which unifies memory and storage technology into a 'storage class memory' with the advantages of both. This would allow, for example, the design of computers where the operating systems and other data do not need to be read out from the hard disk drive to the main memory when it is switched on.

Today three technologies dominate the market – each of them highly specialized in its field of application, dynamic random access memory (DRAM), flash memory, and magnetic hard disk drives. Emerging technologies such as ferroelectric RAM (FeRAM), magnetoresistive RAM, phase-change RAM, and resistive memory have or are about to enter the market. Over the past decades a wide variety of organic and hybrid materials concepts have also been utilized as hysteretic elements in memory cells, including conjugated polymers, small molecules, organic/hybrid materials, chalcogenides, thin-film perovskites, electromechanical switches, ferroelectric organic polymers, *etc.* Depending on the individual functional principle, these materials are integrated into memory elements utilizing a transistor, diode, capacitor or resistor device structure rendering an individual memory cell. Yet, according to the International Technology Roadmap for Semiconductors, in an overall comparison and benchmarking of the specific type of hysteretic materials used in memory cells, it is not only important to judge the material's electrical characteristics (with respect to scalability, read-out time, write time, switching power, retention time and other parameters) but also if the material is compatible with fabrication technology and if the device element fulfils the design constraints for high density integration. Amongst the here-presented candidates, which cover ferroelectric and floating gate organic transistor memory and different resistive memory technologies, the latter one is doubtless the most promising candidate not only to replace the existing technology with an organic counterpart but also to act as storage class memory.

The contributions to this book address and summarize the ongoing development in organic memory technology based on resistive switching and transistor-based memory from the material development, processing as well as from the device operation point of view. The classes of materials presented range from conjugated polymers, to donor–acceptor structures to hybrid material composite concepts to name just a few of them. For all material concepts one finds a comprehensive discussion of the observed electrical switching in the device, reflecting on the fact that despite considerable research effort in this field, the true nature of resistive switching is still uncertain.

Chapter 1 gives a basic and historic overview on existing and emerging memory device concepts based on organic, polymeric and hybrid materials. Existing transistor-based, capacitor-based and resistor-based device concepts are discussed together with the observed operation principles. Chapter 2 focuses on the recent development of resistive memory technologies based on organic and polymeric materials. The chapter covers basic materials design principles to be applied, followed by a comprehensive discussion on the switching mechanism in organic resistor memory elements, followed by a number of application-related examples of organic memory devices. Chapter 3 reports on the use of donor–acceptor structures in resistive memory devices. The chapter highlights the importance of structure to property relationships in understanding the switching mechanism and also discusses multilevel resistive memory devices. Chapter 4 focuses on describing the use of polyimide and functionalized polyimides for resistive memory devices. Different applications of such devices are reviewed together with an in detail discussion on the switching mechanism. Chapter 5 reports on the use of non-conjugated polymers with electroactive chromophore pendants in resistive memory devices. The design of molecules is discussed together with their performance in devices and the observation of switching. Chapter 6 focuses on the use of polymer composites for resistive switching devices. In addition to switching observations in hybrid polymer–nanoparticle systems and multilayer structures, polymer–ionic liquid composites are also discussed. Chapter 7 discusses the design and development of different conjugated polymers for resistive switching elements, including a discussion of the different observed switching mechanisms. Chapter 8 describes the use of a variety of homo- and copolymers bearing donor–acceptor moieties for their use in resistive memory devices. Chapter 9 reviews the use of ferroelectrics, polymer electrets, polymer–molecular hybrids and self-assembled monolayers in non-volatile transistor based memory devices. Chapter 10 discusses the physics and fabrication of floating gate charge storage devices based on organic transistors and possible implications for their use in emerging flexible digital memory electronics. Chapter 11 discusses the use of ferroelectric materials in capacitor-, transistor- or diode-based storage devices elucidating the interplay of polymer orientation, interfacial engineering and device configuration, and memory performance.

References

1. M. Hilbert and P. López, *Science*, 2011, **332**, 60.
2. G. E. Moore, *Electronics*, 1965, 114.

Preface

Organic memory devices have received extensive research interest recently due to their advantages of structural flexibility, processability, light weight and their simple manufacturing processes compared to those employed in inorganic silicon technology. Such devices could have potential applications in flexible electronic components and wearable products. Three types of organic memory devices have been reported in the literature, including resistor-, capacitor- and transistor-type devices. There are several review articles in the field of organic memory but most of them report on a limited subject, such as resistive memory. There is no comprehensive book on tuning different types of memory characteristics through molecular design and morphology manipulation. Thus, I accepted the invitation from the Royal Society of Chemistry to edit this book to provide information on the recent developments of organic electrical memory materials, device fabrication, and switching mechanisms. In order to meet this objective, I have invited the world-leading researchers in this field to write the book chapters, including Prof. En-Tang Kang, Prof. Mitsuru Ueda, Prof. Jian Mei Lu, Prof. Guey-Sheng Liou, Prof. Moonhor Ree, Prof. Jae-Suk Lee, and Prof. Yu-Tai Tao. In this book, different types of charge storage materials, such as organic small molecules, functional polyimides, nonconjugated polymers with pendent chromophores, conjugated polymers, and polymer nanocomposites, are summarized and correlated with their electrical memory characteristics. In addition, various device configurations, including resistor and transistor-type memory, are also discussed in detail along with the defined charge storage mechanisms. The effects of the charge transfer strength, charge trapping/de-trapping ability, surface polarity, interfacial energy barrier, and morphology of the memory active layer on the electrical switching behaviours are also comprehensively discussed in this book. I would like to express my sincere appreciation to the chapter authors for spending their precious time despite their relatively busy schedules. In addition, without the proofreading and administration work of Professor Cheng-Liang Liu and Dr. Hung Chin Wu, I could not finish this book. I hope that this book will not only provide a fundamental understanding of the chemical and physical aspects of organic electrical memory materials but will also spark future development of advanced high density data storage devices.

Wen-Chang Chen

Dedicated to my family: Lora, Claire, and Lucia.

Contents

Chapter 1	**Organic Electronic Memory Devices** *Bin Zhang, Yu Chen, Koon-Gee Neoh, and En-Tang Kang*	**1**
	1.1 Introduction	1
	1.2 Basic Concepts of Electronic Memory	2
	1.3 History of Organic/Polymer Electronic Memory Devices	4
	1.4 Classification of Electrical Memory Devices	7
	1.4.1 Transistor-Type Electronic Memory	7
	1.4.2 Capacitor-Type Electronic Memory	9
	1.4.3 Resistor-Type Electronic Memory	11
	1.5 Types of Organic-Based Electrical Memory Devices	15
	1.5.1 Organic Molecules	15
	1.5.2 Polymeric Materials	23
	1.5.3 Organic–Inorganic Hybrid Materials	34
	1.6 Conclusions and Outlook	42
	References	44
Chapter 2	**Organic Resistor Memory Devices** *Tadanori Kurosawa, Tomoya Higashihara, and Mitsuru Ueda*	**54**
	2.1 Device Structures of Organic Resistor Memory Devices	54
	2.2 Switching Characteristics of Organic Resistor Memory Devices	56
	2.2.1 DRAM Characteristics	56
	2.2.2 SRAM Characteristics	58
	2.2.3 FLASH Characteristics	59
	2.2.4 WORM Characteristics	59

RSC Polymer Chemistry Series No. 18
Electrical Memory Materials and Devices
Edited by Wen-Chang Chen
© The Royal Society of Chemistry 2016
Published by the Royal Society of Chemistry, www.rsc.org

2.3	Materials	62
	2.3.1 Materials Design Principle	62
	2.3.2 Organic Molecules	62
	2.3.3 Polymeric Materials	66
	2.3.4 Organic–Inorganic Hybrid Materials	76
2.4	Mechanism Discussion of Organic Resistor Memory Devices	78
	2.4.1 Filamentary Conduction	78
	2.4.2 Space Charge and Traps	82
	2.4.3 Charge Transfer Effect	85
	2.4.4 The Conformation Effect and Other Effects	87
2.5	Applications of Organic Resistor Memory Devices	88
	2.5.1 Three-Dimensional Devices	90
	2.5.2 Increasing Density by Scale Down	91
	2.5.3 Flexible Memory Devices	92
	2.5.4 Transferrable Memory Devices	93
2.6	Concluding Remarks	94
	References	95

Chapter 3 Donor–Acceptor Organic Molecule Resistor Switching Memory Devices — **101**

Jianmei Lu, Hua Li, and Qing-Feng Xu

3.1	Introduction	101
3.2	Organic Molecules as Active Material in Memory Devices	102
	3.2.1 The Common Electron Donor and Acceptor Groups Used in Molecular Based Memory Devices	102
3.3	Organometallic Complexes as Active Materials in Memory Devices	110
3.4	Organic Molecules for Multilevel Resistive Memory Devices	113
	3.4.1 Azo Benzene Derivative Based Ternary Memory Devices	113
	3.4.2 Other Functional Molecules with Ternary Memory	122
	3.4.3 Molecules with Ternary Memory Performance Through Photoelectric Effects	122
3.5	Structural Effects on Memory Performance	124
	3.5.1 Switching Voltage and ON/OFF Ratio	125
	3.5.2 Memory Types	131
3.6	Conclusion	132
	References	132

Chapter 4 High Performance Polyimides for Resistive Switching Memory Devices — 136
Hung-Ju Yen, Jia-Hao Wu, and Guey-Sheng Liou

    4.1 Introduction to Polyimide-Based Electrical Memory Devices — 136
    4.2 Mechanism — 137
        4.2.1 Charge Transfer — 137
        4.2.2 Space Charge Traps — 140
        4.2.3 Filament Conduction — 140
    4.3 Polyimides for Volatile Memory Devices — 140
        4.3.1 DRAM Properties — 142
        4.3.2 SRAM Properties — 143
    4.4 Polyimides for Non-Volatile Memory Devices — 145
        4.4.1 WORM Properties — 145
        4.4.2 Flash Properties — 148
    4.5 Molecular Design of the Volatility — 151
        4.5.1 Donor Effect — 151
        4.5.2 Acceptor Effect — 156
        4.5.3 Linkage Effect — 158
        4.5.4 Thickness Effect — 160
    4.6 Polyimide–Inorganic Hybrid Materials — 161
    4.7 Flexible Polyimide Electrical Memory Devices — 163
    References — 163

Chapter 5 Nonconjugated Polymers with Electroactive Chromophore Pendants — 167
Moonhor Ree, Yong-Gi Ko, Sungjin Song, and Brian J. Ree

    5.1 Introduction — 167
    5.2 Nonconjugated Polymers with Electron-Donating Chromophore Pendants — 168
        5.2.1 Polymers with Carbazole Pendants: Backbone and Molecular Orbital Effects — 168
        5.2.2 Polymers with Triphenylamine-Based Pendants: Substituent and Molecular Orbital Effects — 174
    5.3 Nonconjugated Polymers with Electron-Accepting Chromophore Pendants — 183
    5.4 Nonconjugated Polymers with Electron-Donor and -Acceptor Chromophore Pendants — 189
    5.5 Stability and Reliability of Electrical Memory in Polymer-Based Devices — 194
    5.6 Memory Mechanisms in Polymer-Based Devices — 197
    5.7 Concluding Remarks — 202
    References — 202

| Chapter 6 | **Polymer Composites for Electrical Memory Device Applications** | **206** |

Cheng-Liang Liu and Wen-Chang Chen

- 6.1 Introduction — 206
- 6.2 Polymer–Organic Molecule Composites — 207
- 6.3 Organic Polymer–Inorganic Nanomaterial Composites — 210
  - 6.3.1 Multistacked Composite with Metallic Nanoparticles Embedded — 210
  - 6.3.2 Organic Polymer–Metal Nanoparticle Hybrids — 212
  - 6.3.3 Organic Polymer–Other Inorganic Nanomaterial Composites — 215
- 6.4 Polymer–Carbon Allotrope Composites — 217
  - 6.4.1 Polymer–Buckyball Cluster Composites — 217
  - 6.4.2 Polymer–Carbon Nanotube Composites — 220
  - 6.4.3 Polymer–Graphene Sheet Composites — 222
- 6.5 Polymer–Ionic Liquid Composites — 223
- 6.6 Conclusion — 224
- References — 224

| Chapter 7 | **Conjugated Polymers for Memory Device Applications** | **233** |

Cheng-Liang Liu and Wen-Chang Chen

- 7.1 Introduction — 233
- 7.2 Fluorene-Based Conjugated Homopolymers and Copolymers — 234
- 7.3 Thiophene-Based Conjugated Polymers — 238
- 7.4 Carbazole-Containing Conjugated Polymers — 240
- 7.5 Conjugated Poly(azomethine)s — 243
- 7.6 Other Intrinsic Conjugated Polymers — 245
- 7.7 Conjugated Polymers Containing Metal Complexes — 246
- 7.8 Conjugated Polyelectrolytes — 251
- 7.9 Conclusion — 253
- References — 253

| Chapter 8 | **Non-Volatile Memory Properties of Donor–Acceptor Block Copolymers** | **256** |

Nam-Goo Kang, Myung-Jin Kim, and Jae-Suk Lee

- 8.1 Introduction — 256
  - 8.1.1 Categories of Electronic Memory — 257
  - 8.1.2 Brief Description of Organic Memory Materials and Devices — 258

Contents   xix

	8.1.3 Block Copolymers and Self-Assembled Nanostructures	260
	8.2 Non-Volatile Memory Based on Well-Defined Polymer Structures	261
	8.2.1 Rod–Coil block copolymers	261
	8.2.2 Well-Defined Homopolymers	268
	8.2.3 Well-Defined Coil–Coil Block Copolymers	275
	8.3 Summary and Outlook	290
	References	291

Chapter 9 Organic Transistor Memory Devices and Materials 295
Chiao-Wei Tseng and Yu-Tai Tao

9.1 Basic Concepts of Organic Field-Effect Transistors 295
9.2 Transistor Memory Devices 297
9.3 Organic Transistor Memory Devices 299
  9.3.1 Organic Transistor Memory Devices Based on Ferroelectrics 300
  9.3.2 Organic Transistor Memory Devices Based on Charge Trapping 303
9.4 Organic Transistor Memory Devices Incorporating Nanoparticles 316
  9.4.1 Inserting NPs into the Gate Dielectric 316
  9.4.2 Incorporating NPs in the Semiconductor 318
  9.4.3 Embedding NPs at the Semiconductor/Dielectric Interface 321
9.5 Conclusion 322
References 324

Chapter 10 Organic Floating Gate Transistor Memory Devices 330
Hung Chin Wu, Ying-Hsuan Chou, Hsuan-Chun Chang, and Wen-Chang Chen

10.1 Nanoparticle Embedded Materials as Charge Storage Layers 330
  10.1.1 Thermal Evaporation 331
  10.1.2 Chemical Assembly 335
  10.1.3 Polymer–Nanoparticle Hybrids 340
10.2 Functional Moiety Embedded Materials as Charge Storage Layers 342
10.3 Switching Mechanism 347
10.4 Flexible Electrical Memory Devices 348
10.5 Conclusion 351
References 351

Chapter 11	Organic Ferroelectric Memory Devices	355
	Hsuan-Chun Chang, Hung-Chin Wu, and Wen-Chang Chen	
	11.1 Introduction	355
	11.2 Materials for Ferroelectricity	357
	11.3 Principles of Organic Ferroelectric Memory Operation	361
	11.3.1 Organic Ferroelectric Capacitors	361
	11.3.2 Organic Ferroelectric Field-Effect Transistors	364
	11.3.3 Organic Ferroelectric Diodes	369
	11.4 Application and Summary of Organic Ferroelectric Memory Devices	373
	References	374
Chapter 12	**Summary and Outlook**	**377**
	Wen-Chang Chen	
	References	380

Subject Index   **381**

CHAPTER 1

Organic Electronic Memory Devices

BIN ZHANG[a,b], YU CHEN*[b], KOON-GEE NEOH[a], AND EN-TANG KANG*[a]

[a]Department of Chemical & Biomolecular Engineering, National University of Singapore, 10 Kent Ridge, 119260, Singapore; [b]Key Lab for Advanced Materials, Institute of Applied Chemistry, East China University of Science and Technology, 130 Meilong Road, Shanghai 200237, China
*E-mail: cheket@nus.edu.sg, chentangyu@yahoo.com

1.1 Introduction

As the performance of digital gadgets for information technology advances, the complexity of data storage devices increases correspondingly. Conventional memory devices are implemented on semiconductor-based integrated circuits, such as transistors and capacitors. In order to achieve greater density of data storage and faster access to information, more components are deliberately packed onto a single chip. The feature size of transistors has decreased from 130 nm in the year 2000 to 32 nm at present.[1,2] Silicon-based semiconductor devices become less stable below 22 nm, and the reliability to store and read individual bits of information will be substantially reduced by severe "cross-talk" issues. Moreover, power consumption and unwanted heat generation are also of increasing concern, and the fidelity of addressing the memory units diminishes correspondingly. Therefore, the current

RSC Polymer Chemistry Series No. 18
Electrical Memory Materials and Devices
Edited by Wen-Chang Chen
© The Royal Society of Chemistry 2016
Published by the Royal Society of Chemistry, www.rsc.org

state-of-the-art memory technologies are no longer capable of fulfilling the requirements for information storage of the near future.[3]

Regarding the aspiration for new data storage technologies, ferroelectric random access memory (FeRAM),[4] magnetoresistive random access memory (MRAM),[5] phase change memory (PCM),[6] and organic/polymer memory have appeared on the scene of the information technology industry.[7-9] Instead of information storage and retrieval by encoding "0" and "1" as the amount of stored charge in the current silicon-based memory devices, the new technologies are based on electrical bistability of materials arising from changes in certain intrinsic properties, such as magnetism, polarity, phase, conformation and conductivity, in response to the applied electric field. The advantages of organic and polymer electronic memory include good processability, molecular design through chemical synthesis, simplicity of device structure, miniaturized dimensions, good scalability, low-cost potential, low-power operation, multiple state properties, 3D stacking capability and large capacity for data storage.[10-16]

Extensive studies toward new organic/polymeric materials and device structures have been carried out to demonstrate their unique memory performances.[17-22] This chapter provides an introduction to the basic concepts and history of electronic memory, followed by a brief description of the structures and switching mechanisms of electrical memory devices classified as transistors, capacitors and resistors. Then, the progress of organic-based memory materials and devices is systematically summarized and discussed. Lastly, the challenges posed to the development of novel organic electrical memory devices are summarized.

1.2 Basic Concepts of Electronic Memory

The basic goal of a memory device is to provide a means for storing and accessing binary digital data sequences of "1's" and "0's", as one of the core functions (primary storage) of modern computers. An electronic memory device is a form of semiconductor storage which is fast in response and compact in size, and can be read and written when coupled with a central processing unit (CPU, a processor). In conventional silicon-based electronic memory, data are stored based on the amount of charge stored in the memory cells. Organic/polymer electronic memory stores data in an entirely different way, for instance, based on different electrical conductivity states (ON and OFF states) in response to an applied electric field. Organic/polymer electronic memory is likely to be an alternative or at least a supplementary technology to conventional semiconductor electronic memory.

According to the storage type of the device, electronic memory can be divided into two primary categories: volatile and non-volatile memory. Volatile memory eventually loses the stored information unless it is provided with a constant power supply or refreshed periodically with a pulse. The most widely used form of primary storage today is volatile memory. As shown in Figure 1.1, electronic memory can be further divided into sub-categories,

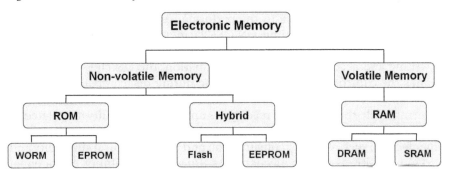

Figure 1.1 Classification of electronic memory devices. Reproduced with permission from ref. 22, © 2008 Elsevier Ltd.

as read only memory (ROM), hybrid memory, and random access memory (RAM). ROM is factory programmable only; data is physically encoded in the circuit and cannot be programmed after fabrication. Hybrid memory allows data to be read and re-written at any time. RAM requires the stored information to be periodically read and re-written, or refreshed, otherwise the data will be lost. Among these types of electronic memory, write-once read-many-times (WORM) memory,[7] hybrid non-volatile and rewritable (flash) memory,[8] static random access memory (SRAM) and dynamic random access memory (DRAM) are the most widely reported polymer memory devices.[23,24]

A WORM memory device can be used to store archival standards, databases and other massive data where information has to be reliably preserved for a long period of time. Conventional CD-Rs, DVD±Rs or programmable-read-only-memory (PROM) devices are examples of WORM memory. Flash memory is another type of non-volatile electronic memory. Different from WORM memory, its stored state can be electrically reprogrammed and it has the ability to write, read, erase and retain the stored state. Thus it is mutable or rewritable in nature. Due to its non-volatility, no power is needed to maintain the information stored in flash memory. DRAM is a type of volatile random access memory that stores each bit of data in a separate capacitor within an integrated circuit. Since real-world capacitors have charge-leaking tendencies, the stored data eventually fade unless the device is refreshed periodically. Because of this periodical refresh requirement, it is a volatile and dynamic memory. The volatility, ultrafast data access time and structural simplicity hold great promise for high density and fast responding performance, making DRAM memory the main memory for most computers. SRAM is another type of volatile memory. The term "static" differentiates it from "dynamic" RAM (DRAM) which must be periodically refreshed. SRAM exhibits data remanence, but it is still volatile and the stored data are eventually lost when the memory remains in the power-off state. SRAM is faster and more reliable than the more common DRAM. Due to its high cost, SRAM is often used only as a memory cache.

Parameters of importance to the performance of a memory cell include switching (write and erase) time, ON/OFF current ratio (or memory window), read cycles, and retention ability. The switching time influences the rate of writing and accessing the stored information, the ON/OFF current ratio defines the control of the misreading rate during device operation, with a higher value being essential for the device to function with minimal misreading error, while the number of read cycles and retention ability are related to the stability and reliability of the memory devices. For practical applications, other factors, such as power consumption and cost, structural simplicity and packing density, as well as mechanical stiffness and flexibility, are of equal importance when designing and fabricating new memory devices.

1.3 History of Organic/Polymer Electronic Memory Devices

Different forms of storage, based on various natural phenomena, have been reported since the 1940s. A computer system usually contains several kinds of storage, each with an individual purpose. In the 1960s, there was a great interest in the electrical properties of amorphous semiconductors and disordered structures, arising from their unusual electrical properties which also make them promising materials for device applications.[25,26]

In 1968, Gregor observed bistable negative resistance in polymer materials and noted that a Pb/polydivinylbenzene/Pb bistable electrical switching device is capable of acting as an information storage device.[27] In 1969, Szymansk et al. reported bistable electrical conductivity phenomena in thin tetracene films sandwiched between metal electrodes.[28] In 1970, Sliva et al. reported that devices based on Saran® wrap, phthalocyanines and polystyrene all exhibited bistable switching behavior.[29] Subsequently, Segui et al. demonstrated reproducible bistable switching in polymer thin films prepared by glow-discharge polymerization.[30] Inspired by these pioneering studies, a wide variety of organic and polymer materials have been explored for threshold and memory switching effects.[31–33] Many of the observed electrical memory effects were due to the formation of filamentary conduction paths, and the performance was not satisfactory for practical applications. Memory switching effects in polymethylmethacrylate, polystyrene, polyethylmethacrylate and polybutylmethacrylate films were ascribed to field-controlled polymer chain ordering and disordering.[34] Memory switching effects in poly(*N*-vinylcarbazole) (PVK) thin films were attributed to trapping–detrapping processes associated with impurities in PVK.[35]

Studies of the transition behavior of some ferroelectric polymers began in the 1980s.[36–38] Thin films of ferroelectric materials can be repeatedly switched between two stable ferroelectric polarization states, and are capable of exhibiting non-volatile memory effects. Polymer films obtained by solution processing techniques were so thick that some devices required operating voltages of at least 30 V. Bune et al. reported a major breakthrough in the fabrication of ferroelectric films by the Langmuir–Blodgett (LB) technique in 1995.[39] The resulting ferroelectric films are as thin as 1 nm and can be switched using a voltage as low as 1 V.[40] Rapid progress in polymer ferroelectric random access memory (FeRAM) as a promising memory technology has since been achieved.[41–43]

An organic transistor memory device using a sexithiophene oligomer as the conductor and an inorganic ferroelectric material as the gate insulator was demonstrated in 2001 by Velu et al.[44] Subsequently, ferroelectric organic and polymer materials have also been utilized as gate insulators in field-effect transistors (OFETs).[45–48] High performance all-organic or polymer transistor memory devices have been demonstrated by Naber et al.[49–51] Transistor memory devices can be faster and more readily integrated with traditional electronics. However, they are not able to meet the high density and low-cost requirements since an additional terminal is required between the gate and the semiconducting channel. A WORM type memory device based on polymer fuses was demonstrated by Forrest and coworkers in 2003.[7] The memory element consists of a thin film p–i–n silicon diode and a conductive polymer fuse, composed of poly(ethylene dioxythiophene) (PEDOT) oxidatively p-doped by poly(styrene sulfonic acid) (PSS).

Bistable electrical switching and memory effects involving charge transfer (CT) complexes were first reported in an electronic device based on a copper (electron donor) and 7,7,8,8-tetracyanoquinodimethane (TCNQ, electron acceptor) complex.[52] Subsequently, a wide variety of organometallic and all-organic CT complexes have also been explored for non-volatile electronic memory applications.[52] Polymer memory devices based on CT effects from doping of a polymer matrix by electron donors, such as 8-hydroxyquinoline (8HQ), tetrathiafulvalene (TTF), polyaniline (PANI), poly-3-hexylthiophene (P3HT), or electron acceptors such as gold nanoparticles, copper metallic filaments and phenyl C_{61}-butyric acid methyl ester (PCBM), have been reported.[8,9,53–55] Carbon nanotubes (CNTs) possess intense π-conjugation and strong electron-withdrawing ability. The CT complexes of CNTs and P3HT, a conjugated copolymer or poly(N-vinylcarbazole) (PVK) have been reported to exhibit a bistable electrical memory effect.[56–58] By utilizing copolymers containing both donor (D) and acceptor (A) moieties in the basic unit, phase separation and ion aggregation could be effectively avoided in a single-component polymer film, resulting in uniform film morphology and improved device performance. D–A polymer-based electrically bistable memory devices have received considerable attention. The molecular design-*cum*-synthesis approach has allowed several polymer electronic memory devices, including flash memory, WORM memory and DRAM to be realized.[59–68]

In order to achieve ultrahigh density memory devices, organic materials with multilevel stable states are highly desirable. In 2004, Pal et al. reported multilevel conductivity and conductance switching in supramolecular structures of Rose Bengal.[69] Subsequently, they observed one low- and three high-conducting states in ultra-thin film devices, and all four accessible states have associated memory effects for data-storage applications.[70]

Multilevel conductance switching in poly[2-methoxy-5-(2′-ethyl-hexyloxy)-1,4-phenylene vinylene] (MEH-PPV) films was first reported by Lauters *et al.* in 2005.[71] They observed that the ITO/MEH-PPV/Al device had the ability to store a continuum of conductance states. These states were non-volatile and could be switched reproducibly by applying appropriate programing biases above a certain threshold voltage. Devices demonstrating multistability where more than two conducting states can be programmed into a single switching element will dramatically increase the amount of data stored per area or volume. Further progress in the development of multilevel organic/polymer memory has been made in recent years.[72–76]

In 1971, Chua proposed a new circuit element, a memristor, which is the fourth passive circuit element beyond the fundamental resistor, capacitor and inductor.[77] A memristor is capable of processing information in the same way as biological systems, mimicking the function of a mammalian synapse, with the ability to learn and memorize new information. According to the redefinition of Chua in 2011, all two-terminal non-volatile resistive switching memory devices are memristors, regardless of the device material or the physical operating mechanism.[78] A polymer memristor was first reported in cobalt(III)-containing conjugated (CP) and non-conjugated (NCP) polymers with an azo-aromatic backbone by Higuchi *et al.* in 2011.[79] Single crystals of a cyclodextrin-based metal–organic framework (MOF) infused with an ionic electrolyte and flanked by silver electrodes can act as memristors.[80] The metal/single-crystal MOF/metal heterostructure can be switched between high and low conductivity states due to the self-limiting oxidative reactions of the metal anode.

The International Technology Roadmap for Semiconductors (ITRS) has identified polymer memory as an emerging memory technology since the year 2005. Figure 1.2 indicates the number of related publications each year

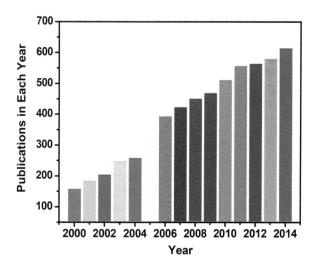

Figure 1.2 Statistics for publications on organic and polymer memory from 2000 to 2014. Data derived from ISI Web of Science.

worldwide since the year 2000. Research work on polymer memory before 2008 was summarized in a comprehensive review by Ling et al.[22] Liu and Chen highlighted recent developments in the field of D–A polymers for resistive switching memory device applications.[20] Chen et al. reviewed the application of electrically, thermally and chemically modified graphene and polymer-functionalized graphene derivatives for switching and information storage applications.[19] Most recently, Huang and coworkers summarized recent progress concerning the use of polymers or polymer composites as active materials for resistive memory devices.[17] The impetus for the research effort in this area arises from the fact that organic/polymer electronic memory devices have been a promising alternative or supplementary device to conventional memory technologies facing the problem of miniaturization from microscale to nanoscale.

1.4 Classification of Electrical Memory Devices

According to the device structure, electronic memory devices can be divided into three primary categories: transistors, capacitors and resistors. With their respective ability to amplify electronic signals, to store charges, and to produce proportional electric currents, electronic memory devices can be constructed from transistors, capacitors and resistors.

1.4.1 Transistor-Type Electronic Memory

Inorganic transistors are widely used in conventional semiconductor memory. For example, a group of six or more field-effect transistors can be integrated to assemble a SRAM cell, while a floating gate transistor can be integrated to assemble a flash memory cell. Organic (including polymer) transistors are also of great potential for memory applications.[81–84] Organic field-effect transistor (OFET) type memory devices have attracted considerable research interest due to their easily integrated structure and the non-destructive reading of a single transistor.[85–87] In addition, the mechanical flexibility of organic materials makes them compatible with plastic substrates for lightweight and flexible device design.

1.4.1.1 Device Structure

The organic transistor inherits its design features from inorganic MOSFET precursors.[88,89] It is composed of three main components: source, drain and gate electrodes, a dielectric insulator layer and an active semiconductor layer, as illustrated in Figure 1.3a.[22] The electrodes can be n- or p-Si, ITO, PEDOT:PSS, TaN, Au, Pt, Al, Cu, Cr or other metals. Among these, Au electrodes are often used for OFETs because the work function of gold is close to the ionization potential of many polymer materials, which leads to an ohmic contact in the device. An OFET memory device consists of at least one

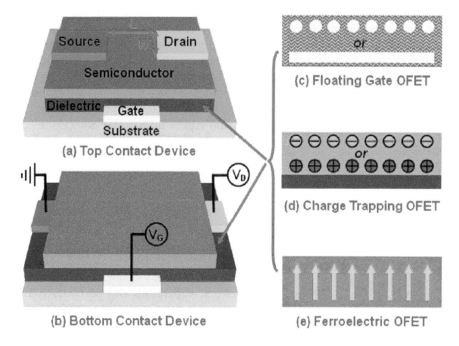

Figure 1.3 OFET configurations: (a) top contact device and (b) bottom contact device with different dielectric layers; (c) floating gate OFET, (d) charge trapping OFET and (e) ferroelectric OFET, exhibiting memory effects. The separation between the source and drain electrodes defines the channel length (L), the width of the electrodes defines the channel width (W). Reproduced with permission from ref. 22, © 2008 Elsevier Ltd.

polymeric material either in its dielectric insulator layer or active semiconductor layer or both. The device is usually supported by a glass, wafer, or plastic substrate. Within the basic MOSFET design, there are two types of device configuration: top contact and bottom contact (Figure 1.3a and b). In the former, the source and drain electrodes are fabricated on a preformed semiconductor layer, whereas the latter is constructed by depositing the organic materials over the contacts.[90]

1.4.1.2 Operation Mechanism

A voltage applied to one pair of the transistor's terminals (either source–drain or source–gate) can change the current flowing through another pair of terminals. The voltage applied between the source and drain is referred to as the source–drain voltage, V_D. The current flowing through the semiconductor film from source to drain is referred to as the source–drain current, I_D. For a lightly doped or undoped semiconductor, the concentration of free charge carriers in the channel is very low. When $V_G = 0$ V, I_D is very low and the transistor is initially in the OFF state. With an increase in V_G, a layer of mobile charges from the source can accumulate at the interface between the semiconductor and the insulator. Due to the increased charge carrier concentration in the semiconductor, I_D increases significantly and turns the transistor to the ON state.[91] The gate threshold voltage (V_{th}) is defined as the voltage necessary to induce mobile charges, that is, the gate voltage at which the square root of the saturation I_D begins to increase substantially. Alternatively, field-effect transistor memory can also be operated at constant V_G and sweeping V_D.

To be a non-volatile transistor for memory applications, the charge must be stored or polarized in domains within the bulk of the dielectric layer, or at interfaces between the gate contact and the semiconductor channel. An additional voltage, *via* charge storage or polarization, is thus introduced between the gate and the semiconductor channel to alter the charge distribution in the transistor. On the basis of charge storage and polarization methods in the dielectric layer or interfaces, OFET memory devices can be divided into three categories: floating gate OFET memory, charge trapping OFET memory, and ferroelectric OFET memory (Figure 1.3c and d).[22]

1.4.2 Capacitor-Type Electronic Memory

Capacitors can store charges on two parallel plate electrodes under an applied electric field. Based on the amount of charge stored in the cell, the bit level (either "0" or "1") can be encoded accordingly. When the medium between the electrodes is merely a dielectric, the stored charge will be lost eventually.[92] Thus, DRAM using a dielectric capacitor is volatile memory, and the information stored in DRAM eventually fades unless the capacitor charge is refreshed periodically. On the other hand, if the medium is ferroelectric in nature, permanent electric polarization can be maintained and longer retention time can be achieved. A ferroelectric material can maintain permanent electric polarization that can be repeatedly switched between two stable states by an external electric field. Thus, memory based on ferroelectric capacitors (FeRAM) is non-volatile memory.[4] FeRAM needs no periodic refreshing and it still retains its data in the case of power failure. Organic and polymeric ferroelectric materials can also be used in DRAM and FeRAM applications.[93-97]

1.4.2.1 Device Structure

Several FeRAM structures, including 1T1C (T – transistor, C – capacitor),[98] 2T2C,[99] 1T2C[100] and others, have been developed. The simplest DRAM and FeRAM cells have similar structures, both utilizing 1T1C as the building components. Figure 1.4a shows an image of a 1T1C FeRAM device, while a schematic circuit diagram of the cell is shown in Figure 1.4b.[98] The plate line (PL) of a FeRAM device has a variable voltage level to enable the switching of the polarization of the ferroelectric capacitor. The upper electrode of

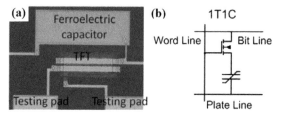

Figure 1.4 (a) An image of a 1T1C FeRAM device, and (b) circuit schematic of a typical 1T1C FeRAM cell. Reproduced with permission from ref. 98, © 2012 Elsevier B.V.

the capacitor is made of Pt, Ir or Ru, and the lower electrode is Pt/Ti.[101] The local interconnect between the access transistor and the storage node of the capacitor is TiN. These materials are refractive and can form conductive oxides, such as IrO_x or RuO_x.

1.4.2.2 Operation Mechanism

Ferroelectric materials exhibit polarization–electric field hysteresis loops. As shown in Figure 1.5, when voltages are applied from 0 V to $+V_{cc}$, the polarization state changes from point A to B to C progressively.[22] Similarly, the polarization state changes progressively from point D to E to F, with the applied voltages increasing from 0 V to $-V_{cc}$. When voltages are applied from $+V_{cc}$ to 0 V and from $-V_{cc}$ to 0 V, the polarization state changes, moving from point C to D and from point F to A, respectively. The amount of polarization charge can be maintained well, without reversing their direction. Under such a situation, zero-voltage remnant polarization states with opposite direction (P_r, points A and D) can be obtained by a large saturation voltage ($\pm V_{cc}$), and can be reversed or switched by the coercive voltage ($\pm V_c$), the minimum value of the voltage necessary to reverse, or switch, the polarization state. Thus, "0" and "1" can be defined as the two stable states, "upward polarization" and "downward polarization", to apply ferroelectric characteristics to electronic memory.[22]

"1" or "0" data can be written to a FeRAM cell by applying a voltage $+V_{cc}$ or $-V_{cc}$ to both electrodes of the ferroelectric capacitor. For instance, to write "0", the word line (WL) is turned on (meaning that the access transistor is on), the bit line (BL) is held at 0 V and the plate line (PL) is cycled from 0 V to V_{cc} to 0 V (Figures 1.4b and 1.5).[22] This polarizes the ferroelectric capacitor in the "0" state. After writing, data is retained even if the selected WL becomes unselected (meaning that the transistor is off). When reading "0" or "1" data from a cell, prior to selecting the WL, the BL must be precharged to 0 V to retain the high-impedance condition. Next, the WL is selected and V_{cc} is applied to the PL (Figures 1.4b and 1.5). For the cell in the "0" state, there will be a minimal voltage change on the BL since the capacitor does

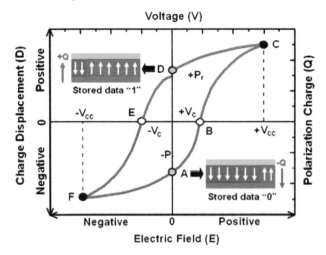

Figure 1.5 Charge displacement–electric field (*D–E*) hysteresis loop and ferroelectric capacitor polarization conditions. Reproduced with permission from ref. 22, © 2008 Elsevier Ltd.

not switch the polarization direction. For the cell in the "1" state, there will be a larger voltage variation on the BL since the capacitor switches the polarization direction and the difference in compensating charge flows onto the BL. When reading the "1" state, the reversal in polarity causes the data to be destroyed, creating a "0" state. Therefore, it is necessary for the capacitor to be re-polarized to the "1" state before closing the WL and moving on to the next operation. To rewrite the "1" state, the BL voltage level is set at V_{cc} and the PL voltage level is set at 0 V. Finally, the WL is turned off and the "1" state is stored again (Figures 1.4b and 1.5).[22] When reading the "0" state, since no reversal of polarity occurs, the datum is not destroyed and the re-writing process is not required.

1.4.3 Resistor-Type Electronic Memory

Devices incorporating switchable resistive materials are generically classified as resistor-type memory, or resistive random access memory (RRAM). Unlike transistor and capacitor memory devices, resistor-type memory does not require a specific cell structure (*e.g.* FET) or to be integrated with the CMOS (complementary metal-oxide-semiconductor) technology. Resistor-type memory devices store data in an entirely different form, for instance, based on different electrical conductivity states (ON and OFF states). Electrical bistability usually arises from changes in the intrinsic properties of materials, such as charge transfer, phase change, conformation change and reduction–oxidation (redox) reaction, in response to an applied voltage or electric field.[22]

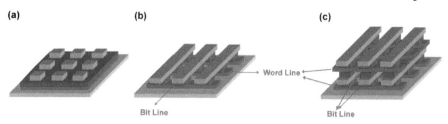

Figure 1.6 Schematic diagram of (a) a 3 × 3 polymer memory device, (b) a 3 (word line) × 3 (bit line) cross-point memory array, and (c) a 3 (layer) × 3 (word line) × 3 (bit line) stacked memory device.

1.4.3.1 Device Structure

Resistor-type electronic memory usually has a simple structure with an organic/polymer thin film sandwiched between two electrodes on a supporting substrate (glass, silicon wafer, plastic or metal foil). The configuration of the top and bottom electrodes can be either symmetric or asymmetric, with aluminum, gold, copper, p- or n-doped silicon, and ITO being the most widely used electrode materials. Test structures usually consist of a spin-coated polymer thin film on the bottom electrode, for instance, ITO, with the top electrodes deposited through a shadow mask *via* thermal evaporation in a vacuum chamber. The area covered by the top electrode forms the active device area. The basic configuration of a test memory device is shown in Figure 1.6a. The individual memory cells can be integrated into a cross-bar (two dimensional or 2D) memory array (Figure 1.6b), and further stacked into three-dimensional (3D) data storage devices (Figure 1.6c). Each cell in the 2D memory array or 3D stacked device can be identified by its unique Cartesian coordinates. Due to the two terminal simple structure and the nanoscale active organic/polymer thin film, high data storage density can be realized in organic/polymer memory.

1.4.3.2 Operation Mechanism

Resistor-type memory is based on the change of the electrical conductivity of materials in response to an applied voltage (electric field). Various mechanisms have been proposed to explain electrical conductance switching in organic/polymer memory devices. Among them, the most widely reported mechanisms include filament conduction, space charges and traps, charge transfer effects, and conformational changes.

1.4.3.2.1 Filament Conduction. Filament conduction is used to describe the high conductivity (ON) state where the current is highly localized in a small fraction of the device area. It is believed that filament conduction is normally associated with physical damage to the device, and thus results in artifact memory effects which are difficult to control and reproduce. Two types of filament conduction, carbon-rich filament and metallic filament conduction, are widely reported in polymer memory.[102] The former is caused by local degradation of polymer films and will increase the mobility of charge carriers only.[30] The latter results from local fusing, migrating or sputtering of electrode metal through the films, leading to an increase in both the charge carrier mobility and concentration.[103]

1.4.3.2.2 Space Charges and Traps.
The intrinsic electrical conductivity of organic and polymer materials is far lower than that of metals. When the electrode–film contact is ohmic, charge carriers will be easily injected from the electrode into the organic thin film and accumulated near the interface to form a space charge buildup. The electrostatic repulsion between individual charges can screen the applied electric field and further limit charge injection into the film.[104] Consequently, hysteresis in the current–voltage (I–V) characteristics of the film is observed. Space charges in materials may arise from several sources, such as electrode injection of charge carriers, ionized dopants in interfacial depletion regions and accumulation of mobile ions at electrode/organic film interfaces. Capacitance–voltage (C–V) characteristics can also show hysteresis arising from space charges.[105] The hysteric behavior, either in I–V or C–V characteristics, can be utilized to create data storage devices. A device can be programmed by applying a voltage pulse to write a state, and read by measuring the device current under a small probe voltage.

When traps are present either in the bulk of the material or at the interface regions, the charge carrier mobility will be significantly reduced. Adsorbed oxygen molecules in organic films,[35] intra-molecular donor–acceptor structures[106] and semiconductor or metal nanoparticles[107] can act as charge trapping centers. As greater numbers of charge carriers are injected with increasing voltage, the traps in the organic thin film are gradually filled.[35] When all of the traps are eventually filled, the newly injected charge carriers will no longer be affected by the fully filled traps. An abrupt increase in the current is observed, and the transition from the OFF to the ON state is related to the level of occupancy of the charge traps. The current is limited by re-excitation (de-trapping) of the trapped carriers in the trap-filled state.[108] Both space charges and traps play an important role in the electronic processes and switching behavior of organic electronic devices.[109]

1.4.3.2.3 Charge Transfer Effects.
A charge transfer (CT) complex is defined as an electron donor–acceptor (D–A) complex, characterized by an electronic transition to an excited state in which a partial transfer of charge occurs from the donor moiety to the acceptor moiety. The conductivity of a CT complex is dependent on the ionic binding between the D–A components.[110] As illustrated in Figure 1.7,[22,111] if the donor is characterized by small size and low ionization potential, a strongly ionic salt forms and a complete transfer of charge (or with the CT degree value, $\delta > 0.7$) occurs from the donor to the acceptor, making the ionic salt insulating. When the donor is very large and has a high ionization potential, a neutral molecular solid

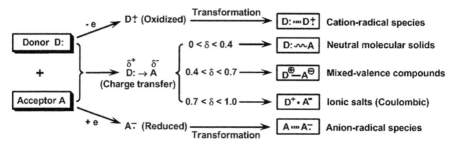

Figure 1.7 Schematic representation of the formation of ion-radical species and charge transfer complexes. Reproduced with permission from ref. 22, © 2008 Elsevier Ltd.

($\delta < 0.4$) forms, which is also insulating. If the donor has intermediate size and ionization potential, it tends to form a weakly ionic salt with the acceptor, which possesses incomplete CT ($0.4 < \delta < 0.7$) and thus is potentially conductive. The formation of a conductive CT complex can be employed to design molecular electronic devices. Many organic CT systems, including organometallic complexes, carbon allotrope (fullerene, carbon nanotubes and graphene)-based polymer complexes, gold nanoparticle–polymer complexes, and single polymers with intra-molecular D–A structures have been explored for memory applications.[112–115]

1.4.3.2.4 Conformational Changes. The spatial conformation of a material can significantly affect the distribution of electron density within the macromolecule and/or the π-conjugation of the system, thus effectively controlling the material's electrical conductivity. Electrical bistability arising from electric field-induced conformational changes has been reported in organic and polymeric materials.[116–118] For instance, carbazole containing polymers are capable of exhibiting field-induced conformational change *via* rotations of randomly oriented carbazole groups to form a more regioregular arrangement for facilitated carrier delocalization and transport.[117] A transition in the molecular conformation of anthracenyl moieties can tune the energy levels of anthracene -containing polymers, and improve the bulk charge recombination.[118] Substituted biphenyl or bipyridine molecules have been inserted as the active material between a pair of metallic electrodes, to control the longitudinal conduction in response to an applied electric field.[119,120] The mechanism of operation is based on the action of an electric field perpendicular to the ring–ring bond on the torsional angle and, as a consequence, the inter-ring conjugation. With the inclusion of suitable substituents on the aromatic rings, the transverse electric field can increase the dihedral angle, and thus change the conformation from one in which electrons can flow freely from side to side, to one in which this flow is hindered. As a result, the conductance of the molecule is varied, providing a means for potential data storage.

To further understand the device transition from the OFF state to the ON state, the current density–voltage data in both states can be fitted to theoretical models. Various mechanisms have been proposed to explain the generation, trapping and transport of charge carriers in organic and polymer memory devices.[121,122]

1.5 Types of Organic-Based Electrical Memory Devices

1.5.1 Organic Molecules

Organic electronic memory devices based on organic molecules were first reported in several acene derivatives including naphthalene, anthracene, tetracene, pentacene, perylene, *p*-quaterphenyl and *p*-quinquephenyl.[123–126] Ma *et al.* reported an organic electrically bistable device with the structure of a single layer of *N*,*N*′-di(naphthalene-l-yl)-*N*,*N*′-diphenyl-benzidine (NPB) embedded between ITO and Ag electrodes.[127] Subsequently, a negative differential resistance (NDR) in organic devices consisting of 9,10-bis-(9,9-diphenyl-9*H*-fluoren-2-yl)-anthracene (DPFA) sandwiched between two electrodes was observed.[128] The device exhibited reproducible NDR and can be electrically switched between the ON state and the OFF state.

Organometallic and all-organic CT complexes have been explored for use in organic memory (Scheme 1.1). Electrical memory phenomena of CT complexes were first reported in a copper and 7,7,8,8-tetracyanoquinodimethane (TCNQ) complex (Cu-TCNQ).[52] Stable and reproducible current-controlled bistable electrical switching was observed in a device with the structure Cu/Cu-TCNQ/Al. Inspired by this discovery, many other organometallic CT complexes with different metals and organic acceptors have been prepared and explored for memory effects over the past few decades.[129–132] A series of all-organic CT complexes were prepared by alternative, mixed or dual deposition in a vacuum chamber. All films of these organic CT complexes exhibited electrically bistable states at room temperature and a short transition time from high to low resistance.[133–137] These CT complexes consisted of methanofullerene 6,6-phenyl C_{61}-butyric acid methyl ester (PCBM) as the organic electron acceptor, and tetrathiafulvalene (TTF) as the organic electron donor. Electrical bistability was demonstrated in devices with a sandwich structure. Devices in the OFF state could be switched to the ON state by applying a 5 V pulse of width shorter than 100 ns, while the ON state could be converted to the OFF state by applying a −9 V pulse of width shorter than 100 ns. Pal *et al.* demonstrated a WORM memory effect in a monolayer of a copper(II) phthalocyanine (CuPc) and fullerenol CT complex.[136] Ma *et al.* observed a reproducible NDR and memory effect in a 1,4-dibenzyl C_{60} (DBC) and zinc phthalocyanine (ZnPc) CT complex.[137] The introduction of DBC enhanced the ON/OFF current ratio and significantly improved the memory stability. The ON/OFF current ratio was up to 2 orders of magnitude. The number of

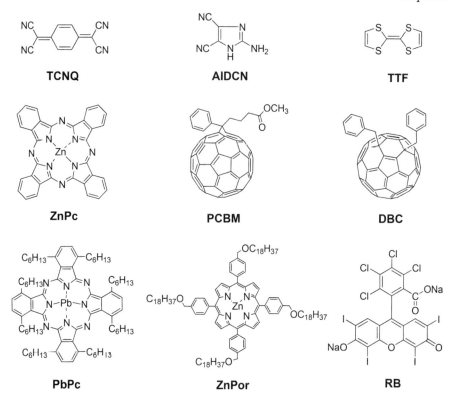

Scheme 1.1 Chemical structures of the molecules used for organic memory devices.

write–read–erase–reread cycles was greater than 10^6, and the retention time reached 10 000 s without current degradation.

Organic dyes, such as phthalocyanines (Pc), porphyrins (Por) and xanthene derivatives have also been explored for electrical memory effects. Ray et al. fabricated bistable memory devices by using 70 nm thick spun films of PbPc molecules sandwiched between an ITO substrate and Al top electrodes.[138] The bistable electrical switching effects were attributed to the existence of a depletion layer at the ITO/PbPc interface and an exponential distribution of the energetics of traps in the non-active region of the PbPc films. Similar phenomena can also be observed for some other phthalocyanine derivatives, such as NiPc,[105] CuPc[139] and ZnPc.[140] Pal et al. observed high electrical conductance switching (ON/OFF ratio = 10^5) in single-layer sandwich structures based on the organic molecule Rose Bengal (RB) at room temperature.[141] A molecular switching device utilizing LB monolayer films containing ZnPor as a redox-active component has also been reported.[142] Devices with the structure metal/ZnPor LB monolayer/metal exhibited outstanding switching diode and tunneling diode behavior at room temperature.

Organic memory devices with a triple-layer structure sandwiched between two outer metal electrodes have been reported.[143] The active layer of the

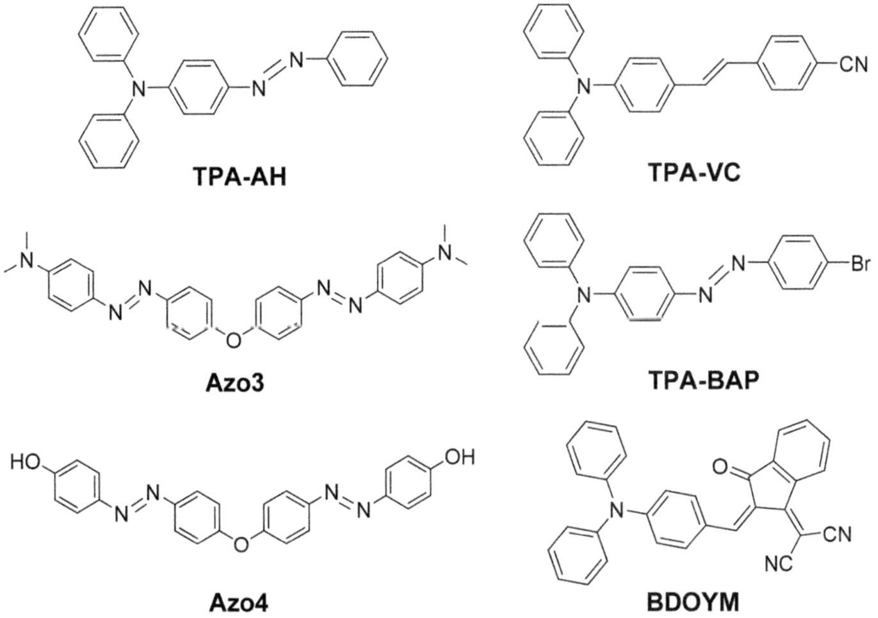

Scheme 1.2 Chemical structures of several typical D–A small molecules reported for binary data-storage devices.

organic bistable device consists of a 2-amino-4,5-imidazoledicarbonitrile (AIDCN)/Al/AIDCN trilayer structure interposed between an anode and a cathode. This device has two distinctive states of conductivity that can be achieved by applying voltage pulses with different polarities. This memory device is non-volatile and rewritable, with an ON/OFF ratio of more than 10^6. The device demonstrates good rewritability characteristics during cycle testing. More than one million write–read–erase cycles were performed on the device without failure. When this organic bistable device is integrated with an organic light-emitting diode, the device can be read out optically.[144]

Small organic molecules containing both an electron donor and an electron acceptor are an important type of material for organic electronic memory devices. As shown in Scheme 1.2, a number of conjugated D–A molecules have been investigated as electrically active materials in binary devices. The triphenylamine-containing D–A molecule (BDOYM) was designed to improve the intrinsic storage performance.[145] By using scanning tunneling microscopy (STM), stable, reliable, reversible data storage was demonstrated.

Subsequently, the effect of different D–A structures on the electronic switching properties of triphenylamine-based molecules was investigated (Figure 1.8).[146] Based on different arrangements of the donor and acceptor units, D–A molecules of two structural types, viz., D–π–A–π–D and A–π–D–π–A were designed. Devices based on the D–π–A–π–D molecule exhibited excellent write–read–erase characteristics with a high ON/OFF ratio of up to 10^6, while the device based on the A–π–D–π–A molecule exhibited irreversible switching behavior and a relatively low ON/OFF ratio. The arrangement

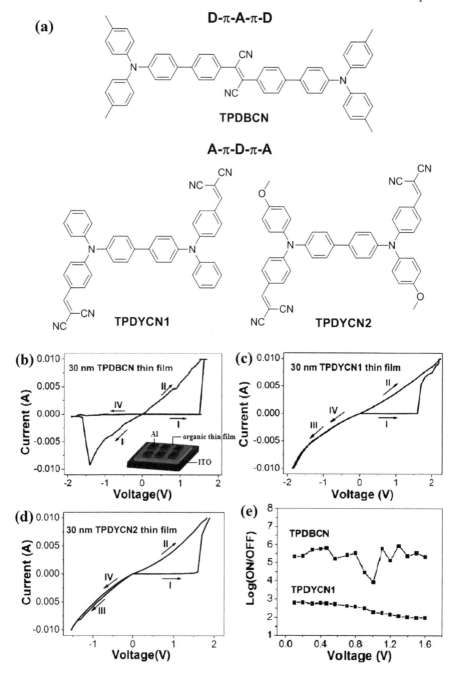

Figure 1.8 (a) Chemical structures of D–A molecules, (b) macroscopic *I–V* characteristics of ITO/TPDBCN/Al, (c) macroscopic *I–V* characteristics of ITO/TPDYCN1/Al, (d) macroscopic *I–V* characteristics of ITO/TPDYCN2/Al, and (e) comparison of the ON/OFF current ratio of (b) and (c). The inset of (b) shows the device structure used for macroscopic *I–V* measurements. Reproduced with permission from ref. 146, © 2010 WILEY-VCH Verlag GmbH & Co. KGaA, Weinheim.

of the donor and acceptor within the molecular backbone plays a key role in the electrical memory behavior of devices.

Lu and coworkers reported a simple DRAM device based on SNACA where the electron acceptor naphthalimide (NA) and electron donor carbazole (CA) were linked by a hydrazone bond (Figure 1.9a).[147] The electrical switching behavior of a Pt/SNACA/Al device is demonstrated in Figure 1.9c. From theoretical calculations, it was found that during the HOMO to LUMO transition, the electron density underwent a minor shift from the electron donor to acceptor in the SNACA backbone, resulting in poor stability of the ON state of the memory device after removal of the electrical field, and thus leading to volatile DRAM characteristics. To alter the distribution of electron density throughout the molecular backbone and thereby tune the corresponding memory type of the device, Lu et al. introduced a π-spacer pyridyl acetylene into the conjugated

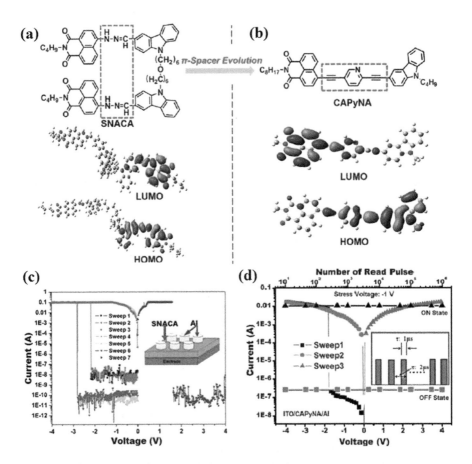

Figure 1.9 Molecular structures, and HOMO and LUMO energy levels of SNACA (a) and CAPyNA (b); current–voltage (I–V) characteristics of the memory device Pt/SNACA/Al (c) and ITO/CAPyNA/Al (d). Reproduced from ref. 147 and 148 with permission from The Royal Society of Chemistry.

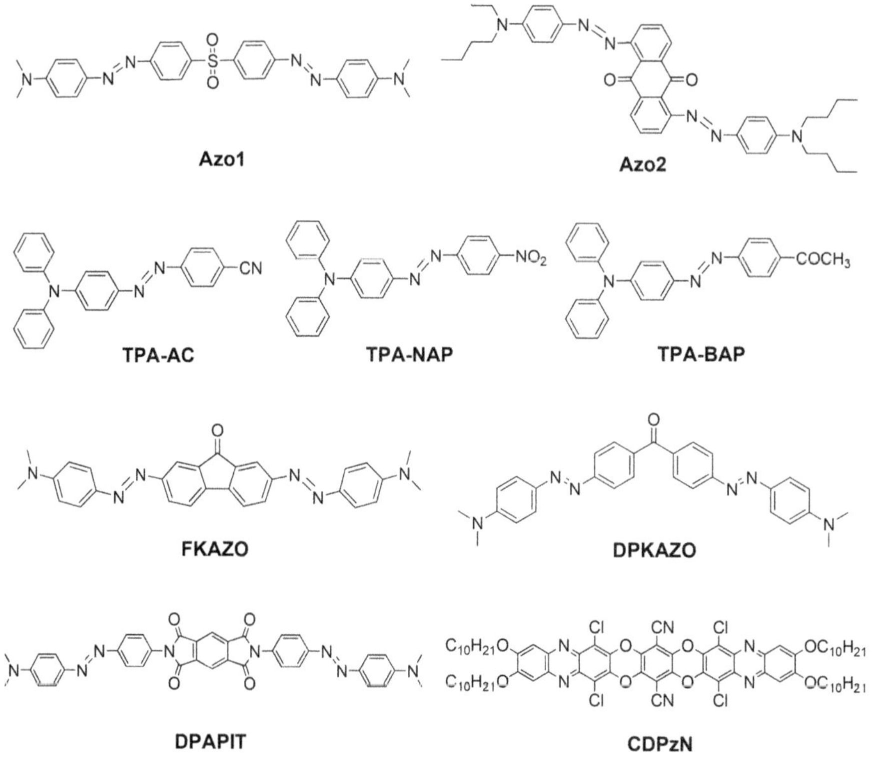

Scheme 1.3 Chemical structures of several typical D–A small molecules reported for ternary data-storage devices.

molecular skeleton to replace the original hydrazone bond between the carbazole unit and the naphthalimide moiety (Figure 1.9b and d).[148]

Small D–A molecules with multilevel stable states can lead to an increased device capacity of 3^n or larger. As illustrated in Scheme 1.3, a series of novel small organic molecules have been designed and synthesized for ternary data-storage devices.[74–76,149–152] Figure 1.10a shows the *I–V* characteristics of an ITO/DPKAZO/Al device.[76] The device demonstrates typical non-volatile ternary WORM memory behavior. As shown in Figure 1.10b, the *I–V* characteristics of the ITO/FKAZO/Al device also exhibit non-volatile ternary WORM characteristics. The two switching threshold voltages of FKAZO are markedly lower than those of DPKAZO (−1.05 and −1.81 V compared to −1.50 and −2.61 V). The low switching threshold voltages are due to the fact that FKAZO has a planar conformation (Figure 1.10c), which can generate a highly ordered arrangement in the film and decrease the hole injection energy barrier. The charge carrier transport process in the D–A molecule FKAZO (Figure 1.10d) was proposed to explain the switching mechanism of the ternary memory device.

Song *et al.* reported a *meta*-conjugated donor–bridge–acceptor (DBA) organic molecule for electronic multilevel storage.[153] This DBA molecule is

Organic Electronic Memory Devices 21

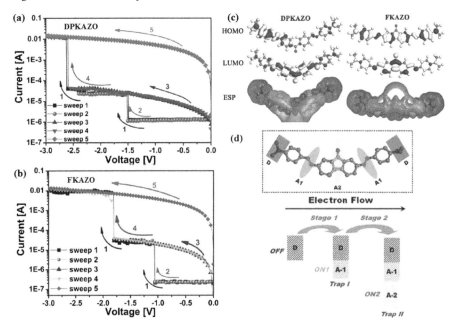

Figure 1.10 Current–voltage (*I*–*V*) characteristics of (a) ITO/DPKAZO/Al device and (b) ITO/FKAZO/Al device. (c) HOMO orbitals, LUMO orbitals, and molecular electrostatic potential (ESP) of DPKAZO and FKAZO from DFT calculations. DPKAZO is twisted and unsymmetrical. FKAZO is planar and symmetrical. (d) Schematic diagram of the charge carrier transport process in an FKAZO-based memory device. Reproduced with permission from ref. 76, © 2012 WILEY-VCH Verlag GmbH & Co. KGaA, Weinheim.

exploited as the storage medium (Figure 1.11a) in which TPA acts as an electron donor and TCF acts as an electron acceptor. The D–A pair is separated by a *meta*-conjugated bridge of a 2,7,12-trisubstituted truxene unit. As shown in Figure 1.11b and c, the device with the structure of ITO/DBA/Al shows good electrically bistable switching behavior between "0" (OFF) and "2" (HC-ON) states in the dark, with a large ON/OFF ratio of more than 10^6. In cooperation with UV light, a new "1" (LC-ON) state is addressable. This "1" state is "written" from the "0" state, by a combination of UV light and a positive voltage of +1.0 V, or from the "2" (HC-ON) state, by a combination of UV light and a negative voltage of −3.4 V.

Certain organic materials and biomacromolecules possess great potential for application in biocompatible, low cost and disposable electronic devices.[154–156] Chen *et al.* used the sericin protein as the functional material for the fabrication of flexible multilevel memory devices.[154] Kundu *et al.* reported the fabrication of transparent bio-memristor devices using natural regenerated silk fibroin protein.[155] As shown in Figure 1.12, by sweeping the dc bias in 4 steps, the ITO/silk/Al device exhibits typical pinch hysteresis-like *I*–*V* characteristics on a linear scale. The observed

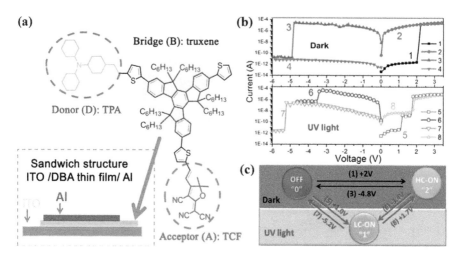

Figure 1.11 (a) Molecular structure of donor–bridge–acceptor compound (DBA). Inset: schematic diagram of ITO/DBA/Al sandwich memory device. (b) I–V characteristics of the memory device. (c) Schematic graph showing the possible transitions and threshold voltage among three states. Reproduced with permission from ref. 153, © 2012, American Chemical Society.

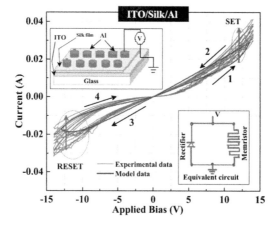

Figure 1.12 I–V characteristics of transparent bio-memristor devices with silk fibroin protein. The corresponding schematic of the fabricated memristor device is shown in the upper inset. The equivalent circuit model consisting of a rectifier in parallel with a memristor is shown in the lower inset. Reproduced with permission from ref. 155, © 2012 WILEY-VCH Verlag GmbH & Co. KGaA, Weinheim.

rectifying nature in the *I–V* characteristics suggests that the silk fibroin can make a rectifier contact with either electrode. Thus the bipolar memristive switching with rectifying characteristics of the silk fibroin memristor was demonstrated.

1.5.2 Polymeric Materials

The molecular structure of polymeric materials can be tailored using electron donors and acceptors of different strengths, spacer moieties with different steric effects, and electroactive pendant groups to induce different switching behaviors for electronic memory applications.

1.5.2.1 Functional Polyimides

Functional polyimides (PIs) are one of the most attractive polymeric materials for organic electrical memory applications due to their good solution processability, high thermal stability and mechanical strength.[18] In functional PIs, phthalimide acts as the electron acceptor, and electron donors (triphenylamine or carbazole moieties) are introduced to form a D–A structure. An electric field-induced CT state can be formed, which is the main mechanism responsible for the memory behavior.[157] Electronic memory devices based on soluble PIs were first reported in 2006.[59] As shown in Figure 1.13a, the functional TP6F-PI contains triphenylamine as electron donor and phthalimide as the acceptor. The hexafluoroisopropylidene (6F) group plays an important role in increasing the solubility of the PIs due to its bulkiness and low surface energy. TP6F-PI exhibited excellent thermal stability, with a 10% weight-loss

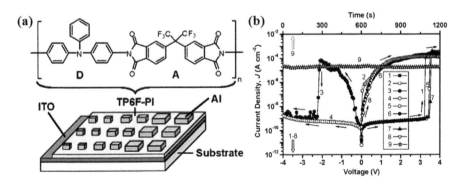

Figure 1.13 (a) Molecular structure (top) of functional PI (TP6F-PI) and schematic diagram (bottom) of single-layer memory device. (b) *J–V* characteristics of a 0.16 mm² Al/TP6F-PI/ITO device. The ON state was maintained by refreshing at 1 V every 5 s. Sweeps 1, 2, 5, 6, 7, and 8: 0 to +4 V (with power off for 1 min for sweeps 6 and 7). Sweeps 3 and 4: 0 to −4 V. Reproduced with permission from ref. 59, © 2006, American Chemical Society.

temperature of 524 °C and a glass transition temperature of 316 °C. The Al/TP6F-PI/ITO device exhibits dynamic random access memory (DRAM) behavior with an ON/OFF current ratio of up to 10^5 (Figure 1.13b). Field-induced charge transfer from triphenylamine to phthalimide is considered to dominate the switching behavior. Since 2006, a large number of memory effects, including volatile and non-volatile switching behavior, have been observed in a number of functional PIs.[158–167]

Flexible dual functional devices with multi-colored electrochromic and volatile memory characteristics have been fabricated from a solution-processable TPA-containing PI (Figure 1.14a).[166] Introduction of highly electron-donating starburst TPA moieties into the polymer main chain not only stabilizes its radical cations but also leads to good solubility and

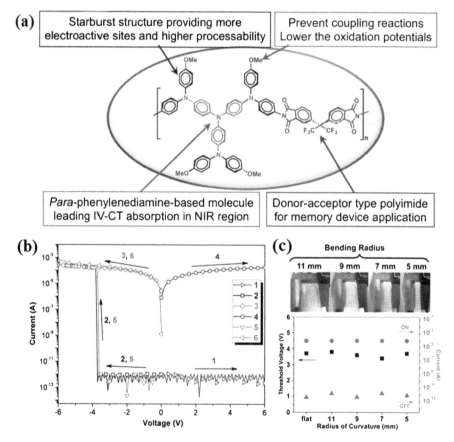

Figure 1.14 (a) Design strategy for starburst triarylamine-based polyimide 9Ph-6FDA. (b) *I–V* characteristics, and (c) appearance in various bent states, and variation of current and threshold voltage with different bending radii of the ITO/9Ph-6FDA/Al flexible memory device. Reproduced with permission from ref. 166, © 2013 WILEY-VCH Verlag GmbH & Co. KGaA, Weinheim.

film-formation properties of the PI. As shown by the *I–V* characteristics of Figure 1.14b, the device can switch from the OFF state to the ON state during the negative sweep. The fifth sweep was conducted after turning off the power for about 3 min. It was found that the ON state had relaxed to the steady OFF state without an erasing process. The device could be reprogrammed to the ON state again at a threshold voltage of −3.6 V. Thus the ITO/9Ph-6FDA/Al flexible memory device exhibits SRAM memory behavior. Reliable and reproducible switching memory behavior of this flexible memory device (Figure 1.14c) can be maintained under mechanical bending stress. Furthermore, the polymer device showed electrochromism with a high contrast ratio, high coloration efficiency, low switching time, and outstanding stability for long-term electrochromic operation in both the visible and near-infrared regions. Ree *et al.* reported the fabrication and operation of an electrically programmable non-volatile memory device based on a thermally and dimensionally stable PI containing carbazole moieties (6F-HAB-CBZ PI).[167] The fabricated Al/6F-HAB-CBZ/Al devices exhibited excellent unipolar switching behavior, and could be repeatedly written, read, and erased.

1.5.2.2 Non-Conjugated Polymers with Pendants

Non-conjugated polymers with pendent electroactive donors, acceptors and chromophores are another kind of polymer material favorable for electronic memory.[168–172] Chen *et al.* developed D–A random copolymers P(VTPA$_x$-BOXD$_y$) and P(CNVTPA$_x$BOXD$_y$) containing pendent electron-donating TPA and electron-accepting 1,3,4-oxadiazole units for memory device applications (Figure 1.15).[66] The tunable switching behavior (SRAM or DRAM) was explored by using different ratios of the pendent TPA donor and the BOXD acceptor. Similarly, the electrical switching behavior based on pendent polymers containing electron-donating carbazole units (VPK) and electron-withdrawing

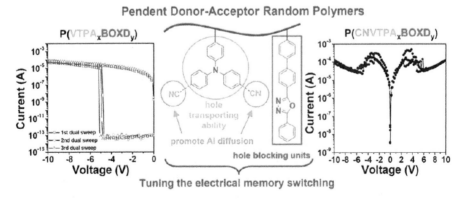

Figure 1.15 Donor–acceptor random copolymers with pendent triphenylamine and 1,3,4-oxidazole for memory device applications. Reproduced with permission from ref. 66, © 2011, American Chemical Society.

oxadiazole-containing units (OXD or BOXD) can be tuned using the donor/acceptor ratio or acceptor trapping ability.[168] The electrical I–V characteristics change between the diode, the volatile memory, and the insulator, depending on the relative donor/acceptor ratios. The unstable ON state in the P(VPK$_8$OXD$_2$) or P(VPK$_8$BOXD$_2$) device was due to shallow hole traps with spontaneous back transfer of charge carriers when the electric field was removed to produce volatile SRAM behavior.

Novel random and block copolymers with controlled morphologies and electrical properties have been developed to prevent aggregation and thus improve the reproducibility and stability of memory devices.[169,171] Lee et al. synthesized D–A type block copolymers of poly(9-(4-vinylphenyl)-carbazole)-b-poly(2-vinylpyridine) (denoted PVPCz-b-P2VP).[169] By adjusting the block ratios of PVPCz and P2VP, various morphologies, ranging from spherical to lamellar, as well as different electrical behaviors including metallic, bistable switching and insulating, were observed.

For the Disperse Red 1-functionalized PVK copolymer (PVDR), in which carbazole entities serve as electron donors, and DR1 moieties serve as electron acceptors,[68] both inter- and intrachain electron transfer can occur easily. PVDR nanoaggregates self-assemble via π–π stacking interactions of the carbazole groups in the polymer system after adding a solution of PVDR in N,N'-dimethylformamide (DMF) to dichloromethane (DCM). The ITO/PVDR/Al device fabricated from the pre-assembled PVDR film exhibited a typical WORM memory effect.

The use of azobenzene moieties as spacers between electron donor and acceptor groups in the pendant chains of vinyl-based polymers has been investigated.[173-177] The reversibility of the ON state was found to be dependent on the terminal moiety of the azobenzene chromophores.[173] Polymers with electron-accepting terminal moieties (–Br or –NO$_2$) in the pendent azobenzene exhibited WORM memory behavior, while those with electron-donating terminal moieties (–OCH$_3$) showed rewritable memory behavior. The switching threshold voltage of the devices varied almost linearly with the functional azobenzene moiety content in the copolymer, which was attributed to the reduced energy barrier between the HOMO of the copolymer film and the ITO work function.[174] The ON/OFF ratios of the WORM devices (10^4–10^6) were also higher than those of the rewritable devices (10^3–10^4). Two poly(N-vinylcarbazole) derivatives with pendant donor–trap–acceptor (D–T–A) structures, PVK-AZO-2CN and PVK-AZO-NO$_2$, have been prepared for memory devices.[175,176] The Al/polymer/ITO devices all exhibited WORM memory effects with a low switching threshold voltage and a high ON/OFF current ratio, with the Al/PVK-AZO-2CN/ITO device with stronger electron acceptors (2CN) exhibiting an even higher ON/OFF ratio. The charge transfer and trapping processes were further stabilized by the conformational relaxation of the total energy of the D–T–A system through donor–acceptor electrostatic interaction. When the terminal acceptor moieties were removed from the azobenzene chromophores, pure insulators were obtained.

1.5.2.3 Conjugated Polymers

Considerable effort has been devoted to develop novel conjugated polymers for information and communication technology.[178–180] The incorporation of different electron acceptors into conjugated polymer donors significantly affects the memory properties, while the induced trapping environment or charge transfer channel determines the volatility of the memory device.[20] D–A type conjugated polymers (Scheme 1.4) have been utilized to fabricate different types of memory device, such as volatile DRAM and SRAM devices, and non-volatile WORM and Flash devices.[181–186] A device with the sandwich structure ITO/PFOxPy/Al could write, read, erase and refresh its electronic states, fulfilling the functionality of a DRAM device. Its memory behavior was attributed to space charge and de-trapping effects.

A device based on a fluorene conjugated polymer with electron-accepting oxidiazole (PCFO) exhibited WORM memory effects,[67] with a switching threshold voltage of −2.3 V and an ON/OFF current ratio of 10^5. Another polyfluorene-based copolymer (PFTPACN) containing electron-rich TPA and electron-poor cyano substituents on the side chain also exhibited typical WORM memory characteristics.[184] In addition, non-volatile rewritable memory behavior was observed in a device based on conjugated poly[9,9-bis(4-diphenylaminophenyl)-2,7-fluorene] donors covalently bridged with Disperse Red 1 acceptors (DR1-PDPAF-DR1).[181]

Conjugated polymers with good solution processability have been used to fabricate flexible memory devices. A conjugated polymer with fluorene and thiophene donors in the main chain and the phenanthro[9,10-*d*]-imidazole (PFT-PI) acceptor in the side chain was employed as the active layer for a flexible device.[183] The fabricated PEN/Al/PFT-PI/Al flexible device exhibited non-volatile rewritable memory behavior. The *J–V* characteristics did not change significantly before and after continuous bending stress. Subsequently, the tunable electrical switching characteristics of the vinylene-based conjugate polymers, PVC-PI, PVT-PI, and PVTPA-PI, consisting of different donors of carbazole (C), thiophene (T) and triphenylamine (TPA) and the acceptor phenanthro[9,10-*d*] imidazole (PI) were demonstrated.[185] The donor structure affects the polymer conformation, the D–A electrical interaction, and the LUMO energy levels for stabilizing the charge separation. The PEN/Al/PVC-PI/Al flexible device exhibited SRAM behavior while the PVTPA-PI device exhibited WORM behavior. However, the PVT-PI device only exhibited diode-like electrical behavior. The PVC-PI and PVTPA-PI flexible memory devices could operate at low voltages (less than <2.5 V) with high ON/OFF current ratios (>10^4) and exhibited excellent durability in repeated bending tests.

Benzodithiophene (BDT)-based conjugated polymers that exhibit good performance in both organic field-effect transistors and solar cells have also been used in polymer memory.[186] A series of BDT-based D–D and D–A conjugated polymers have been prepared (Figure 1.16). Reliable and reproducible WORM memory properties were demonstrated in ITO/CPs/Al flexible memory devices. Conjugated polymers with different backbones and side

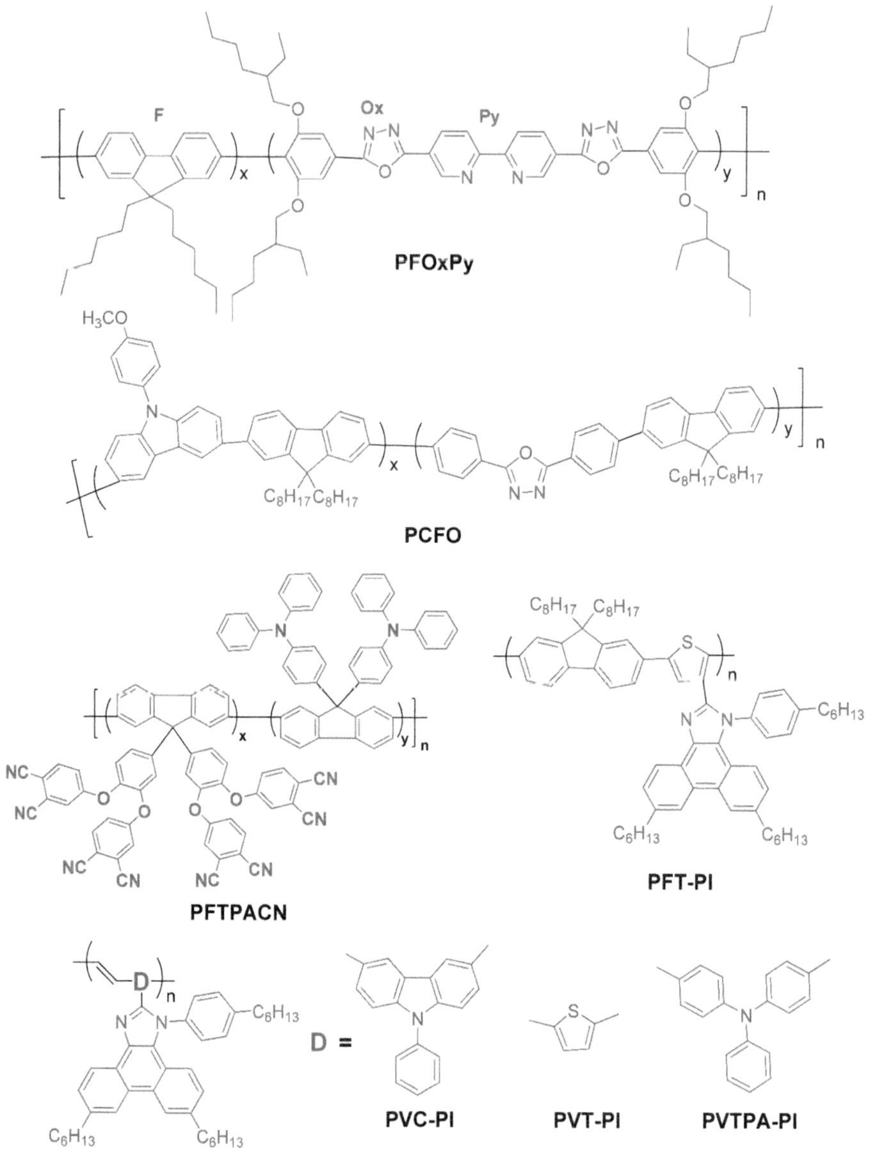

Scheme 1.4 Chemical structures of some D–A type conjugated polymers for memory devices.

chains exhibited different stabilities of their charge-separated states. Conjugated polyelectrolytes,[187] poly(3,4-ethylenedioxythiophene):poly(styrenesulfonate) (PEDOT:PSS)[188,189] and doped conjugated polymers[73,190] have also been used as switching media for flexible devices. McCreery et al. fabricated redox-based polymer memristors using polythiophene as the electroactive component and ethylviologen perchlorate as the redox counter-electrode

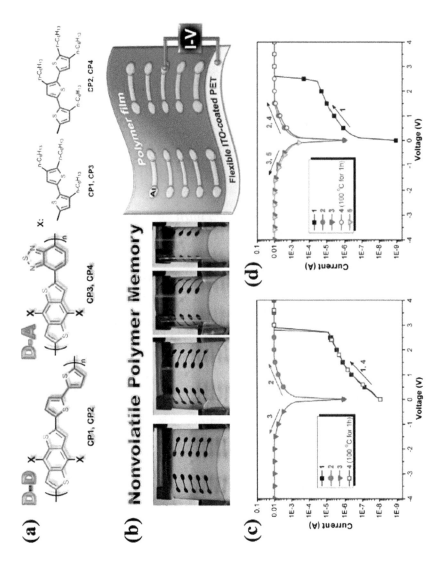

Figure 1.16 (a) Chemical structures of BDT-based conjugated polymers, (b) configuration of flexible memory device. (c) *I-V* characteristics of ITO/CP2/Al memory device. (d) *I-V* characteristics of ITO/CP4/Al memory device. Reproduced from ref. 186 with permission from The Royal Society of Chemistry.

material.[190] The devices exhibited switching immediately after fabrication and did not require the "electroforming" step required in many types of memory device. The conductivity of the polythiophene layer could be reversibly switched between high and low conductance states with a "write–erase" (W–E) bias, with only minor degradation of response after 200 W–E cycles.

A poly(Schiff base) is a conjugated polymer with imine groups (C=N) in its backbone. Li *et al.* fabricated an organic resistive switching memory device based on protonic-acid-doped polyazomethine (PA-TsOH).[73] The resistance of this memory device can be controlled gradually and exactly *via* manipulation of the doping level of the PA:TsOH system. The uniformity in the device performance is superior to that reported for RRAM devices, suggesting that controllable protonic doping is an effective way to optimize the uniformity of the resistive switching behavior. By applying different RESET voltages to the Pt/PA-TsOH/Pt devices, the memory device can be programmed into four resistive states consecutively for more than 700 switching cycles, increasing the data storage capacity exponentially.

1.5.2.4 Polymers Chemically Modified with Fullerenes or Graphene

Carbon nanomaterials such as fullerenes, graphene and their derivatives have exhibited good performance in optoelectronic devices. PVK with covalently attached fullerene (PVK–C_{60}),[58] in which the carbazole group serves as the electron donor and hole-transporting moiety, and C_{60} serves as the electron acceptor, has been synthesized. The fabricated ITO/PVK–C_{60}/Al device exhibited non-volatile rewritable memory behavior. Ree *et al.* also attached a fullerene acceptor covalently to a carbazole containing polymer in a controlled manner.[191] With C_{60} acting as a charge trap, poly(2-(*N*-carbazolyl) ethyl methacrylate) end-capped with fullerene (PCzMA-C_{60}) exhibited both bipolar and unipolar rewritable memory characteristics. By introducing 10 wt% PCBM into poly-4-methoxytriphenylamine,[61] Liou *et al.* demonstrated a WORM memory device with lower switching-ON voltage (0.9 V) and much higher ON/OFF ratio (10^9) than those of the blended composites.

Chemical functionalization also plays a key role in tailoring the structure, processability, and physicochemical and electronic properties of GO nanosheets. By using the "grafting to" or "grafting from" method, a variety of electroactive polymers have been covalently incorporated onto the surface as well as the edge of GO nanosheets to fabricate graphene-based polymer memory devices.[113–115,192–195] Triphenylamine (TPA)-based conjugated polyazomethine, covalently grafted to GO (TPAPAM–GO),[113] has been used directly for fabricating memory devices (Figure 1.17). With efficient hole injection, high carrier mobilities and the low ionization potential of the TPAPAM polymer chains, the devices exhibited bistable electrical switching and non-volatile rewritable memory effects, with a small switch-on voltage of about −1 V and an ON/OFF current ratio of more than 10^3. The CT state of TPAPAM–GO is

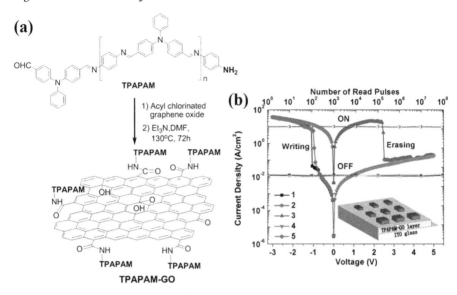

Figure 1.17 (a) Molecular structure of conjugated polymer functionalized graphene oxide TPAPAM–GO; (b) J–V characteristics of Al/TPAPAM–GO/ITO device. Reproduced with permission from ref. 113, © 2010 WILEY-VCH Verlag GmbH & Co. KGaA, Weinheim.

effectively stabilized by electron delocalization in the graphene nanosheets, which leads to the non-volatile nature of the memory device.

A PANI-functionalized GO derivative (GO–PANI) was prepared *via in-situ* oxidative graft polymerization of aniline on GO nanosheets.[195] After growing PANI from the surface of GO, spindle-shaped PANI nanofibers with large length-to-diameter ratios appear to surround the GO nanosheets and act as a tunneling barrier for electrons moving from one GO sheet to another. The Al/GO–PANI/ITO device exhibited typical bistable electrical switching and non-volatile rewritable memory effects, which can be attributed to electric field-induced charge transfer between the PANI chains and the GO sheets. Covalent functionalization of GO with electroactive polymers is thus an effective and versatile approach to tuning the electronic properties of GO. In addition, the conductance of PANI can be further manipulated by ionic doping with mineral as well as organic acids, demonstrating great potential for the development of novel ionic electronics.

1.5.2.5 Polymers Containing Metal Complexes

When introduced into polymer backbones or pendants, the electrical affinity of transition-metal complexes can improve the stability of conductive states, and their reversible redox properties can make a polymer memory device more suitable for practical applications.[196] Among various transition-metal complexes, ferrocene (Fe^{2+}) is of particular interest because of its well-studied

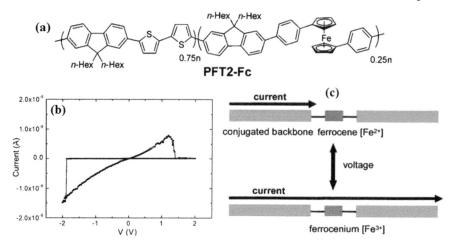

Figure 1.18 (a) Chemical structure of the ferrocene containing polymer, PFT2-Fc. (b) *I–V* characteristics of PFT2-Fc device. (c) Postulated mechanism for memory behavior. Reproduced with permission from ref. 197, © 2007, American Chemical Society.

reversible redox behavior, and its stable oxidized form ferrocenium (Fe^{3+}), which renders the possibility of non-volatility for memory applications. Choi *et al.* synthesized a new conjugated polymer (PFT2-Fc) with redox-active functionality, ferrocene, in the polymer main chain (Figure 1.18).[197] The fabricated Al/LiF/PFT2-Fc/ITO device exhibited non-volatile rewritable memory behavior. Its switching mechanism involves ferrocene acting as a voltage-dependent *in-situ* dopant during the redox process to result in the enhancement of conductivity of the polymer film.

Ling *et al.* coupled 1.3 mol% of a rare-earth metal complex (Eu-complex) with a carbazole donor to synthesize the D–A type functional copolymers, PKEu (Scheme 1.5).[112] The as-fabricated transparent and uniform film avoided the phenomena of ion aggregation and phase separation commonly found in mixed or doped systems. The Al/PKEu/ITO device exhibited non-volatile rewritable memory behavior with an ON/OFF ratio as high as 10^4. Transient current density *versus* time measurements indicated that the device transition occurred in less than 20 μs. They further introduced 1,3,4-oxadiazole moieties (electron acceptors) into the polymer chains, to obtain a new triblock copolymer, PCzOxEu.[198] The device based on PCzOxEu exhibited much better rewritable memory performance, with an ON/OFF current ratio up to 10^5, a switching response time of ~1.5 μs, more than 10^6 read cycles and a retention time of more than 8 h.

Non-volatile flash memory devices using polyfluorenes containing Ir(III) complexes in the main chain (iamP1–iamP3) as the active material have been reported.[60] The fluorene moieties act as electron donors and the Ir(III) complex units as the electron acceptors. Charge transfer and traps in the polymers are probably responsible for the conductance-switching behavior

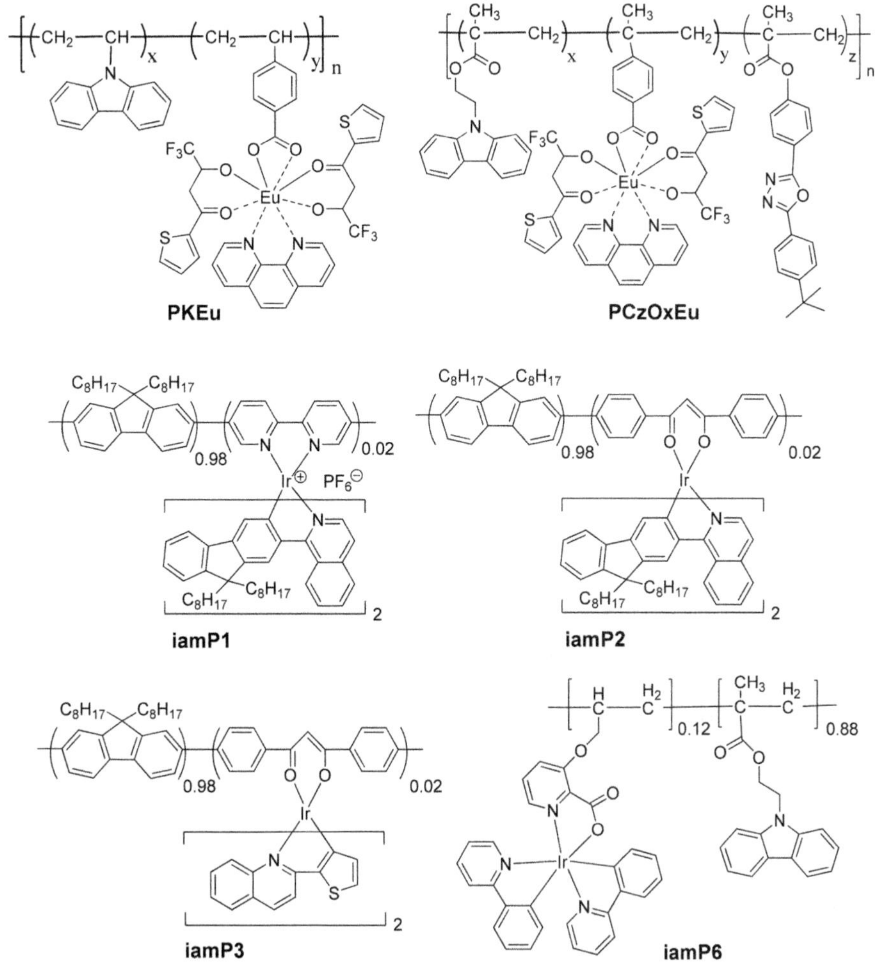

Scheme 1.5 Chemical structures of some polymers containing metal complexes for memory devices.

and the memory effect. Polycarbazole and polyfluorene containing cationic Ir(III) complexes in their side chains also exhibited electrical bistability, and non-volatile flash memory devices based on them have been realized.[199] Polymer iamP6 comprises flexible spacers (O=C–O–C–C units) bridging the electroactive pendent chromophore carbazoles to permit conformational change under an electric field and a donor–acceptor system containing carbazole as donor and an Ir(III) complex as acceptor to permit charge transfer. The ITO/iamP6/Al device exhibited excellent ternary memory performance, including low reading, writing and erasing voltages, and good stability for the three states.

Higuchi *et al.* studied the memory behavior of a Co(III)-containing conjugated (CP) and non-conjugated polymer (NCP) with an azo-aromatic

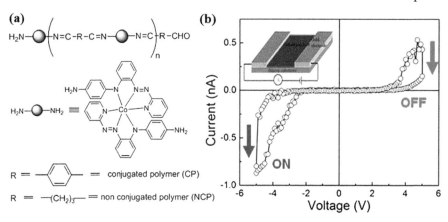

Figure 1.19 (a) Chemical structure of the Co(III) polymer of an azo aromatic ligand; (b) I–V characteristics of the CP device. Reproduced with permission from ref. 79, © 2011, American Chemical Society.

backbone (Figure 1.19).[79] The I–V characteristics of the CP device traced in the voltage range ±5 V showed the current to jump from the initial low level to a high level at −5 V, as well as another jump from the high level to the initial level at +5 V. This bistable behavior with a "pinch-off" in the I–V characteristics is a typical feature of a memristive device. The CP device could be operated at different voltage levels with different current responses, therefore providing multilevel storage capability to replace the existing binary computing devices.

1.5.3 Organic–Inorganic Hybrid Materials

Generally, organic–inorganic hybrid materials are composed of organic layers containing fullerenes, carbon nanotubes, graphene, metal nanoparticles, semiconductor nanoparticles or inorganic quantum dots (QDs).

1.5.3.1 Organic–Carbon Allotrope Hybrid Materials

Fullerene and its derivatives possess high electron-withdrawing ability, and are able to capture up to six electrons. For organic electronic memory applications, they have been widely used as electron acceptors to form CT complexes with polymer-containing electron donors, such as thiophene, fluorene, carbazole and aniline derivatives.[200–204] A hybrid material-based electrical memory device with the configuration rGO/P3HT:PCBM/Al exhibited electrically bistable behavior and the WORM memory effect, with an ON/OFF ratio of more than 10^4.[200] The memory effect was attributed to the polarization of PCBM domains and the formation of a localized internal electrical field among the adjacent domains.

Chen *et al.* reported the fabrication of a memory device with the structure of poly[4-(9,9-dihexylfloren-2-yl)styrene]-*block*-poly(2-vinylpyridine)

(P(St-Fl)-*b*-P2V):PCBM composites sandwiched between the top Al electrode and the bottom ITO electrode.[201] The switching mechanism could be explained by a charge injection dominated thermionic emission model (OFF state) and a charge transfer dominated Poole–Frenkel (PF) emission model (ON state).

In the absence of electron donor moieties in hybrid materials, C_{60} molecules can be used as the charge storage medium.[205-208] By simply changing the concentration of C_{60} in the insulating polystyrene (PS) matrix, the device with the structure Al/PS:C_{60}/Al exhibits three distinctly different electrical behaviors, namely a true insulator (C_{60}%(wt) = 1.0–5.0%), a rewritable memory device (C_{60}%(wt) = 5.0–7.5%) with an ON/OFF current ratio larger than 10^4 and a WORM type memory device (C_{60}%(wt) = 7.5–20%). Memory effects have also been observed in a memory device with C_{60} molecules dispersed in a poly(vinyl phenol) (PVP) matrix. This simple device exhibits distinct electrical hysteresis with low and high conductivity states. Charge transfer and retention in the C_{60} molecules are proposed to account for the electrical switching.

Polyimide (PI) has been used as a dielectric and mechanical-support material in the electronics industry due to its thermal stability, good chemical resistance, and excellent mechanical properties. Kim *et al.* fabricated organic non-volatile memory devices on PET flexible substrates using PI:PCBM hybrid materials as active layers.[209] These flexible memory devices showed successful rewritable switching properties even under bent conditions. Three-dimensional (3D) stacking of memory devices provides a way to achieve a great increase in memory cell density. A 3D-stacked 8 × 8 cross-bar array of polymer resistive memory devices with the PI:PCBM hybrid material as the memory element has been fabricated using a spin-coating process.[210] The memory cells in each layer exhibited excellent non-volatile memory performance, and they can be independently written, read and erased with a high ON/OFF ratio, good endurance and good retention capability.

Organic/polymer:PCBM hybrid materials have been integrated into one diode–one resistor (1D–1R) arrays to demonstrate practical implementation of organic memory.[211,212] For instance, Kim *et al.* reported the first 64-bit organic flexible memory cell array in a 1D–1R architecture with photopatternable hybrid memory (PS:PCBM) and diode (P3HT) systems.[212] Figure 1.20 shows the integrated structure of the PEN/Al/PS:PCBM/Al/P3HT/Au/Al 1D–1R cell. A P3HT/Bis-FB-N3 solution was chosen as the diode material, and a PS/PCBM/Bis-FB-N3 mixture was used for the unipolar memory component. The 1D–1R cell array can avoid cross-talk problems and encode letters based on the standard ASCII character code.

Controllable electrical conductance switching and non-volatile memory effects have been observed in Al/PVK–CNT/ITO sandwich structures.[57] Unique device behavior, including (i) insulator behavior, (ii) bistable electrical conductance switching effects (rewritable memory and WORM memory effects), and (iii) conductor behavior, can be realized in an ITO/PVK–CNT/Al sandwich structure by increasing the CNT content in the PVK–CNT composite film. The turn-on voltage of the bistable devices decreased with the increase in CNT content of the composite film.

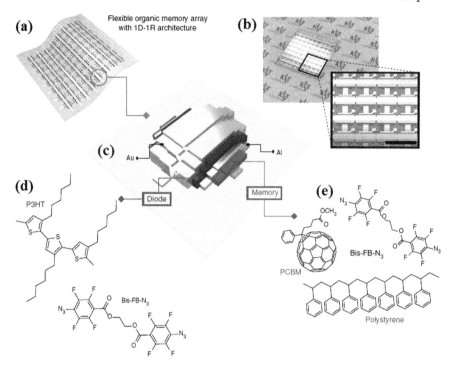

Figure 1.20 (a) Illustration of as-fabricated 1D–1R organic resistive memory cell array on a flexible PEN substrate and (b) optical image. Scale bar: 5 mm. (c) Schematic of unit cell of 1D–1R cell (Al/PS:PCBM/Al/P3HT/Au/Al). (d) Chemical structure of P3HT and Bis-FB-N_3 for ultraviolet patternable diode layer. (e) Chemical structure of PS, PCBM and Bis-FB-N_3 for ultraviolet patternable memory layer. Reproduced with permission from ref. 212, © 2013, Rights Managed by Nature Publishing Group.

Park et al. developed resistive-type non-volatile polymer memory devices for high temperature operation using hybrid nanocomposites of the conjugated block copolymer poly(styrene-*block-para*phenylene) (PS-*b*-PPP) and single-walled carbon nanotubes (SWNTs) by simple solution blending.[213] With certain SWNT concentration regimes, the nanocomposites exhibited bipolar non-volatile rewritable memory characteristics. High temperature operation of the device was realized at up to 100 °C without significant degradation of the memory performance.

Kim et al. fabricated mechanically flexible, multilevel switching resistive memory by solution casting of PS–chemically-doped multiwalled CNT composites.[214] The controlled work function and high dispersibility of the substitutionally doped CNTs significantly improved the resistive memory performance of the PS:CNT composite devices. As shown in Figure 1.21, devices consisting of quadruple layers of Al/PS:BCNT/PS:NCNT/Al were prepared on polyimide substrates (BCNT = boron-doped CNT, NCNT = nitrogen-doped CNT). The flexible device with a quadruple layer exhibited

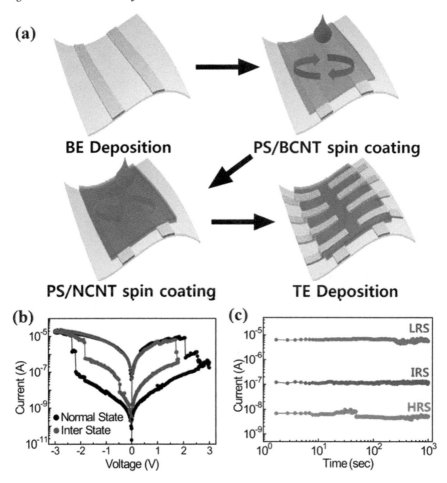

Figure 1.21 (a) A schematic illustration of the fabrication process of a multistate resistive memory device. (b) *I–V* and (c) retention characteristics of multistate devices. Reproduced with permission from ref. 214, © 2012, American Chemical Society.

stable triple state memory characteristics (*i.e.*, low resistance state (LRS), high resistance state (HRS) and interstate (IRS)), unlike triple layer devices consisting of Al/PS:BCNT or PS:NCNT/Al. The ON/OFF ratios for LRS/IRS and IRS/HRS were 50 and 15, respectively. The switching mechanism of the CNT:PS composite memory devices follows a charge storage mechanism by trapping and de-trapping.

Graphene nanosheets have been of particular interest for application in organic–inorganic hybrid material-based memory devices.[215–221] Figure 1.22 shows a schematic structure of the hydrogen bonding between poly(styrene-*block*-4-vinylpyridine) (PS-*b*-P4VP) and GO sheets.[217] The ITO/PS-*b*-P4VP:7 wt% GO composite/Al device exhibits a WORM memory effect with an ON/OFF ratio of 10^5 at −1.0 V. The switching mechanism was attributed to the

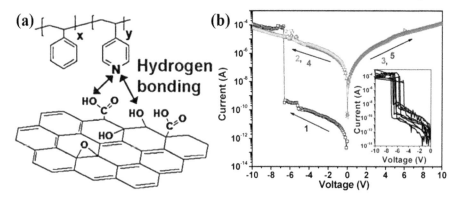

Figure 1.22 (a) Schematic structures of PS-*b*-P4VP and GO composites. (b) *I*–*V* characteristics of a 7 wt% GO composite device. Reproduced from ref. 217 with permission from The Royal Society of Chemistry.

charge trapping environment operating across the PS-*b*-P4VP/GO interface and due to the GO intrinsic defects. Controlling the physical interaction of the copolymers and functional GO nanosheets can generate a well-dispersed charge storage composite device for future flexible information technology.

An interconnected, one-dimensional/two-dimensional lamellar film was produced through encapsulation of self-assembled tetracene-derived organic wires by GO nanosheets.[215] By combining the hole transporting ability of the p-type tetracene semiconductor and the charge trapping properties of GO, the memory devices exhibited reproducible and non-volatile electrical bistability. Zhang *et al.* demonstrated resistive memory effects in a device with the structure ITO/PVK:graphene/Al.[218] Devices with 2 wt% graphene in the composite exhibited the WORM memory effect, while devices with 4 wt% graphene showed rewritable memory effects. The conductance switching effects of the composites can be attributed to electron trapping in the graphene nanosheets of the electron-donating/hole-transporting PVK matrix.

Figure 1.23 shows a schematic diagram of an Al/polymer/graphene/polymer/ITO memory device, which was fabricated by laminating two glass substrates coated with patterned electrodes and spacers.[216] Devices based on the Al/PS/graphene/PS/ITO sandwich structure exhibit WORM memory behavior, while devices with the structure Al/PVK/graphene/PVK/ITO show volatile memory behavior. The distinct electrical behavior of the two devices is probably related to the different depths of charge traps formed between the graphene sheets and polymer matrix, and provides a new route to tailor the memory effects of hybrid bistable devices.

Li *et al.* systematically investigated the rewritable memory effects of devices based on polyimide (PI) and graphene oxide (GO) with the structure Ag/PI/GO:PI/PI/ITO.[220] The stacked layer structure for the GO–PI film indicates that the GO nanosheets are well-packed during the spin coating process. The *I*–*V* curves for the as-fabricated device exhibit multilevel resistive-switching properties under various reset voltages. Multilevel conduction states arose from the varying filling degrees of traps in the active layer at different reset bias.

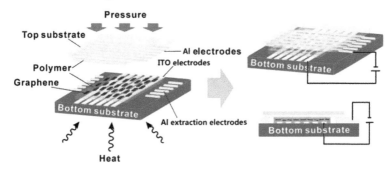

Figure 1.23 Schematic diagram of hybrid bistable memory devices fabricated utilizing graphene sheets sandwiched between polymer layers. Reproduced with permission from ref. 216, © 2011 Elsevier B.V.

1.5.3.2 Organic–Inorganic Nanocomposites

Hybrid electronic memory devices have been reported in some organic composites containing metal nanoparticles (NPs), quantum dots (QDs) and metal oxide NPs.[222–227] Yang et al. fabricated a memory device with the structure of a Au NP and 8-hydroxyquinoline-containing PS film sandwiched between two metal electrodes.[8] Electronic transitions were attributed to electric field-induced charge transfer between the Au NPs and 8-hydroxyquinoline. Au NPs can also be introduced into electroactive polymers such as P3HT,[54] PANI[224] and PVK[226] to realize memory behavior. A rewritable memory effect was demonstrated in a hybrid memory device with an active layer consisting of P3HT and Au NPs capped with 1-dodecanethiol sandwiched between two metal electrodes.[54] The device was fabricated through a simple solution processing technique and exhibited remarkable electrically bistable behavior.

In such electroactive polymer–metal nanoparticles hybrid systems, the electroactive polymers usually act as electron donors while the metal nanoparticles act as electron acceptors; the electrical bistability is related to electric field-induced charge transfer between the two components. Thus, the compatibility of the two components is crucial for optimal device performance. Yang et al. reported a non-volatile rewritable memory device based on polyaniline (PANI) nanofibers decorated with Au NPs.[55] The active polymer layer was created by growing nanometer sized gold particles within the PANI nanofibers using a redox reaction with chloroauric acid. The solution syntheses of PANI nanofibers with different sizes of autoreduced Au NPs were also investigated and the relationship between Au NP size and bistable memory response was evaluated (Figure 1.24).[224] The performance of devices made from four different solutions, with the Au NPs in four distinct size ranges is shown in Figure 1.24g.

Metal NP–polymer hybrids have also been used in organic polymer floating-gate transistor memory.[82,85] Organic memory devices consisting of P3HT electrospun nanofiber transistors functionalized with surface-modified Au NPs have been reported.[82] Figure 1.25 illustrates a prototypical

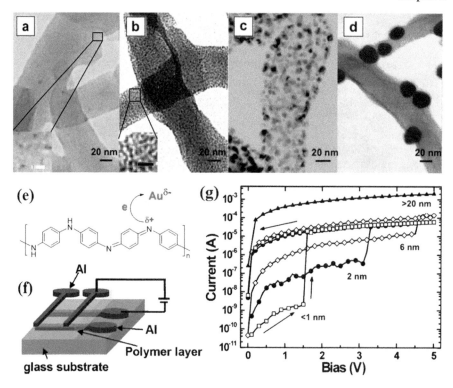

Figure 1.24 TEM images of autoreduced gold nanoparticles of (a) <1, (b) 2, (c) 6, and (d) >20 nm grown on polyaniline nanofibers. (e) Schematic structure of PANI-nanofiber/Au NP material after application of +3 V. An increase in charge transfer from PANI to the Au NPs is believed to occur. (f) The structure of the PANI nanofiber/Au NP bistable memory device. (g) Characteristic I–V scans for autoreduced Au NPs/PANI nanofiber hybrid memory devices. Reproduced with permission from ref. 224, © 2011, American Chemical Society.

electrospun P3HT:Au hybrid nanofiber-based transistor memory device on a PEN substrate with a bottom-gate top-contact configuration using thermally evaporated Au as back gate. The device remains reliable and stable even under bending conditions (radius: 5–30 mm) or after 1000 repetitive bending cycles. The high performance P3HT:Au hybrid nanofiber-based transistor memory devices allow their integration in future flexible logic circuits.

Other kinds of inorganic nanomaterials, such as ZnO, TiO_2 and MoS_2, have also been introduced into the organic polymer layer for hybrid memory devices. For instance, Kim and co-workers investigated the electrical properties and carrier transport mechanisms of memory devices fabricated by utilizing ZnO NPs embedded in PVK or PMMA layers as active materials.[223,225] Kim *et al.* fabricated an 8 × 8 array structured device using TiO_2 NPs embedded in a PVK film.[228] By varying the concentration of TiO_2 NPs, the ON/OFF

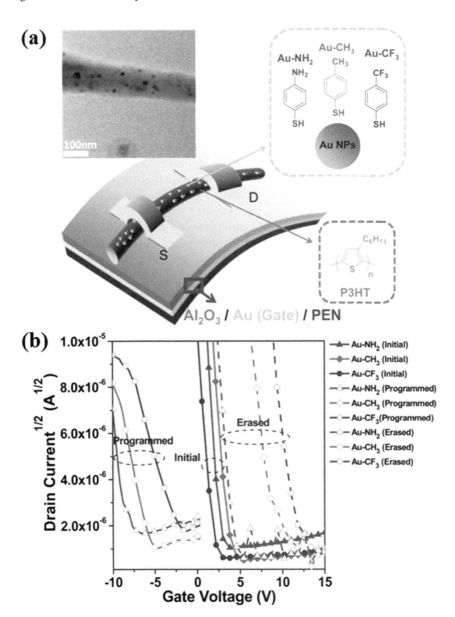

Figure 1.25 (a) Schematic configuration of hybrid nanofiber-based transistor memory devices, chemical structures of P3HT and surface-modified Au NPs, and representative plan-view TEM measurements of a P3HT:Au hybrid nanofiber. (b) Plot of $I_{ds}^{1/2}$ versus V_G curves of the P3HT:Au NP nanofiber-based transistor memory devices (programmed state: $V_G = -5$ V, 1 ms; erased state: $V_G = 5$ V, 1 ms; at a fixed $V_D = -5$ V). Reproduced with permission from ref. 82, © 2013 WILEY-VCH Verlag GmbH & Co. KGaA, Weinheim.

ratio of the devices could be modulated. More recently, Liou *et al.* fabricated memory devices using hydroxy-containing fluorene-based polyimide hybrids with tunable TiO$_2$ contents.[229] The hybrid memory devices with different TiO$_2$ concentrations from 0 wt% to 50 wt% exhibited tunable memory properties from DRAM to SRAM to WORM with an ON/OFF current ratio as high as 10^8. The unique energy level and quantum confinement effects also render MoS$_2$ as a charge trapping component for memory applications. Zhang *et al.* developed a facile method for exfoliation and dispersion of MoS$_2$ with the aid of polyvinylpyrrolidone (PVP).[230] The flexible device fabricated by a solution processing technique with the configuration rGO/PVP:MoS$_2$/Al exhibited electrical bistability and a non-volatile rewritable memory effect.

Yolk–shell type nanospheres, consisting of a movable Au nanocore in the hollow cavity of a hairy electroactive polymer shell (Au@air@PTEMA-*g*-P3HT hybrid nanorattles, PTEMA = poly(2-(thiophen-3-yl)ethyl methacrylate), P3HT = poly(3-hexylthiophene)) can be readily dispersed in toluene and uniformly integrated into polystyrene (PS) thin films (Figure 1.26).[231] By controlling the nanorattle content in the composite film, the corresponding devices are capable of exhibiting unique device behaviors, including insulating behavior (with 5 wt% nanorattles), WORM (with 10 wt% nanorattles) and rewritable memory effects (with 25 wt% nanorattles), and conductive behavior (with 50 wt% nanorattles). In the structure of the Au@air@PTEMA-*g*-P3HT hybrid nanorattles, the PTEMA-*g*-P3HT shell can act as an electron donor, with the Au nanocore as an electron acceptor. The switching mechanism of the hybrid devices can probably be attributed to electric field-induced charge transfer from the P3HT shell to the Au nanocore.

Kim *et al.* investigated the electrical bistability and operating mechanism of an organic–inorganic hybrid device consisting of CdSe–ZnS core–shell nanoparticles embedded in a PVK layer.[232] The devices exhibited non-volatile electrically bistable behavior. Similar electrical behavior was also observed in memory devices involving CuInS$_2$(CIS)–ZnS core–shell type QDs blended with a PVK layer.[233] Pal *et al.* investigated the bistability of a memory device based on core–shell CdS:PB (phloxine B) hybrid nanoparticles.[234] The device based on core–shell nanoparticles exhibited a higher ON/OFF ratio and a lower switching threshold voltage than those based on the individual components.

1.6 Conclusions and Outlook

Research on novel memory materials, device structures and mechanisms is currently at a rapid growth stage, since it is recognized that organic-based electrical memory devices may be an alternative or supplemental technology to the conventional memory technologies facing the problem of miniaturization from micro- to nanoscale. By using organic/polymeric materials as storage media, flexible and miniaturized memory devices with simple structure can be fabricated with particular ease through solution processing. More importantly, the electronic properties of organic/polymeric materials, and thus the device performance, can be tailored through molecular design *cum* chemical synthesis. This chapter provided an introduction to the basic

Organic Electronic Memory Devices 43

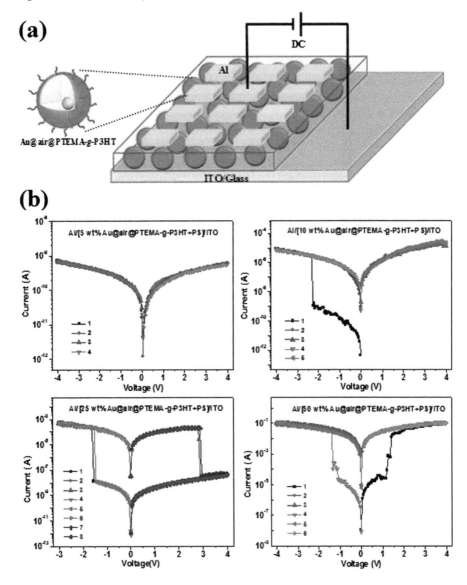

Figure 1.26 (a) A schematic diagram of the Al/Au@air@PTEMA-*g*-P3HT+PS/ITO switching device. (b) Typical *I–V* characteristics of the devices with different weight ratios of Au@air@PTEMA-*g*-P3HT nanorattles. Reproduced from ref. 231 with permission from The Royal Society of Chemistry.

concepts, history, device structures and switching effects associated with organic/polymer memory, and systematically summarized recent advances in organic/polymer memory materials and device performance.

Although significant advances have been made in the field of organic/polymer electronic memory over the past decade, it is still at the early stage of development in comparison to commercially viable inorganic devices. The immediate challenges facing organic electronic memory devices are the fabrication of organic thin film devices with reproducible switching and transport properties, and their integration into addressable memory arrays. The relationships among the materials structure, device parameters and electrical transition phenomena should be further explored to understand the solid-state physics and electronics. Refinements in materials design and preparation methods, and improvements in device fabrication, characterization and integration techniques will be needed to advance organic memory technology. In spite of the great challenges faced by memory devices, organic/polymer electronic memory proves to be one of the most promising electronic technologies of the 21st century.

References

1. R. F. Service, *Science*, 2003, **302**, 556.
2. T. Mikolajick, N. Nagel, S. Riedel, T. Mueller and K. H. Kuesters, *Mater. Sci.–Pol.*, 2007, **25**, 33.
3. R. Campano, *Nanotechnology*, 2001, **12**, 85.
4. N. Setter, D. Damjanovic, L. Eng, G. Fox, S. Gevorgian, S. Hong, A. Kingon, H. Kohlstedt, N. Y. Park, G. B. Stephenson, I. Stolitchnov, A. K. Taganstev, D. V. Taylor, T. Yamada and S. Streiffer, *J. Appl. Phys.*, 2006, **100**, 051606.
5. J. De Boeck, W. Van Roy, J. Das, V. Motsnyi, Z. Liu, L. Lagae, H. Boeve, K. Dessein and G. Borghs, *Semicond. Sci. Technol.*, 2002, **17**, 342.
6. S. Hudgens and B. Johnson, *MRS Bull.*, 2004, **29**, 829.
7. S. Moller, C. Perlov, W. Jackson, C. Taussig and S. R. Forrest, *Nature*, 2003, **426**, 166.
8. J. Y. Ouyang, C. W. Chu, C. R. Szmanda, L. P. Ma and Y. Yang, *Nat. Mater.*, 2004, **3**, 918.
9. C. W. Chu, J. Ouyang, H. H. Tseng and Y. Yang, *Adv. Mater.*, 2005, **17**, 1440.
10. F. M. Raymo, *Adv. Mater.*, 2002, **14**, 401.
11. M. A. Reed, C. Zhou, C. J. Muller, T. P. Burgin and J. M. Tour, *Science*, 1997, **278**, 252.
12. Y. Yang, J. Ouyang, L. P. Ma, R. J. H. Tseng and C. W. Chu, *Adv. Funct. Mater.*, 2006, **16**, 1001.
13. Y. Yang, L. P. Ma and J. H. Wu, *MRS Bull.*, 2004, **29**, 833.
14. R. F. Service, *Science*, 2001, **293**, 1746.
15. J. C. Scott, *Science*, 2004, **304**, 62.
16. A. Stikemann, *Technol. Rev.*, 2002, **105**, 31.
17. W.-P. Lin, S.-J. Liu, T. Gong, Q. Zhao and W. Huang, *Adv. Mater.*, 2014, **26**, 570.
18. T. Kurosawa, T. Higashihara and M. Ueda, *Polym. Chem.*, 2013, **4**, 16.
19. Y. Chen, B. Zhang, G. Liu, X. Zhuang and E.-T. Kang, *Chem. Soc. Rev.*, 2012, **41**, 4688.
20. C.-L. Liu and W.-C. Chen, *Polym. Chem.*, 2011, **2**, 2169.

21. B. Cho, S. Song, Y. Ji, T.-W. Kim and T. Lee, *Adv. Funct. Mater.*, 2011, **21**, 2806.
22. Q.-D. Ling, D.-J. Liaw, C. Zhu, D. S.-H. Chan, E.-T. Kang and K.-G. Neoh, *Prog. Polym. Sci.*, 2008, **33**, 917.
23. Y.-L. Liu, K.-L. Wang, G.-S. Huang, C.-X. Zhu, E.-S. Tok, K.-G. Neoh and E.-T. Kang, *Chem. Mater.*, 2009, **21**, 3391.
24. Q.-D. Ling, Y. Song, S.-L. Lim, E. Y.-H. Teo, Y.-P. Tan, C. Zhu, D. S. H. Chan, D.-L. Kwong, E.-T. Kang and K.-G. Neoh, *Angew. Chem., Int. Ed.*, 2006, **45**, 2947.
25. N. F. Mott, *Adv. Phys.*, 1967, **16**, 49.
26. S. R. Ovshinsk, *Phys. Rev. Lett.*, 1968, **21**, 1450.
27. L. V. Gregor, *IBM J. Res. Dev.*, 1968, **12**, 140.
28. A. Szymansk, D. C. Larson and M. M. Labes, *Appl. Phys. Lett.*, 1969, **14**, 88.
29. P. O. Sliva, G. Dir and C. Griffiths, *J. Non-Cryst. Solids*, 1970, **2**, 316.
30. Y. Segui, B. Ai and H. Carchano, *J. Appl. Phys.*, 1976, **47**, 140.
31. K. Antonowi, L. Cacha and J. Turlo, *Carbon*, 1973, **11**, 1.
32. J. Gazso, *Thin Solid Films*, 1974, **21**, 43.
33. L. F. Pender and R. J. Fleming, *J. Appl. Phys.*, 1975, **46**, 3426.
34. H. K. Henisch and W. R. Smith, *Appl. Phys. Lett.*, 1974, **24**, 589.
35. Y. Sadaoka and Y. Sakai, *J. Chem. Soc., Faraday Trans.*, 1976, **72**, 1911.
36. T. Yagi, M. Tatemoto and J. Sako, *Polym. J.*, 1980, **12**, 209–223.
37. M. Siradjuddin, V. K. Raju and P. J. Reddy, *Phys. Status Solidi A*, 1984, **81**, K37.
38. N. Yamauchi, *Jpn. J. Appl. Phys., Part 1*, 1986, **25**, 590.
39. A. Bune, S. Ducharme, V. Fridkin, L. Blinov, S. Palto, N. Petukhova and S. Yudin, *Appl. Phys. Lett.*, 1995, **67**, 3975.
40. T. J. Reece, S. Ducharme, A. V. Sorokin and M. Poulsen, *Appl. Phys. Lett.*, 2003, **82**, 142.
41. R. Weiss, *Electron. Des.*, 2001, **49**, 56.
42. S. J. Kang, Y. J. Park, I. Bae, K. J. Kim, H.-C. Kim, S. Bauer, E. L. Thomas and C. Park, *Adv. Funct. Mater.*, 2009, **19**, 2812.
43. J. Y. Son, S. Ryu, Y.-C. Park, Y.-T. Lim, Y.-S. Shin, Y.-H. Shin and H. M. Jang, *ACS Nano*, 2010, **4**, 7315.
44. G. Velu, C. Legrand, O. Tharaud, A. Chapoton, D. Remiens and G. Horowitz, *Appl. Phys. Lett.*, 2001, **79**, 659.
45. H. E. Katz, X. M. Hong, A. Dodabalapur and R. Sarpeshkar, *J. Appl. Phys.*, 2002, **91**, 1572.
46. M. Mushrush, A. Facchetti, M. Lefenfeld, H. E. Katz and T. J. Marks, *J. Am. Chem. Soc.*, 2003, **125**, 9414.
47. S. Dutta and K. S. Narayan, *Adv. Mater.*, 2004, **16**, 2151.
48. S. H. Lim, A. C. Rastogi and S. B. Desu, *J. Appl. Phys.*, 2004, **96**, 5673.
49. R. C. G. Naber, C. Tanase, P. W. M. Blom, G. H. Gelinck, A. W. Marsman, F. J. Touwslager, S. Setayesh and D. M. De Leeuw, *Nat. Mater.*, 2005, **4**, 243.
50. R. Schroeder, L. A. Majewski and M. Grell, *Adv. Mater.*, 2004, **16**, 633.

51. M. Halik, H. Klauk, U. Zschieschang, G. Schmid, C. Dehm, M. Schutz, S. Maisch, F. Effenberger, M. Brunnbauer and F. Stellacci, *Nature*, 2004, **431**, 963.
52. R. S. Potember, T. O. Poehler and D. O. Cowan, *Appl. Phys. Lett.*, 1979, **34**, 405.
53. L. P. Ma, S. Pyo, J. Ouyang, Q. F. Xu and Y. Yang, *Appl. Phys. Lett.*, 2003, **82**, 1419.
54. A. Prakash, J. Ouyang, J.-L. Lin and Y. Yang, *J. Appl. Phys.*, 2006, **100**, 054309.
55. R. J. Tseng, J. X. Huang, J. Ouyang, R. B. Kaner and Y. Yang, *Nano Lett.*, 2005, **5**, 1077.
56. B. Pradhan, S. K. Batabyal and A. J. Pal, *J. Phys. Chem. B*, 2006, **110**, 8274.
57. G. Liu, Q.-D. Ling, E. Y. H. Teo, C.-X. Zhu, D. S.-H. Chan, K.-G. Neoh and E.-T. Kang, *ACS Nano*, 2009, **3**, 1929.
58. Q.-D. Ling, S.-L. Lim, Y. Song, C.-X. Zhu, D. S.-H. Chan, E.-T. Kang and K.-G. Neoh, *Langmuir*, 2007, **23**, 312.
59. Q.-D. Ling, F.-C. Chang, Y. Song, C.-X. Zhu, D.-J. Liaw, D. S.-H. Chan, E.-T. Kang and K.-G. Neoh, *J. Am. Chem. Soc.*, 2006, **128**, 8732.
60. S.-J. Liu, Z.-H. Lin, Q. Zhao, Y. Ma, H.-F. Shi, M.-D. Yi, Q.-D. Ling, Q.-L. Fan, C.-X. Zhu, E.-T. Kang and W. Huang, *Adv. Funct. Mater.*, 2011, **21**, 979.
61. C.-J. Chen, J.-H. Wu and G.-S. Liou, *Chem. Commun.*, 2014, **50**, 4335.
62. J.-H. Wu, H.-J. Yen, Y.-C. Hu and G.-S. Liou, *Chem. Commun.*, 2014, **50**, 4915.
63. C.-J. Chen, Y.-C. Hu and G.-S. Liou, *Chem. Commun.*, 2013, **49**, 2536.
64. Y.-G. Ko, W. Kwon, H.-J. Yen, C.-W. Chang, D. M. Kim, K. Kim, S. G. Hahm, T. J. Lee, G.-S. Liou and M. Ree, *Macromolecules*, 2012, **45**, 3749.
65. Y.-K. Fang, C.-L. Liu, C. Li, C.-J. Lin, R. Mezzenga and W.-C. Chen, *Adv. Funct. Mater.*, 2010, **20**, 3012.
66. Y.-K. Fang, C.-L. Liu, G.-Y. Yang, P.-C. Chen and W.-C. Chen, *Macromolecules*, 2011, **44**, 2604.
67. X.-D. Zhuang, Y. Chen, B.-X. Li, D.-G. Ma, B. Zhang and Y. Li, *Chem. Mater.*, 2010, **22**, 4455.
68. X.-D. Zhuang, Y. Chen, G. Liu, B. Zhang, K.-G. Neoh, E.-T. Kang, C.-X. Zhu, Y.-X. Li and L.-J. Niu, *Adv. Funct. Mater.*, 2010, **20**, 2916.
69. A. Bandyopadhyay and A. J. Pal, *Appl. Phys. Lett.*, 2004, **84**, 999.
70. B. Mukherjee and A. J. Pal, *Appl. Phys. Lett.*, 2004, **85**, 2116.
71. M. Lauters, B. McCarthy, D. Sarid and G. E. Jabbour, *Appl. Phys. Lett.*, 2005, **87**, 231105.
72. S.-J. Liu, P. Wang, Q. Zhao, H.-Y. Yang, J. Wong, H.-B. Sun, X.-C. Dong, W.-P. Lin and W. Huang, *Adv. Mater.*, 2012, **24**, 2901.
73. B. Hu, X. Zhu, X. Chen, L. Pan, S. Peng, Y. Wu, J. Shang, G. Liu, Q. Yan and R.-W. Li, *J. Am. Chem. Soc.*, 2012, **134**, 17408.
74. P.-Y. Gu, F. Zhou, J. Gao, G. Li, C. Wang, Q.-F. Xu, Q. Zhang and J.-M. Lu, *J. Am. Chem. Soc.*, 2013, **135**, 14086.
75. H. Li, Q. Xu, N. Li, R. Sun, J. Ge, J. Lu, H. Gu and F. Yan, *J. Am. Chem. Soc.*, 2010, **132**, 5542.

76. S. Miao, H. Li, Q. Xu, Y. Li, S. Ji, N. Li, L. Wang, J. Zheng and J. Lu, *Adv. Mater.*, 2012, **24**, 6210.
77. L. O. Chua, *IEEE Trans. Circuit Theory*, 1971, **18**, 507.
78. L. Chua, *Appl. Phys. A: Mater. Sci. Process.*, 2011, **102**, 765.
79. A. Bandyopadhyay, S. Sahu and M. Higuchi, *J. Am. Chem. Soc.*, 2011, **133**, 1168.
80. S. M. Yoon, S. C. Warren and B. A. Grzybowski, *Angew. Chem., Int. Ed.*, 2014, **53**, 4437.
81. H. Jiang, H. Zhao, K. K. Zhang, X. Chen, C. Kloc and W. Hu, *Adv. Mater.*, 2011, **23**, 5075.
82. H.-C. Chang, C.-L. Liu and W.-C. Chen, *Adv. Funct. Mater.*, 2013, **23**, 4960.
83. Y.-C. Chiu, C.-L. Liu, W.-Y. Lee, Y. Chen, T. Kakuchi and W.-C. Chen, *NPG Asia Mater.*, 2013, **5**, e35.
84. Y.-H. Chou, Y.-C. Chiu and W.-C. Chen, *Chem. Commun.*, 2014, **50**, 3217.
85. H.-C. Chang, C.-L. Liu and W.-C. Chen, *ACS Appl. Mater. Interfaces*, 2013, **5**, 13180.
86. P.-Z. Jian, Y.-C. Chiu, H.-S. Sun, T.-Y. Chen, W.-C. Chen and S.-H. Tung, *ACS Appl. Mater. Interfaces*, 2014, **6**, 5506.
87. A.-D. Yu, W.-Y. Tung, Y.-C. Chiu, C.-C. Chueh, G.-S. Liou and W.-C. Chen, *Macromol. Rapid Commun.*, 2014, **35**, 1039.
88. A. L. Briseno, S. C. B. Mannsfeld, M. M. Ling, S. Liu, R. J. Tseng, C. Reese, M. E. Roberts, Y. Yang, F. Wudl and Z. Bao, *Nature*, 2006, **444**, 913.
89. C. R. Newman, C. D. Frisbie, D. A. da Silva, J. L. Bredas, P. C. Ewbank and K. R. Mann, *Chem. Mater.*, 2004, **16**, 4436.
90. Y. M. Sun, Y. Q. Liu and D. B. Zhu, *J. Mater. Chem.*, 2005, **15**, 53.
91. K.-J. Baeg, Y.-Y. Noh, J. Ghim, S.-J. Kang, H. Lee and D.-Y. Kim, *Adv. Mater.*, 2006, **18**, 3179.
92. J. C. Scott and L. D. Bozano, *Adv. Mater.*, 2007, **19**, 1452.
93. K. Asadi, D. M. De Leeuw, B. De Boer and P. W. M. Blom, *Nat. Mater.*, 2008, **7**, 547.
94. Z. Hu, M. Tian, B. Nysten and A. M. Jonas, *Nat. Mater.*, 2009, **8**, 62–67.
95. A. K. Tripathi, A. J. J. M. van Breemen, J. Shen, Q. Gao, M. G. Ivan, K. Reimann, E. R. Meinders and G. H. Gelinck, *Adv. Mater.*, 2011, **23**, 4146.
96. M. A. Khan, U. S. Bhansali and H. N. Alshareef, *Adv. Mater.*, 2012, **24**, 2165.
97. M. A. Khan, U. S. Bhansali, M. N. Almadhoun, I. N. Odeh, D. Cha and H. N. Alshareef, *Adv. Funct. Mater.*, 2014, **24**, 1372.
98. D. Mao, I. Mejia, A. L. Salas-Villasenor, M. Singh, H. Stiegler, B. E. Gnade and M. A. Quevedo-Lopez, *Org. Electron.*, 2013, **14**, 505.
99. J. T. Evans and R. Womack, *IEEE J. Solid-State Circuits*, 1988, **23**, 1171.
100. S. M. Yoon and H. Ishiwara, *IEEE Trans. Electron Devices*, 2001, **48**, 2002.
101. H. Kohlstedt, Y. Mustafa, A. Gerber, A. Petraru, M. Fitsilis, R. Meyer, U. Bottger and R. Waser, *Microelectron. Eng.*, 2005, **80**, 296.
102. W. L. Kwan, B. Lei, Y. Shao and Y. Yang, *Curr. Appl. Phys.*, 2010, **10**, E50.
103. W. Hwang and K. C. Kao, *J. Chem. Phys.*, 1973, **58**, 3521.

104. A. J. Campbell, D. D. C. Bradley and D. G. Lidzey, *J. Appl. Phys.*, 1997, **82**, 6326.
105. H. S. Majumdar, A. Bandyopadhyay and A. J. Pal, *Org. Electron.*, 2003, **4**, 39.
106. L. Li, Q.-D. Ling, S.-L. Lim, Y.-P. Tan, C. Zhu, D. S. H. Chan, E.-T. Kang and K.-G. Neoh, *Org. Electron.*, 2007, **8**, 401.
107. L. D. Bozano, B. W. Kean, M. Beinhoff, K. R. Carter, P. M. Rice and J. C. Scott, *Adv. Funct. Mater.*, 2005, **15**, 1933.
108. T. Ouisse and O. Stephan, *Org. Electron.*, 2004, **5**, 251.
109. D. M. Taylor, *IEEE Trans. Dielectr. Electr. Insul.*, 2006, **13**, 1063.
110. A. Dei, D. Gatteschi, C. Sangregorio and L. Sorace, *Acc. Chem. Res.*, 2004, **37**, 827.
111. J. B. Torrance, *Acc. Chem. Res.*, 1979, **12**, 79.
112. Q. D. Ling, Y. Song, S. J. Ding, C. X. Zhu, D. S. H. Chan, D. L. Kwong, E. T. Kang and K. G. Neoh, *Adv. Mater.*, 2005, **17**, 455.
113. X.-D. Zhuang, Y. Chen, G. Liu, P.-P. Li, C.-X. Zhu, E.-T. Kang, K.-G. Neoh, B. Zhang, J.-H. Zhu and Y.-X. Li, *Adv. Mater.*, 2010, **22**, 1731.
114. B. Zhang, G. Liu, Y. Chen, L.-J. Zeng, C.-X. Zhu, K.-G. Neoh, C. Wang and E.-T. Kang, *Chem.–Eur. J.*, 2011, **17**, 13646.
115. B. Zhang, Y.-L. Liu, Y. Chen, K.-G. Neoh, Y.-X. Li, C.-X. Zhu, E.-S. Tok and E.-T. Kang, *Chem.–Eur. J.*, 2011, **17**, 10304–10311.
116. D. He, H. Zhuang, H. Liu, H. Liu, H. Li and J. Lu, *J. Mater. Chem. C*, 2013, **1**, 7883.
117. S. L. Lim, Q. Ling, E. Y. H. Teo, C. X. Zhu, D. S. H. Chan, E.-T. Kang and K. G. Neoh, *Chem. Mater.*, 2007, **19**, 5148.
118. D. Ma, M. Aguiar, J. A. Freire and I. A. Hummelgen, *Adv. Mater.*, 2000, **12**, 1063.
119. I. Cacelli, A. Ferretti, M. Girlanda and M. Macucci, *Chem. Phys.*, 2007, **333**, 26.
120. I. Cacelli, A. Feretti, M. Girlanda and M. Macucci, *Chem. Phys.*, 2006, **320**, 84.
121. W. Y. Wang, T. Lee and M. A. Reed, *Phys. Rev. B*, 2003, **68**, 035416.
122. C. A. Mills, D. M. Taylor, A. Riul and A. P. Lee, *J. Appl. Phys.*, 2002, **91**, 5182.
123. A. R. Elsharkawi and K. C. Kao, *J. Phys. Chem. Solids*, 1977, **38**, 95.
124. H. Kasica, W. Wlodarski, H. Kurczewska and A. Szymanski, *Thin Solid Films*, 1975, **30**, 325–333.
125. D. Tondelier, K. Lmimouni, D. Vuillaume, C. Fery and G. Haas, *Appl. Phys. Lett.*, 2004, **85**, 5763.
126. J. Swiatek, *Thin Solid Films*, 1979, **61**, 321.
127. J. S. Chen and D. G. Ma, *Appl. Phys. Lett.*, 2005, **87**, 023505.
128. J. Chen, L. Xu, J. Lin, Y. Geng, L. Wang and D. Ma, *Appl. Phys. Lett.*, 2006, **89**, 083514.
129. T. Erlbacher, M. P. M. Jank, H. Ryssel, L. Frey, R. Engl, A. Walter, R. Sezi and C. Dehm, *J. Electrochem. Soc.*, 2008, **155**, H693.
130. Q. Zhang, L. Z. Kong, Q. J. Zhang, W. J. Wang and Z. Y. Hua, *Solid State Commun.*, 2004, **130**, 799.

131. Z. Q. Xue, M. Ouyang, K. Z. Wang, H. X. Zhang and C. H. Huang, *Thin Solid Films*, 1996, **288**, 296–299.
132. Z. Y. Hua, W. Xu, G. R. Chen, X. J. Yan, X. L. Mo and Q. Zhang, *Synth. Met.*, 2003, **137**, 1531.
133. H. J. Gao, K. Sohlberg, Z. Q. Xue, H. Y. Chen, S. M. Hou, L. P. Ma, X. W. Fang, S. J. Pang and S. J. Pennycook, *Phys. Rev. Lett.*, 2000, **84**, 1780.
134. J. C. Li, Z. Q. Xue, W. M. Liu, S. M. Hou, X. L. Li and X. Y. Zhao, *Phys. Lett. A*, 2000, **266**, 441.
135. Y. Iwasa, T. Koda, Y. Tokura, S. Koshihara, N. Iwasawa and G. Saito, *Appl. Phys. Lett.*, 1989, **55**, 2111.
136. B. Mukherjee and A. J. Pal, *Chem. Mater.*, 2007, **19**, 1382.
137. J. Lin, M. Zheng, J. Chen, X. Gao and D. Ma, *Inorg. Chem.*, 2007, **46**, 341.
138. B. Mukherjee, A. K. Ray, A. K. Sharma, M. J. Cook and I. Chambrier, *J. Appl. Phys.*, 2008, **103**, 074507.
139. C.-H. Tu, Y.-S. Lai and D.-L. Kwong, *Appl. Phys. Lett.*, 2006, **89**, 062105.
140. L. P. Ma, Q. F. Xu and Y. Yang, *Appl. Phys. Lett.*, 2004, **84**, 4908.
141. A. Bandyopadhyay and A. J. Pal, *Appl. Phys. Lett.*, 2003, **82**, 1215.
142. J. R. Koo, H. S. Lee, Y. Ha, Y. H. Choi and Y. K. Kim, *Thin Solid Films*, 2003, **438**, 123.
143. L. P. Ma, J. Liu and Y. Yang, *Appl. Phys. Lett.*, 2002, **80**, 2997.
144. L. P. Ma, J. Liu, S. M. Pyo and Y. Yang, *Appl. Phys. Lett.*, 2002, **80**, 362.
145. Y. Shang, Y. Wen, S. Li, S. Du, X. He, L. Cai, Y. Li, L. Yang, H. Gao and Y. Song, *J. Am. Chem. Soc.*, 2007, **129**, 11674.
146. Y. Ma, X. Cao, G. Li, Y. Wen, Y. Yang, J. Wang, S. Du, L. Yang, H. Gao and Y. Song, *Adv. Funct. Mater.*, 2010, **20**, 803.
147. H. Li, Z. Jin, N. Li, Q. Xu, H. Gu, J. Lu, X. Xia and L. Wang, *J. Mater. Chem.*, 2011, **21**, 5860.
148. G. Wang, S. Miao, Q. Zhang, H. Liu, H. Li, N. Li, Q. Xu, J. Lu and L. Wang, *Chem. Commun.*, 2013, **49**, 9470.
149. S. Miao, H. Li, Q. Xu, N. Li, J. Zheng, R. Sun, J. Lu and C. M. Li, *J. Mater. Chem.*, 2012, **22**, 16582.
150. Y. Zhang, H. Zhuang, Y. Yang, X. Xu, Q. Bao, N. Li, H. Li, Q. Xu, J. Lu and L. Wang, *J. Phys. Chem. C*, 2012, **116**, 22832.
151. S. Miao, Y. Zhu, H. Zhuang, X. Xu, H. Li, R. Sun, N. Li, S. Ji and J. Lu, *J. Mater. Chem.*, 2013, **1**, 2320.
152. H. Zhuang, Q. Zhang, Y. Zhu, X. Xu, H. Liu, N. Li, Q. Xu, H. Li, J. Lu and L. Wang, *J. Mater. Chem. C*, 2013, **1**, 3816.
153. C. Ye, Q. Peng, M. Li, J. Luo, Z. Tang, J. Pei, J. Chen, Z. Shuai, L. Jiang and Y. Song, *J. Am. Chem. Soc.*, 2012, **134**, 20053.
154. H. Wang, F. Meng, Y. Cai, L. Zheng, Y. Li, Y. Liu, Y. Jiang, X. Wang and X. Chen, *Adv. Mater.*, 2013, **25**, 5498.
155. M. K. Hota, M. K. Bera, B. Kundu, S. C. Kundu and C. K. Maiti, *Adv. Funct. Mater.*, 2012, **22**, 4493.
156. F. Meng, L. Jiang, K. Zheng, C. F. Goh, S. Lim, H. H. Hng, J. Ma, F. Boey and X. Chen, *Small*, 2011, **7**, 3016.

157. D. M. Kim, S. Park, T. J. Lee, S. G. Hahm, K. Kim, J. C. Kim, W. Kwon and M. Ree, *Langmuir*, 2009, **25**, 11713.
158. S. G. Hahm, S. Choi, S.-H. Hong, T. J. Lee, S. Park, D. M. Kim, J. C. Kim, W. Kwon, K. Kim, M.-J. Kim, O. Kim and M. Ree, *J. Mater. Chem.*, 2009, **19**, 2207.
159. N.-H. You, C.-C. Chueh, C.-L. Liu, M. Ueda and W.-C. Chen, *Macromolecules*, 2009, **42**, 4456.
160. Y. Li, H. Xu, X. Tao, K. Qian, S. Fu, Y. Shen and S. Ding, *J. Mater. Chem.*, 2011, **21**, 1810.
161. Y.-Q. Li, R.-C. Fang, A.-M. Zheng, Y.-Y. Chu, X. Tao, H.-H. Xu, S.-J. Ding and Y.-Z. Shen, *J. Mater. Chem.*, 2011, **21**, 15643.
162. S. Park, K. Kim, D. M. Kim, W. Kwon, J. Choi and M. Ree, *ACS Appl. Mater. Interfaces*, 2011, **3**, 765.
163. G. Tian, D. Wu, S. Qi, Z. Wu and X. Wang, *Macromol. Rapid Commun.*, 2011, **32**, 384.
164. Y. Liu, Y. Zhang, Q. Lan, S. Liu, Z. Qin, L. Chen, C. Zhao, Z. Chi, J. Xu and J. Economy, *Chem. Mater.*, 2012, **24**, 1212.
165. C.-J. Chen, H.-J. Yen, Y.-C. Hu and G.-S. Liou, *J. Mater. Chem. C*, 2013, **1**, 7623.
166. H.-J. Yen, C.-J. Chen and G.-S. Liou, *Adv. Funct. Mater.*, 2013, **23**, 5307.
167. S. G. Hahm, S. Choi, S.-H. Hong, T. J. Lee, S. Park, D. M. Kim, W.-S. Kwon, K. Kim, O. Kim and M. Ree, *Adv. Funct. Mater.*, 2008, **18**, 3276.
168. Y.-K. Fang, C.-L. Liu and W.-C. Chen, *J. Mater. Chem.*, 2011, **21**, 4778.
169. N.-G. Kang, B. Cho, B.-G. Kang, S. Song, T. Lee and J.-S. Lee, *Adv. Mater.*, 2012, **24**, 385.
170. G. Wen, Z. Ren, D. Sun, T. Zhang, L. Liu and S. Yan, *Adv. Funct. Mater.*, 2014, **24**, 3446.
171. B. Ahn, D. M. Kim, J.-C. Hsu, Y.-G. Ko, T. J. Shin, J. Kim, W.-C. Chen and M. Ree, *ACS Macro Lett.*, 2013, **2**, 555.
172. F.-L. Ye, C.-J. Lu, H.-X. Chen, Y.-H. Zhang, N.-J. Li, L.-H. Wang, H. Li, Q.-F. Xu and J.-M. Lu, *Polym. Chem.*, 2014, **5**, 752.
173. S. L. Lim, N.-J. Li, J.-M. Lu, Q.-D. Ling, C. X. Zhu, E.-T. Kang and K. G. Neoh, *ACS Appl. Mater. Interfaces*, 2009, **1**, 60.
174. N. Fan, H. Liu, Q. Zhou, H. Zhuang, Y. Li, H. Li, Q. Xu, N. Li and J. Lu, *J. Mater. Chem.*, 2012, **22**, 19957.
175. G. Liu, B. Zhang, Y. Chen, C.-X. Zhu, L. Zeng, D. S.-H. Chan, K.-G. Neoh, J. Chen and E.-T. Kang, *J. Mater. Chem.*, 2011, **21**, 6027.
176. B. Zhang, G. Liu, Y. Chen, C. Wang, K.-G. Neoh, T. Bai and E.-T. Kang, *ChemPlusChem*, 2012, **77**, 74.
177. H. Zhuang, X. Xu, Y. Liu, Q. Zhou, X. Xu, H. Li, Q. Xu, N. Li, J. Lu and L. Wang, *J. Phys. Chem. C*, 2012, **116**, 25546.
178. A. Ajayaghosh, *Chem. Soc. Rev.*, 2003, **32**, 181.
179. W. Zhang, C. Wang, G. Liu, J. Wang, Y. Chen and R.-W. Li, *Chem. Commun.*, 2014, **50**, 11496.
180. W. Zhang, C. Wang, G. Liu, X. Zhu, X. Chen, L. Pan, H. Tan, W. Xue, Z. Ji, J. Wang, Y. Chen and R.-W. Li, *Chem. Commun.*, 2014, **50**, 11856.

181. Q.-D. Ling, E.-T. Kang, K.-G. Neoh, Y. Chen, X.-D. Zhuang, C. Zhu and D. S. H. Chan, *Appl. Phys. Lett.*, 2008, **92**, 143302.
182. P. Wang, S.-J. Liu, Z.-H. Lin, X.-C. Dong, Q. Zhao, W.-P. Lin, M.-D. Yi, S.-H. Ye, C.-X. Zhu and W. Huang, *J. Mater. Chem.*, 2012, **22**, 9576.
183. H.-C. Wu, A.-D. Yu, W.-Y. Lee, C.-L. Liu and W.-C. Chen, *Chem. Commun.*, 2012, **48**, 9135.
184. L.-J. Zeng, G. Liu, B. Zhang, J. Chen, Y. Chen and E.-T. Kang, *Polym. J.*, 2012, **44**, 257.
185. H.-C. Wu, C.-L. Liu and W.-C. Chen, *Polym. Chem.*, 2013, **4**, 5261.
186. H.-J. Yen, H. Tsai, C.-Y. Kuo, W. Nie, A. D. Mohite, G. Gupta, J. Wang, J.-H. Wu, G.-S. Liou and H.-L. Wang, *J. Mater. Chem. C*, 2014, **2**, 4374.
187. B. Cho, J.-M. Yun, S. Song, Y. Ji, D.-Y. Kim and T. Lee, *Adv. Funct. Mater.*, 2011, **21**, 3976.
188. J. M. Son, W. S. Song, C. H. Yoo, D. Y. Yun and T. W. Kim, *Appl. Phys. Lett.*, 2012, **100**, 183303.
189. U. S. Bhansali, M. A. Khan, D. Cha, M. N. AlMadhoun, R. Li, L. Chen, A. Amassian, I. N. Odeh and H. N. Alshareef, *ACS Nano*, 2013, **7**, 10518–10524.
190. R. Kumar, R. G. Pillai, N. Pekas, Y. Wu and R. L. McCreery, *J. Am. Chem. Soc.*, 2012, **134**, 14869.
191. S. G. Hahm, N.-G. Kang, W. Kwon, K. Kim, Y.-G. Ko, S. Ahn, B.-G. Kang, T. Chang, J.-S. Lee and M. Ree, *Adv. Mater.*, 2012, **24**, 1062.
192. G. Liu, X. Zhuang, Y. Chen, B. Zhang, J. Zhu, C.-X. Zhu, K.-G. Neoh and E.-T. Kang, *Appl. Phys. Lett.*, 2009, **95**, 253301.
193. B. Zhang, Y. Chen, L. Xu, L. Zeng, Y. He, E.-T. Kang and J. Zhang, *J. Polym. Sci., Part A: Polym. Chem.*, 2011, **49**, 2043.
194. B. Zhang, Y. Chen, G. Liu, L.-Q. Xu, J. Chen, C.-X. Zhu, K.-G. Neoh and E.-T. Kang, *J. Polym. Sci., Part A: Polym. Chem.*, 2012, **50**, 378.
195. B. Zhang, Y. Chen, Y. Ren, L.-Q. Xu, G. Liu, E.-T. Kang, C. Wang, C.-X. Zhu and K.-G. Neoh, *Chem.–Eur. J.*, 2013, **19**, 6265.
196. S.-J. Liu, Y. Chen, W.-J. Xu, Q. Zhao and W. Huang, *Macromol. Rapid Commun.*, 2012, **33**, 461.
197. T.-L. Choi, K.-H. Lee, W.-J. Joo, S. Lee, T.-W. Lee and M. Y. Chae, *J. Am. Chem. Soc.*, 2007, **129**, 9842.
198. Q.-D. Ling, W. Wang, Y. Song, C.-X. Zhu, D. S.-H. Chan, E.-T. Kang and K.-G. Neoh, *J. Phys. Chem. B*, 2006, **110**, 23995.
199. S.-J. Liu, W.-P. Lin, M.-D. Yi, W.-J. Xu, C. Tang, Q. Zhao, S.-H. Ye, X.-M. Liu and W. Huang, *J. Mater. Chem.*, 2012, **22**, 22964.
200. J. Liu, Z. Yin, X. Cao, F. Zhao, A. Lin, L. Xie, Q. Fan, F. Boey, H. Zhang and W. Huang, *ACS Nano*, 2010, **4**, 3987.
201. J.-C. Hsu, C.-L. Liu, W.-C. Chen, K. Sugiyama and A. Hirao, *Macromol. Rapid Commun.*, 2011, **32**, 528.
202. S.-L. Lian, C.-L. Liu and W.-C. Chen, *ACS Appl. Mater. Interfaces*, 2011, **3**, 4504.
203. S. Gao, C. Song, C. Chen, F. Zeng and F. Pan, *J. Phys. Chem. C*, 2012, **116**, 17955.

204. C.-J. Chen, Y.-C. Hu and G.-S. Liou, *Chem. Commun.*, 2013, **49**, 2804.
205. H. S. Majumdar, J. K. Baral, R. Osterbacka, O. Ikkala and H. Stubb, *Org. Electron.*, 2005, **6**, 188.
206. S. Paul, A. Kanwal and M. Chhowalla, *Nanotechnology*, 2006, **17**, 145.
207. H. Jo, J. Ko, J. A. Lim, H. J. Chang and Y. S. Kim, *Macromol. Rapid Commun.*, 2013, **34**, 355.
208. S. Qi, H. Iida, L. Liu, S. Irle, W. Hu and E. Yashima, *Angew. Chem., Int. Ed.*, 2013, **52**, 1049.
209. Y. Ji, B. Cho, S. Song, T.-W. Kim, M. Choe, Y. H. Kahng and T. Lee, *Adv. Mater.*, 2010, **22**, 3071.
210. S. Song, B. Cho, T.-W. Kim, Y. Ji, M. Jo, G. Wang, M. Choe, Y. H. Kahng, H. Hwang and T. Lee, *Adv. Mater.*, 2010, **22**, 5048.
211. T.-W. Kim, D. F. Zeigler, O. Acton, H.-L. Yip, H. Ma and A. K. Y. Jen, *Adv. Mater.*, 2012, **24**, 828.
212. Y. Ji, D. F. Zeigler, D. S. Lee, H. Choi, A. K. Y. Jen, H. C. Ko and T.-W. Kim, *Nat. Commun.*, 2013, **4**, 2707.
213. S. K. Hwang, J. R. Choi, I. Bae, I. Hwang, S. M. Cho, J. Huh and C. Park, *Small*, 2013, **9**, 831.
214. S. K. Hwang, J. M. Lee, S. Kim, J. S. Park, H. I. Park, C. W. Ahn, K. J. Lee, T. Lee and S. O. Kim, *Nano Lett.*, 2012, **12**, 2217.
215. S. Wang, K. K. Manga, M. Zhao, Q. Bao and K. P. Loh, *Small*, 2011, **7**, 2372.
216. C. Wu, F. Li, T. Guo and T. W. Kim, *Org. Electron.*, 2012, **13**, 178.
217. A.-D. Yu, C.-L. Liu and W.-C. Chen, *Chem. Commun.*, 2012, **48**, 383.
218. Q. Zhang, J. Pan, X. Yi, L. Li and S. Shang, *Org. Electron.*, 2012, **13**, 1289.
219. D. I. Son, T. W. Kim, J. H. Shim, J. H. Jung, D. U. Lee, J. M. Lee, W. II Park and W. K. Choi, *Nano Lett.*, 2010, **10**, 2441.
220. C. Wu, F. Li, Y. Zhang, T. Guo and T. Chen, *Appl. Phys. Lett.*, 2011, **99**, 042108.
221. G. L. Li, G. Liu, M. Li, D. Wan, K. G. Neoh and E. T. Kang, *J. Phys. Chem. C*, 2010, **114**, 12742.
222. J. Ouyang and Y. Yang, *Appl. Phys. Lett.*, 2010, **96**, 063506.
223. D. I. Son, C. H. You, J. H. Jung and T. W. Kim, *Appl. Phys. Lett.*, 2010, **97**, 013304.
224. C. O. Baker, B. Shedd, R. J. Tseng, A. A. Martinez-Morales, C. S. Ozkan, M. Ozkan, Y. Yang and R. B. Kanert, *ACS Nano*, 2011, **5**, 3469.
225. D. I. Son, D. H. Oh, J. H. Jung and T. W. Kim, *J. Nanosci. Nanotechnol.*, 2011, **11**, 711.
226. D. I. Son, D. H. Park, J. Bin Kim, J.-W. Choi, T. W. Kim, B. Angadi, Y. Yi and W. K. Choi, *J. Phys. Chem. C*, 2011, **115**, 2341.
227. S. I. White, P. M. Vora, J. M. Kikkawa and K. I. Winey, *Adv. Funct. Mater.*, 2011, **21**, 233.
228. B. Cho, T.-W. Kim, M. Choe, G. Wang, S. Song and T. Lee, *Org. Electron.*, 2009, **10**, 473.
229. C.-L. Tsai, C.-J. Chen, P.-H. Wang, J.-J. Lin and G.-S. Liou, *Polym. Chem.*, 2013, **4**, 4570.

230. J. Liu, Z. Zeng, X. Cao, G. Lu, L.-H. Wang, Q.-L. Fan, W. Huang and H. Zhang, *Small*, 2012, **8**, 3517.
231. T. Cai, B. Zhang, Y. Chen, C. Wang, C. X. Zhu, K.-G. Neoh and E.-T. Kang, *Chem.–Eur. J.*, 2014, **20**, 2723.
232. F. Li, D.-I. Son, S.-M. Seo, H.-M. Cha, H.-J. Kim, B.-J. Kim, J. H. Jung and T. W. Kim, *Appl. Phys. Lett.*, 2007, **91**, 122111.
233. J. H. Shim, J. H. Jung, M. H. Lee, T. W. Kim, D. I. Son, A. N. Han and S. W. Kim, *Org. Electron.*, 2011, **12**, 1566.
234. B. C. Das and A. J. Pal, *ACS Nano*, 2008, **2**, 1930.

CHAPTER 2

Organic Resistor Memory Devices

TADANORI KUROSAWA[a], TOMOYA HIGASHIHARA[b], AND MITSURU UEDA*[c]

[a]Department of Chemical Engineering, Stanford University, 443 Via Ortega, Stanford, CA 94305-4125, The United States of America; [b]Graduate School of Science and Engineering, Yamagata University, 4-3-16 Jonan, Yonezawa, Yamagata 992-8510, Japan; [c]Department of Chemistry, Kanagawa University, 3-27-1 Rokkakubashi, Yokohama, Kanagawa 221-8686, Japan
*E-mail: mueda@kanagawa-u.ac.jp

2.1 Device Structures of Organic Resistor Memory Devices

In order to store data, two (or more) distinct electronic states, which are assigned as "0" and "1" or "ON" and "OFF", respectively, are required to be generated in a memory device. Figure 2.1 describes the typical internal structure of a current semiconducting dynamic random access memory (DRAM) device. Inside the memory device, the memory element is located at each intersection of grid-shaped electrical wires which are known as word lines and bit lines. Since this memory element can create two distinct electronic states, the memory device can store data by recognizing the patterns of "0s" and "1s". Furthermore, the memory element is composed of a fine electronic circuit, including a complementary metal oxide semiconductor (CMOS) transistor and capacitor (C). In this electronic circuit, "0" and "1" corresponds to the "discharged"

RSC Polymer Chemistry Series No. 18
Electrical Memory Materials and Devices
Edited by Wen-Chang Chen
© The Royal Society of Chemistry 2016
Published by the Royal Society of Chemistry, www.rsc.org

and "charged" states of the C, respectively. A variety of these semiconducting memory devices can be accessed by tailoring the electronic circuit. Over the past few decades, the capacity of semiconducting memory devices has drastically increased, while their size has dramatically decreased. The combination of increased capacity and reduced size has been achieved by scaling down the size of the electronic circuit of the memory element using conventional lithography technologies. However, while the remarkable growth in information society has enabled us to deal with enormous amounts of data, such techniques are facing several issues such as physical limitations on the resolution of lithography patterns, high processing costs, *etc.*, which will need to be addressed in order to meet the requirements for large data storage. Therefore, studies of alternatives for these current devices are a significant topic.

In comparison to the current semiconducting memory devices, an organic resistor memory device stores data in a completely different fashion. While the current memory devices store data based on the presence of charge, an organic resistor memory device assigns the "0" and "1" or "ON" and "OFF" to the "low conductivity state" and "high conductivity state", respectively.[1-5] This leads to the fact that the organic material itself can be the memory element, which in current semiconducting memory devices generally consists of a minute and complicated electric circuit. Therefore, the device structure can be simplified and the device can be fabricated in several simple steps. Figure 2.2 shows the basic fabrication process for an organic resistor memory device in three steps. First, the bottom electrode is placed on the substrate. Secondly, the active organic layer is deposited. Finally, the top electrode is placed on the active layer.

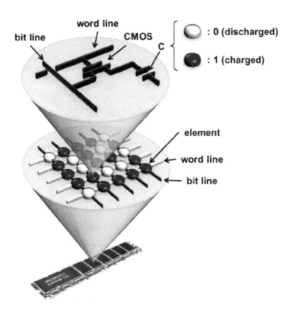

Figure 2.1   Typical structure of DRAM device.

2.2 Switching Characteristics of Organic Resistor Memory Devices

As shown in Figure 2.3, resistor memory can be classified into two types, *i.e.*, volatile memory and non-volatile memory.[2] Volatile memory is a type of memory which cannot hold stored data without an external electronic power supply, that is, written data will disappear after turning off the power. On the other hand, non-volatile memory can retain stored data without a power supply, that is, written data will not disappear unless or even after an additional procedure is carried out. For instance, dynamic random access memory (DRAM) and static random access memory (SRAM), both used in current computers, are types of volatile memory. On the other hand, write once read many (WORM) type memory represented by the compact disc (CD), digital versatile disc (DVD), and Blue-ray Disc is non-volatile. FLASH-type memory, such as the universal serial bus (USB) flash drive, secure digital (SD) memory card, and solid state drive (SSD) are also non-volatile.

In this section, the switching characteristics of organic resistor memory devices together with their typical current–voltage (I–V) or current density–voltage (J–V) curves will be introduced. These characteristics can generally be assigned to various existing current memory types. In addition, other important features, such as stability testing, the effect of read cycles on the ON and OFF states, *etc.*, will be explained where appropriate.

2.2.1 DRAM Characteristics

In 2011, Chen and Ueda's group reported a block copolymer composed of poly(3-hexylthiophene) (P3HT) and poly(3-phenoxymethylthiophene) (P3PT) in which a certain block ratio (P3HT$_{102}$-*b*-P3PT$_{37}$) resulted in bistable

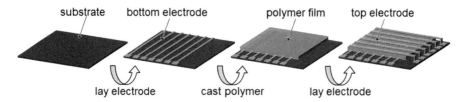

Figure 2.2 A schematic diagram of the fabrication process of an organic memory device. Reprinted with permission from ref. 3. Copyright 2013, The Royal Society of Chemistry.

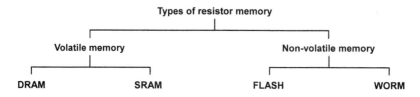

Figure 2.3 Classification of resistor memory devices.

electrical switching properties (Figure 2.4).[6] The memory properties were evaluated using the *I–V* curve measured by a semiconductor analyzer of an indium tin oxide (ITO)/polymer/Al sandwich device, which is a typical structure for polymer memory devices (inset of Figure 2.4a). In the 1st sweep from 0 to −8 V, the current density is low in the range of 10^{-14}–10^{-12} A until the voltage reaches −2.8 V. However, a sharp increase is observed at around −3 V after which the current density is over 10^{-6} A cm^{-2}, indicating a switching behavior from the OFF state to the ON state. The electric transition from the OFF state to the ON state corresponds to the "writing" process of the current memory device. During the subsequent sweep from 0 to −8 V, the ON state can be maintained at any voltage value, which means that the distinct conductivity states (OFF and ON) can be read at low voltage values (*e.g.*, below −2 V). However, by turning off the power for 1 min, the switched ON state returns to the initial OFF state without any erasing process (see the 3rd sweep), which means that the device cannot sustain the ON state without the power supply and behaves as a volatile memory device. The volatile ON state can be sustained by a refreshing voltage pulse of −1 V every 10 s (plot 5 in Figure 2.4b). Based on the described writing, reading, rewriting, and refreshing abilities, the memory property of P3HT$_{102}$-*b*-P3PT$_{37}$ is determined to be volatile DRAM.

Stability and read cycle tests are both essential to evaluate the performance of a device. As shown in Figure 2.5, no degradation in current density is observed for either the ON or OFF states under a constant stress of −1 V, indicating the high stability of the device based on P3HT$_{102}$-*b*-P3PT$_{37}$. In addition, no resistance degradation was observed for both states after more than 10^8 read cycles indicating that the device is insensitive to the read pulses.[6]

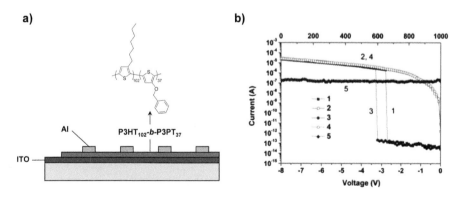

Figure 2.4 (a) Chemical structure of the conjugated block copolymer P3HT$_{102}$-*b*-P3PT$_{37}$. The inset shows a schematic of the memory device. (b) Typical *I–V* characteristics of the ITO/P3HT$_{102}$-*b*-P3PT$_{37}$/Al device in the ON and OFF states whilst maintaining the ON state by refreshing at −1 V every 10 s. Adapted with permission from ref. 6. Copyright 2011, The Royal Society of Chemistry.

2.2.2 SRAM Characteristics

Kang et al. reported the memory properties using a functional polyimide called P(BPPO)-PI (Figure 2.6).[7] The memory properties characterized from the current density–voltage (J–V) curve (Figure 2.6b) of an ITO/P(BPPO)-PI/Al sandwich device are described below. During the 1st positive sweep from 0 to 4 V, a switching behavior from the OFF state to the ON state, corresponding to a "writing" process, is observed at around 2.3 V, and the induced

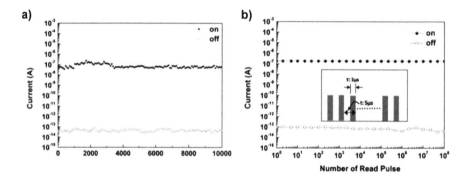

Figure 2.5 (a) Stability of the device based on P3HT$_{102}$-b-P3PT$_{37}$ in either ON or OFF state under constant stress (−1 V) and the ON/OFF current ratio for the same sweep. (b) Effect of read cycles on the ON and OFF states. The insert shows the pulses used for these measurements. Adapted with permission from ref. 6. Copyright 2011, The Royal Society of Chemistry.

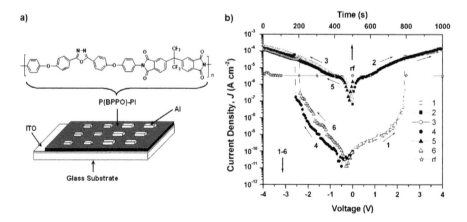

Figure 2.6 (a) Chemical structure of P(BPPO)-PI and schematic diagram of the memory device. (b) J–V characteristics of the device based on P(BPPO)-PI. The sequence and direction of each sweep are indicated by the respective number and arrow. The fourth and sixth sweeps were conducted about 4 min after turning off the power. The ON state was sustained by a refreshing voltage pulse (−1 V) every 5 s, as shown by the "rf" trace. Adapted with permission from ref. 7. Copyright 2009, American Chemical Society.

ON state can be maintained during the subsequent 2nd sweep. Different to PFOxPy, the ON state cannot be reset to the initial OFF state by sweeping the voltage in the reverse direction (3rd sweep), which means that P(BPPO)-PI is not erasable. Another difference between PFOxPy and P(BPPO)-PI is that the ON state of P(BPPO)-PI can be maintained for about 4 min after turning off the power supply. However, the ON state eventually returns to the initial OFF state, which means that the ON state is "remanent" yet volatile. After the ON state has relaxed to the initial OFF state, it can be reswitched to the ON state by reapplying a negative voltage (the 4th sweep), and can remain in the ON state during the subsequent negative voltage sweep (5th sweep). Also, the ON state can be sustained by a refreshing voltage pulse of −1 V of 1 ms duration every 5 s (the rf plots in Figure 2.6b). Based on these results (the rewritable, refreshable, and also the "remanent" yet volatile nature), P(BPPO)-PI can be considered as SRAM, which realizes low power consumption in real applications.

2.2.3 FLASH Characteristics

Poly(N-vinylcarbazole) with fullerene (C_{60}) covalently attached to its backbone (PVK-C_{60}) has been reported for application as non-volatile memory (Figure 2.7a).[8] The J–V curve of the device with the configuration ITO/PVK-C_{60}/Al is shown in Figure 2.7b. An abrupt increase in the current around −3 V is observed during the 2nd voltage sweep corresponding to the "writing" process. The device remains in a high conductivity (ON) state during the subsequent 4th sweep and also after turning off the power, indicating non-volatile behavior. As the voltage is swept in the opposite direction, a sudden decrease in the current around 3.2 V, corresponding to the "erasing" process, is observed (5th sweep) and the device remains in this low conductivity (OFF) state during the subsequent 6th sweep. The ON and OFF states can be reproduced by reapplying the switching threshold voltages (8th to 10th sweeps). By combining the switching phenomena of the device and its volatility, the device based on PVK-C_{60} can be used as non-volatile FLASH-type memory.

Although the I–V (or J–V) curves of DRAM and FLASH memory look identical to each other, there is a big difference between these two which is based on the fact that the FLASH type requires a turn-off voltage to turn the device OFF while DRAM only needs a power supply to stop. Therefore, a write–read–erase–read (WRER) cycle test is performed to evaluate the stability of a FLASH-type memory device. In Figure 2.8, a typical WRER cycle test of a polyimide, namely APTT-6FDA, which also exhibits FLASH memory characteristics, is shown.[9] As can be seen, no degradation in either the ON or OFF state (Figure 2.8b) is observed which indicates the high stability of the device over the WRER cycle.

2.2.4 WORM Characteristics

In 2010, Kang and co-workers reported a functional polyimide (OXTA-PIa) for non-volatile memory applications (Figure 2.9a).[10] The memory effect of the ITO/OXTA-PIa/Al device demonstrated by the J–V curve (Figure 2.9b)

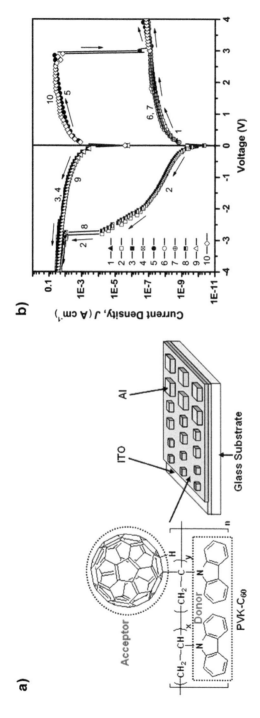

Figure 2.7 (a) Chemical structure of PVK-C_{60} and schematic diagram of its polymer memory device. (b) J–V characteristics of the device. The sweep sequence and direction are indicated by the numbers and arrows, respectively. The 4th and 7th sweeps were conducted after the power had been turned off. Sweep 8–10 were obtained after 20 write–read–erase–read switching cycles. Adapted with permission from ref. 8. Copyright 2007, American Chemical Society.

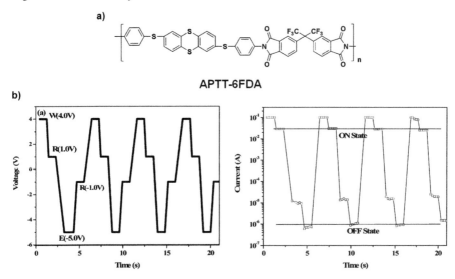

Figure 2.8 (a) Structure of APTT-6FDA. (b) Applied voltage sequence (left) and typical current response to WRER cycles (right). Adapted with permission from ref. 9. Copyright 2009, American Chemical Society.

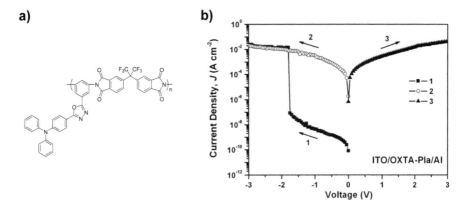

Figure 2.9 (a) Chemical structure of OXTA-PIa. (b) J–V characteristics of a 0.4 × 0.4 mm² ITO/AZTA-PIa/Al device with an initial negative electrical sweep. The sequence and direction of each sweep are indicated by the respective number and arrow. Adapted with permission from ref. 10. Copyright 2010, American Chemical Society.

is described below. During the initial voltage sweep, a "writing" process is observed around −1.8 V. The ON state can be maintained during the subsequent 2nd sweep and also after turning off the electrical power supply, which indicates non-volatile behavior. Moreover, unlike PVK-C$_{60}$, the device cannot return to the initial OFF state even after applying a reverse voltage (3rd sweep). The non-erasable non-volatile switching behavior leads to the fact that the memory property of OXTA-PIa can be determined to be WORM type.

2.3 Materials

Over the past decades, all types of memory characteristics, from volatile memory properties (DRAM and SRAM) to non-volatile memory properties (FLASH and WORM types), have been reproduced in resistive memory systems.[1-5] Control of these memory properties has been achieved, to some extent, by tuning the chemical structure. Therefore, understanding the relationship between the chemical structure and memory properties is a subject of utmost importance in the development of memory materials. We now introduce some of the reported materials, organic molecules, polymeric materials, and organic–inorganic hybrid materials.

2.3.1 Materials Design Principle

Since an organic resistor memory device stores data by utilizing the conductivity response, and the organic material itself functions as the memory element, the general design principle for the employed materials is to generate distinct electronic states in the presence of an electric field. It is difficult to generalize the design concept which will provide such a property since a large number of materials with various mechanisms have been reported.[1-5] However, there are several common features with regard to the structure of these materials. The first major common point is that almost every material consists of a hole transporting or/and electron transporting moiety. In other words, they are composed of an electron donating unit (D) or/and electron acceptor unit (A). The second point is that nearly half of these materials are reported to show semiconducting properties to some extent. The third point in common is that many materials have insulating or trapping sites, which is opposite to the second point.

2.3.2 Organic Molecules

In 1974, Ratner and Aviram discussed the use of a single organic molecule in a simple rectifier which was the first proposal for an organic molecule memory device.[11] It was concluded that for a suitably constructed single organic molecule, current will begin to flow through the π system from the anode to cathode resulting in switching behavior. Up to now, as shown in Scheme 2.1, the development of small molecule memory materials has been eagerly carried out using polymeric materials.[1,2,12] The lack of a molecular weight distribution and the relative ease in obtaining high purity small molecules are thought to be advantageous for polymeric materials in order to elucidate the effect of the molecular structure on the resulting memory properties.

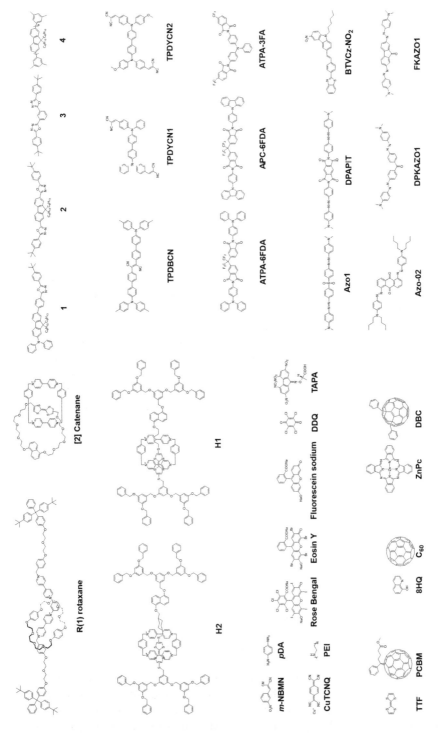

Scheme 2.1 Chemical structures of organic molecules showing memory properties.

One of the early studies of applying a small organic molecule to a memory application was performed by utilizing a monolayer of redox-active rotaxane molecules having two bipyridinium units in the dumbbell component (R(1) rotaxane; Scheme 2.1) in 1999 by Collier and Wong's group.[13] Subsequently, Stoddart and coworkers group reported bistable switching in a single monolayer of bistable [2]catenane consisting of π-electron deficient cyclobis(paraquat-p-phenylene) (CBPQT^{4+}) and π-electron-rich recognition sites such as tetrathiafulvalene (TTF) and 1,5-dioxynaphthalene (NP) anchored by phospholipid counterions (Scheme 2.1).[14] A device with the configuration n-type polycrystalline silicon/single monolayer of [2]catenane/metal top electrode (Ti/Al) exhibits bistable I–V characteristics with turn-on and turn-off voltages of +2 V and −2 V, respectively. Recently, a rotaxane molecule composed of identical ring and recognition sites to the above [2]catenane has been revealed to show bistable switching characteristics (H1 and H2; Scheme 2.1).[15,16] A scanning probe microscopy (SPM) technique was applied to these rotaxane molecules to demonstrate nanometer scale recording. Crystalline organic complexes are also known to show bistable memory phenomena. For instance, when a 1:1 mixed powder of *m*-nitrobenzal malononitrile (*m*-NBMN) and *p*-phenylenediamine (*p*DA) (Scheme 2.1) was deposited on a highly ordered pyrolytic graphite substrate, a nanometer scale recording can be performed by a scanning tunneling microscope.[17,18]

In terms of a single component molecule, Rose Bengal (Scheme 2.1) is a unique molecule whose switching characteristics are known to be dominated by two mechanisms depending on the device configuration.[19,20] When Rose Bengal is embedded in a polyelectrolyte, namely poly(allylamine hydrochloride) (PAH), *via* a layer-by-layer electrostatic self-assembly (ESA) technique, the switching of the device is driven by the conformational transition which will be described later.[19] On the other hand, when Rose Bengal is directly deposited on an electrode, due to the high packing density of the directly deposited film and strong intermolecular interactions, no conformational change or rotation takes place.[20] When the same backbone with a different number of acceptor group molecules, such as Rose Bengal, eosin Y and fluorescein sodium (Scheme 2.1), was studied in a similar manner, the ON/OFF ratio increased (fluorescein sodium < eosin Y < Rose Bengal) with the increasing number of A groups (fluorescein sodium < eosin Y < Rose Bengal).[20] In addition, single planar molecules possessing oxidation–reduction states, such as 2,3-dichloro-5,6-dicyano-1,4-benzoquinone (DDQ) and (+)-2-(2,4,5,7-tetranitro-9-fluorenylideneaminoxy)propionic acid (TAPA), which will not go through a conformational transition, also exhibit bistable switching characteristics. These results suggest that the driving force of the switching behavior of these directly deposited films can be explained by conjugation restoration through electro-reduction.[20]

Taking the materials design principle into account, an organometallic CT complex would obviously exhibit switching characteristics. In fact, copper tetracyanoquinodimethane (CuTCNQ), a well-known CT complex, was found to show bistable switching characteristics when present in a polyethyleneimine

(PEI) film (Scheme 2.1).[21] The device based on this film with the configuration Au/CuTCNQ–PEI/ITO exhibited reproducible switching and unsymmetrical I–V characteristics under an electric field (Figure 2.10). The high conductivity state switches to the low conductivity state at a threshold voltage of 8 V, and this threshold voltage decreases during subsequent sweeps, with the film eventually showing only the low conductivity state. However, the film recovers to the original high conductivity state after removing the film from the electric field for several minutes. Although the initial switching behavior (ON to OFF) is opposite to the conventional one (OFF to ON), as long as bistable switching occurs, the behavior observed in the CuTCNQ system will also be determined to be a memory characteristic. Considering the volatility, this memory characteristic will correspond to DRAM or SRAM.

A bimolecular CT complex of 6,6-phenyl C_{61}-butyric acid methyl ester (PCBM) and TTF dispersed in a polystyrene matrix also shows bistable memory characteristics (Scheme 2.1).[22] In this pair, TTF serves as D while PCBM serves as A. The memory device based on this CT complex switches from the OFF state to the ON state by application of a voltage higher than the threshold. Once the device reaches the ON state, it remains in this state for a prolonged period of time, indicating non-volatile behavior. However, the OFF state can be recovered by simply applying a higher bias. Therefore, the rewritable non-volatile memory characteristic of the PCBM–TTF CT complex is considered to be FLASH-type memory. In a similar manner, bimolecular CT complexes, fullerene C_{60}–8-hydroxyquinoline (8HQ)[23] and 1,4-dibenzyl C_{60} (DBC)–zinc phthalocyanine (ZnPc),[24] both dispersed in polystyrene, were reported to show bistable switching characteristics (Scheme 2.1). Petty and coworkers reported that small organic molecules containing a fluorene structure (compounds **1** and **2**; Scheme 2.1) show bistable switching characteristics.[25] The facts that polyfluorene (compound **4**; Scheme 2.1) shows a much higher ON/OFF ratio and that a similar structure with no fluorene (compound **3**; Scheme 2.1) does not show any switching behavior both indicate the importance of the fluorene unit.

While most research studies have focused on exploring different types of molecules with better performance for data storage, Song *et al.* investigated the

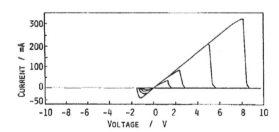

Figure 2.10 Switching characteristics of Au/Cu-TCNQ-PEI/ITO cell. Sweeping velocity: 1 V s^{-1}. The positive quadrant corresponds to a positive bias on the top Au electrode. Reprinted with permission from ref. 21. Copyright 1992, American Institute of Physics.

role of the donor and acceptor unit arrangement.[26] Three different molecules with similar donor and acceptor units in two structural types, namely D–π–A–π–D (TPDBCN), and A–π–D–π–A (TPDYCN1 and TPDYCN2) (Scheme 2.1), were synthesized and their memory performances were evaluated. The device based on the D–π–A–π–D molecule exhibited excellent write–read–erase characteristics with a high ON/OFF ratio of around 10^4–10^6. On the other hand, devices based on the A–π–D–π–A molecules showed irreversible switching behaviors with a relatively low ON/OFF ratio of 10^1–10^3. This study clearly shows that the arrangement of the functional unit affects the final device properties.

A more significant arrangement effect was observed in imide compounds (Scheme 2.1).[27] Devices based on 4,4′-hexafluoroisopropylidenebis[4-(N,N-diphenylamino)phenyl phthalimide] (ATPA-6FDA) and 4,4′-hexafluoroisopropylidenebis[4-(carbazo-9-yl)phenyl phthalimide] (APC-6FDA), whose structures are considered to be D–A–D, revealed reversible negative differential resistance (NDR) characteristics with ON/OFF ratios of 10^4 and 10^2, respectively, and excellent stability. On the other hand, a device prepared from an A–D–A structured imide compound, namely N,N′-[(phenylimino)di-4,1-phenylene]bis(5-trifluoromethyl phthalimide) (APTA-3FA), showed only insulating properties and no switching behavior before annealing the film (Figure 2.11). However, after annealing the device based on APTA-3FA, a clear switching behavior was observed with an ON/OFF ratio of 10^{10}. Since the switching is dominated by the injection and transportation of holes, a film containing a higher amount of the D moiety than the A moiety is preferred and this is considered to be the very reason for the difference in the structural effect on the resulting memory behavior, i.e., D–A–D shows memory behavior while A–D–A only shows insulating properties.[27] The switching phenomenon observed in the thermally treated film of APTA-3FA is thought to result from the diffusion of aluminum atoms from the electrode, since a terrace crystal morphology is formed after the thermal treatment and the boundaries provide an effective pathway (Figure 2.11).

One of the main features of small molecule memory materials is their ternary data-storage capability which was first reported by Lu and Gu's group.[28] The I–V curve of a device based on an azobenzene derivative (Azo1, Scheme 2.1) exhibits two-step transitions which indicate the ternary memory behavior (Figure 2.12). Therefore, devices based on this material have the potential to increase their data-storage capacity from 2^n to 3^n. The basic design principle which gives rise to the ternary switching behavior is to have shallow and deep traps in a single molecule. Based on this design concept, various organic compounds have been synthesized and found to exhibit ternary memory characteristics (Scheme 2.1).[29-33]

2.3.3 Polymeric Materials

Polymeric materials have several strong advantages compared to small molecules in terms of device fabrication and stability due to their excellent film forming ability and mechanical and thermal stabilities. In this section, we

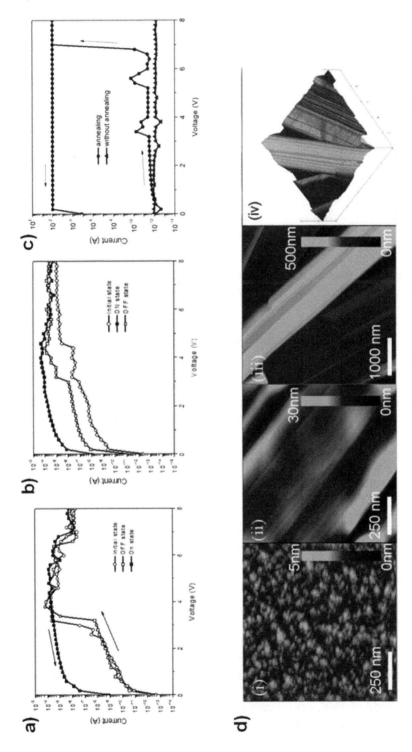

Figure 2.11 *I–V* characteristics of (a) APTA-6FDA, (b) APC-6FDA, and (c) ATPA-3FA. (d) AFM height image of ATPA-3FA: (i) without thermal treatment; (ii and iii) with thermal annealing at 180 °C for 30 min at different scales; (iv) 3-dimensional image of the surface with thermal annealing. Adapted with permission from ref. 27. Copyright 2011, American Chemical Society.

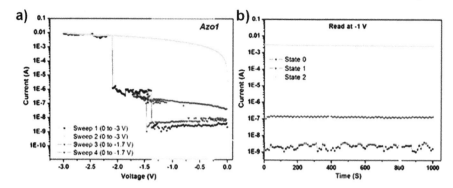

Figure 2.12 (a) *I–V* characteristics of a memory device fabricated with Azo1. (b) Stability of the device in three states under a constant "read" voltage of −1 V. Adapted with permission from ref. 28. Copyright 2010, American Chemical Society.

will introduce various kinds of polymeric materials by classifying them into three categories: (1) polymers with specific pendant groups; (2) conjugated polymers; (3) polyimides.

2.3.3.1 Polymers with Specific Pendant Groups

Poly(*N*-vinylcarbazole) (PVK; Scheme 2.2) and its derivatives are the most well-known and studied polymers with specific pendant groups for memory applications. The switching phenomenon of PVK under an electric field was first observed in 1976 by Sadaoka and Sakai.[34] However, due to the high conductivity of PVK in the initial state which originates from the close packing of the carbazole group,[35,36] the switching behavior was not sufficient enough for memory applications.[37,38] By moving the carbazole unit further from the backbone using an alkyl spacer (PCz; Scheme 2.2), Zhu *et al.* managed to reduce the initial conductivity and achieved an ON/OFF ratio greater than 10^6 of which the memory characteristic was WORM type.[38]

By changing the spacer unit between the backbone and carbazole unit (PBCz; Scheme 2.2), the resulting memory properties changed from nonvolatile WORM type to volatile SRAM type.[37] It was also found that an aromatic spacer would result in a similar effect to the above mentioned alkyl spacer.[39] Chen and coworkers introduced a biphenylethene spacer between the backbone and carbazole unit (PBC; Scheme 2.2) which resulted in WORM-type characteristics. In the same report, they also copolymerized PVK with PBCz (PVK-PBC; Scheme 2.2) which dramatically changed the memory properties from WORM to FLASH type. It was concluded that the introduction of a small vinylcarbazole unit between the neighboring bulky units provided greater conformational freedom for rotation of the pendant unit. Different from spacer embedded PVK, nonplanar phenylfluorene partially attached PVK (PVK-PF)[40] and PVK with fullerene C_{60} attached to the backbone (PVK-C_{60})[8] showed FLASH-type memory characteristics (Scheme 2.2).

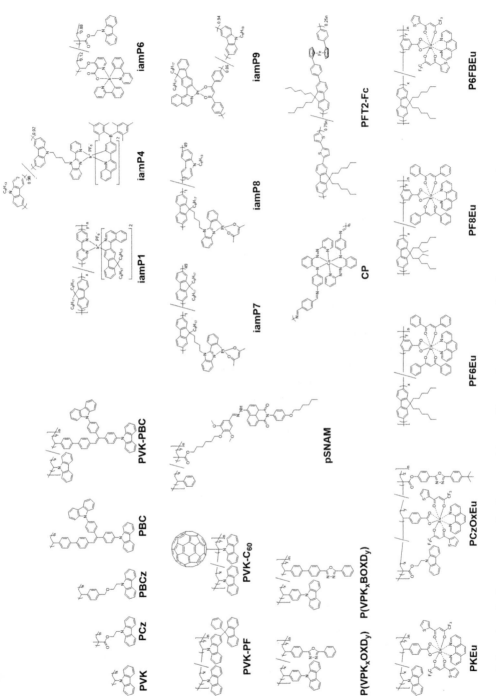

Scheme 2.2 Chemical structures of polymers with specific pendant groups showing memory properties.

Similar to the PVK series, polymers with D and A pendant groups are widely studied for application to memory materials. For instance, Chen and coworkers reported a series of random copolymers (P(VPK$_x$OXD$_y$) and P(VPK$_x$BOXD$_y$); Scheme 2.2) with pendant D and A units based on phenylcarbazole and oxadiazole, respectively, which showed volatile SRAM characteristics.[41] The same group also reported a similar type of random copolymer (P(VTPA$_x$BOXD$_y$)) composed of a triphenylamine-based D pendant and an oxadiazole-based A pendant which exhibited DRAM or SRAM properties depending on the ratio of the donor and acceptor pendant groups.[42] Lu *et al.* investigated the effect of a highly electron deficient 1,8-naphthalimide, commonly used in optoelectronics[43] or cell imaging applications,[44] on memory characteristics.[45] When 1,8-naphthalimide was embedded in a random copolymer system with styrene (pSNAM; Scheme 2.2), DRAM characteristics were observed with the linearly shifting threshold voltage depending on the 1,8-naphthalimide unit content. Polymers with a pendant metal complex are also known to be one of the major categories that exhibit memory characteristics. The first discovery of bistable switching behavior of a rare earth metal complex was in 2003 when porphyrin molecules chelating Eu were demonstrated to exhibit a memory effect.[46] Based on this phenomenon, Kang and coworkers designed a copolymer (PKEu; Scheme 2.2) that contains an Eu complex for polymeric memory device applications.[47] Since this report, a large number of metal complexes containing not only Eu,[48-51] but also Ir,[52-54] Pt,[55,56] Co,[57] and Fe[58] have been reported (Scheme 2.2). In general, these metal complexes behave as the A units. Therefore, the memory characteristics of these complexes are totally different from those of the polymers with a similar structure without the metal element. For example, while PCz only exhibits bistable switching characteristics,[38] iamP6 (Scheme 2.2), which consists of a PCz and Ir complex, shows ternary memory properties.[54]

2.3.3.2 Conjugated Polymers

In terms of organic electronic applications, conjugated polymers are considered to be suitable candidates for polymeric memory materials since their semiconducting nature has been intensively studied, and a wide variety of synthetic methods have been established in order to obtain a well-defined structure. Hence, a large number of memory devices based on conjugated polymers has been reported. Early observations of memory phenomena in conjugated polymers were reported in 1996 for poly(*p*-phenylenevinylene) (PPV; Scheme 2.3) in polymer light emitting diodes.[59] In a vacuum, anomalous *I–V* characteristics and switching phenomena were found for a sandwich structure of metal/polymer/metal, while this behavior could not be observed in air. Later, it was reported that the electrical bistability of PPV had an associated memory behavior, such as read only memory (ROM) or random access memory (RAM) depending on the operation.[60] It was also shown that a device based on an oriented PPV film had a device current of both high and low conductivity states which was two orders of magnitude higher compared to

Organic Resistor Memory Devices

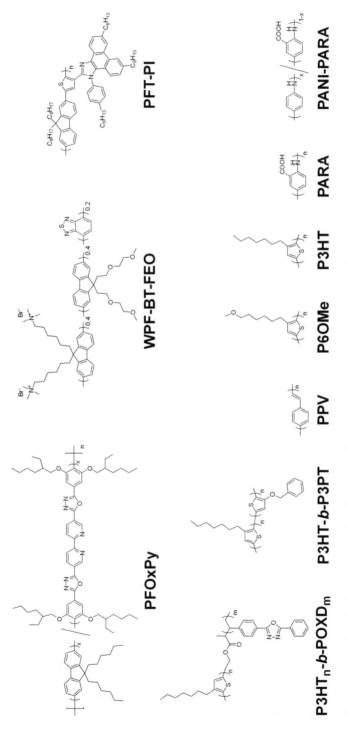

Scheme 2.3 Chemical structures of conjugated polymers showing memory characteristics.

a device based on an unoriented film.[60] Kang et al. revealed that conjugated polyfluorene even without Eu[49,50] would have volatile DRAM type memory characteristics when copolymerized with oxadiazole (Ox) and pyridine (Py) groups (PFOxPy; Scheme 2.3).[61] A conjugated polyelectrolyte composed of polyfluorene (WPF-BT-FEO; Scheme 2.3) is also reported to exhibit switching behavior. This polymer was a milestone for observing the filamentary switching mechanism which will be described in a later section.[62] In 2012, Chen et al. reported a conjugated poly(fluorene-thiophene) with a phenanthro[9,10 d]imidazole acceptor tether (PFT-PI; Scheme 2.3) that could act as a flexible non-volatile FLASH memory device.[63] Among the known conjugated polymers, polythiophene has been the most famous and most studied standard conjugated polymer since its early reports in the 1980s.[64,65] It was reported in 2002 that a device based on poly(3-(6-methoxyhexyl)thiophene) (P6OMe; Scheme 2.3) shows hysteresis-type behavior and can be used as a data storage device.[66] P3HT (Scheme 2.3), a well-known standard material used in photovoltaic applications, has also been used in rewritable memory devices.[67-69] A unique memory device, namely a bulk heterojunction memory device adopting P3HT:PCBM, a well-known and studied combination of materials for application in polymer solar cell devices, has been reported to exhibit WORM-type memory properties.[70] One of the major features of the polythiophene derivative is the ability to carry out precision polymerization via the Kumada Catalyst-Transfer Polymerization (KCTP).[71] Chen et al. reported a D–A type block copolymer composed of P3HT and poly(2-phenyl-5-(4-vinylphenyl)-1,3,4-oxazole) (POXD) synthesized by a combination of KCTP and atom transfer radical polymerization (ATRP) which exhibited memory characteristics in a specific block ratio (P3HT$_{44}$-b-POXD$_{18}$; Scheme 2.3).[72] Subsequently, Chen and Ueda's group reported a block copolymer composed of P3HT and P3PT in which a certain block ratio (P3HT$_{102}$-b-P3PT$_{37}$ and P3HT$_{52}$-b-P3PT$_{39}$; Scheme 2.3) exhibited DRAM characteristics.[6] Other than the above described conjugated polymers, polyaniline derivatives, such as poly(o-anthranilic acid) (PARA) and poly(o-anthranilic acid-co-aniline) (PANI-PARA) (Scheme 2.3), have also shown good non-volatile memory performance.[73] Their memory devices are switchable in a very low voltage range (−0.5 V), and have extremely fast switching response (shorter than 80 ns) with a very high ON/OFF current ratio of 10^5. In addition, these devices have very stable ON- and OFF-states in ambient air without any degradation for a very long time (up to one year) and can be repeatedly written, read and erased.[74]

2.3.3.3 Polyimides

Considering the heat resistance for the device fabrication process and extreme use in real applications, due to their outstanding thermal stability and mechanical strength, polyimides are the most attractive polymers for memory applications. Over the years, a significant number of reports regarding their memory applications have been made and this number is still increasing

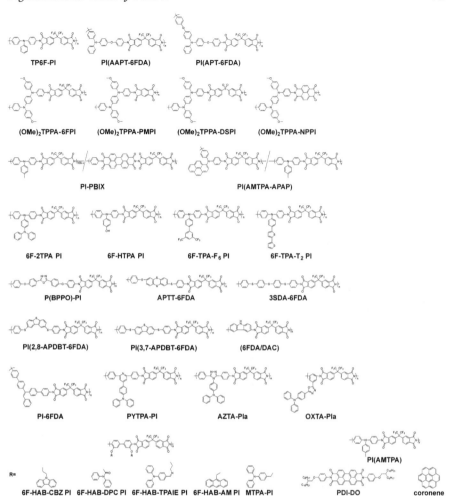

Scheme 2.4 Chemical structures of polyimides showing memory characteristics.

today. The very first report on a polyimide (TP6F-PI; Scheme 2.4) which exhibits DRAM characteristics was published in 2006 by Kang's group.[75] One of the most important points in this report is the design concept of introducing the hexafluoroisopropylidene (6F; Scheme 2.4) group in order to enhance the solubility of the polyimide, which is typically a highly chemically resistive material, and to provide solution processability. The fact that more than 80% of the reported polyimides for memory applications contain the 6F group supports this importance. The successful development of TP6F-PI led to an expansion of the development of polyimide memory materials (Scheme 2.4). Most memory characteristics are determined by the D structure which varies among all the D moieties such as triarylamine (TP6F-PI,[75] PI(AAPT-6FDA), PI(APT-6FDA),[76]

(OMe)$_2$TPPA-6FPI,[77] -PMPI, -DSPI, -NPPI,[78] PI-PBIX,[79] PI(AMTPA-APAP),[80] 6F-2TPA PI,[81] 6F-HTPA PI,[82] 6F-TPA-F$_6$ PI, 6F-TPA-T$_2$ PI[83]), oxadiazole (P(BPPO)-PI),[7] thianthrene (APTT-6FDA),[9] phenylenethioether (3SDA-6FDA),[9] dibenzothiophene (PI(2,8-APDBT-6FDA), PI(3,7-APDBT-6FDA)),[84] carbazole (6FDA/DAC),[85] triphenylethylene (PI-6FDA),[86] triphenylamine connected *via* pyridine (PYTPA-PI),[87] triazole (AZTA-Pia),[88] and oxadiazole (OXTA-Pia).[10] In addition, polyimides having D moieties in the pendant position (6F-HAB-CBZ PI,[89] 6F-HAB-DPC PI,[90] 6F-HAB-TPAIE PI,[91] 6F-HAB-AM PI,[92] MTPA-PI[93]) are also reported to exhibit memory characteristics. The switching mechanism of polyimide memory is thought to be dominated by two major mechanisms which are the charge transfer effect and filamentary conduction. As shown in Scheme 2.4, regardless of the underlying mechanism, all types of memory characteristics can be covered. Among these polyimides, several studies have focused on manipulating the resulting memory characteristics by tuning the chemical structures. For instance, in 2010, Ueda *et al.* reported two triphenylamine-based polyimides exhibiting different memory properties depending on the number of phenoxy linkages.[76] The polyimide possessing the mono-phenoxy linkage (PI(AAPT-6FDA); Scheme 2.4) showed DRAM properties, while with dual-phenoxy linkages (PI(APT-6FDA); Scheme 2.4), it exhibits SRAM characteristics. This is considered to be the result of different conformations coming from the amount of flexible phenoxy linkages. The (OMe)$_2$TPPA series reported by Liou and coworkers also shows a proof of concept for tailoring the memory characteristics by modification of the chemical structures.[77,78] When the triphenylamine-based D moiety ((OMe)$_2$TPPA) is connected to the 6F moiety ((OMe)$_2$TPPA-6FPI; Scheme 2.4) the resulting memory properties would be DRAM type,[77] while connecting a phthalimide instead of a 6F moiety ((OMe)$_2$TPPA-PMPI; Scheme 2.4) would result in SRAM characteristics.[78] Further changing the A unit to diphenylsulfonetetracarboxylic diimide ((OMe)$_2$TPPA-DSPI) and naphthalene tetracarboxylic diimide ((OMe)$_2$TPPA-NPPI) by changing the dipole moment and lowest unoccupied molecular orbital (LUMO) energy levels, respectively, induces non-volatile WORM-type memory characteristics.[78] Recently, Ueda's group established a new design concept of tuning the memory properties by incorporating a large conjugation unit into the backbone.[79,80] For a polyimide system containing perylene bisimide (PI-PBIX; Scheme 2.4), the memory characteristics of the copolyimides containing none or a small amount of the perylene bisimide unit (x = 0, 1, 2.5) showed volatile memory properties, while copolyimides containing a larger amount of the perylene bisimide unit (x = 5, 10) showed non-volatile WORM-type memory characteristics.[79] As for a copolyimide system possessing the pyrene moiety (PI(AMTPA-APAP); Scheme 2.4), the memory properties can be tuned from volatile DRAM to non-volatile FLASH type to non-erasable WORM type with the corresponding pyrene contents of 0, 5, and 100 mol%, respectively (Figure 2.13).[80] The interesting thing about this copolyimide is that when 10 mol% of the pyrene moiety is introduced, 30% of the fabricated cells exhibited FLASH-type memory characteristics, while the remaining 70% were WORM type (Figure 2.13c). This indicated that the

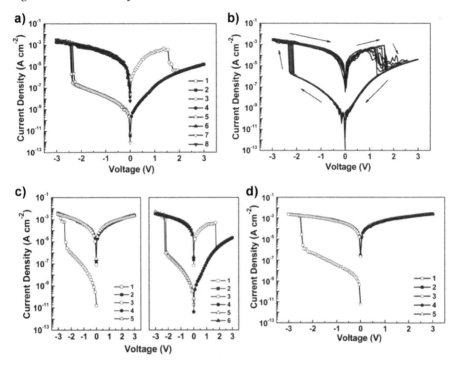

Figure 2.13 *J–V* characteristics of (a) PI(AMTPA), (b) PI(AMTPA95-APAP5), (c) PI(AMTPA90-APAP10), and (d) PI(APAP) device. Adapted with permission from ref. 80. Copyright 2012, The Royal Society of Chemistry.

critical content of the pyrene unit required to tune the memory from FLASH to WORM type is around 10% and this tuning is sensitive to the existing amount of pyrene moieties. These tunable memory characteristics in the copolyimide system result from the stabilization of the radical anion or cation in the charge transfer state by the resonance effect of the large conjugation unit (perylene and pyrene).

More recently, the same group discovered that the large conjugation unit does not necessarily need to be connected to the backbone of the polyimide.[94] When a polyimide (PI(AMTPA); Scheme 2.4) exhibiting volatile DRAM characteristics was blended with coronene (Scheme 2.4), the resulting memory characteristics transitioned to non-volatile FLASH type and non-erasable WORM type with a coronene content of 3 and 15 wt%, respectively. Similar memory property transitions were observed for a blend film of PI(AMTPA) and PDI-DO (Scheme 2.4) with a much lower blending content of PDI-DO, that is 1 wt% for the FLASH type and 10 wt% for the WORM type. The effects of coronene and PDI-DO on the memory transition are similar to those of pyrene in PI(AMTPA-APAP)[80] and perylene in PI-PBIX,[79] respectively. An interesting behavior of a polyimide consisting of a triphenylamine-based D and 6F as an A (6F-2TPA PI; Scheme 2.4) was

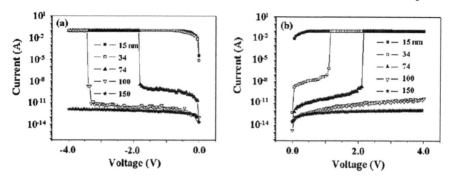

Figure 2.14 *I–V* curves of a Au/6F-2TPA PI/Al sandwich device using various thickness (15–150 nm) films; the voltage was swept in both (a) negative and (b) positive directions. Adapted with permission from ref. 81. Copyright 2009, IOP Publishing Ltd.

reported by Lee's group.[81] The device using a 15 nm 6F-2TPA PI film showed no memory characteristics, while non-volatile WORM-type characteristics were observed using a 34 nm 6F-2TPA PI film and sweeping in the positive direction. When the film thickness was increased to 74 nm, WORM-type properties were observed in both voltage sweep directions and further increasing the thickness to 100 nm resulted in DRAM-type behavior only in the negative voltage sweep direction. Note that for the 34 and 100 nm thick films, no WORM and DRAM behaviors were observed during the sweep in the opposite direction (Figure 2.14).

2.3.4 Organic–Inorganic Hybrid Materials

Several polymer (organic)–inorganic hybrid materials have been reported for application in memory devices. The advantage of hybrid materials is that these materials show mechanical properties similar to regular organic materials while possessing excellent functionality coming from the inorganic counterpart. These composites generally consist of an inorganic compound, matrix, capping agent, and D compound as shown in Scheme 2.5. In 2004, Yang and coworkers reported programmable electrical bistability in a polymer–inorganic hybrid composite film consisting of polystyrene (matrix), Au nanoparticles capped with 1-dodecanethiol (inorganic with capping agent), and 8HQ (D) (Scheme 2.5).[95] The device with the configuration Al/composite/Al exhibited non-volatile FLASH-type memory characteristics. In this composite, the switching behavior is attributed to the charge transfer effect between the 8HQ and Au nanoparticles, while polystyrene is thought to be inert to this electrical transition. Also, the insulating 1-dodecanethiol surrounding the Au nanoparticles prevents the recombination of the charges after removing the electric field, and as a result, non-volatile memory characteristics are observed. When 8HQ was replaced with 9,10-dimethylanthracene

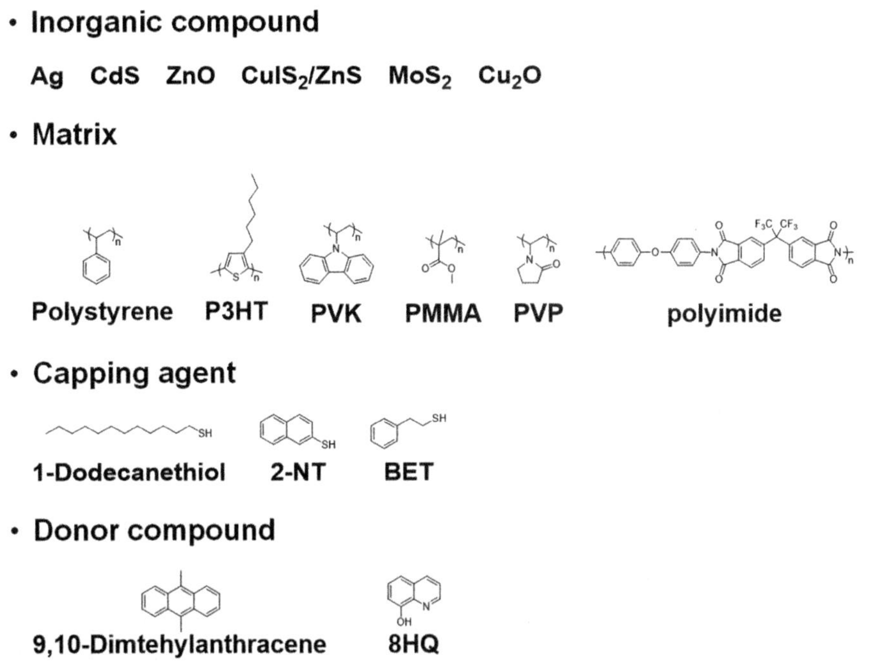

Organic Resistor Memory Devices

- **Inorganic compound**

  Ag   CdS   ZnO   CuIS$_2$/ZnS   MoS$_2$ Cu$_2$O

- **Matrix**

  Polystyrene   P3HT   PVK   PMMA   PVP   polyimide

- **Capping agent**

  1-Dodecanethiol   2-NT   BET

- **Donor compound**

  9,10-Dimtehylanthracene   8HQ

Scheme 2.5 Chemical structures of organic–inorganic hybrid materials showing memory characteristics.

(Scheme 2.5), similar memory properties with a slightly shifted threshold voltage were observed.[96] However, by changing the capping unit from a saturated alkylthiol (1-dodecanethiol) to an aromatic compound, the resulting memory characteristics drastically changed. For instance, a device based on a polystyrene composite film with Au nanoparticles capped with 2-naphthalenethiol (2-NT; Scheme 2.5) showed non-volatile WORM-type memory characteristics.[96] The electrical transition (switching) is also explained by the field induced charge transfer between the D (2-NT) and A (Au nanoparticles). Such WORM-type memory properties can also be observed by capping the Au nanoparticles with 2-benzeneethanethiol (BET; Scheme 2.5). Since the conjugation of 2-BET is lower than that of 2-NT, the ON/OFF ratio is reduced by two orders of magnitude.[96] By comparing these composites, it is obvious that the combination of an electron donating compound and Au nanoparticles would give rise to memory characteristics in the composite. Therefore, Yang *et al.* decided to simultaneously use P3HT (Scheme 2.5) as the matrix and electron donating agent which would simplify the device structure.[97] Also, reducing the number of components in the composite would help in understanding the switching behavior. As a result, a device based on this composite with the configuration Al/Au-nanoparticle:P3HT/Al showed bistable switching behavior that can be accomplished a thousand times with an ON/OFF ratio of about 10^3. Similar to P3HT, PVK (Scheme 2.5), known as a good hole transporting polymer with electron donating ability as shown in the previous section, has also been utilized in Au nanoparticle composite systems.[98] The switching behavior observed in this composite was determined to be FLASH-type memory. For the inorganic counterpart, besides Au, various inorganic materials, such as Ag,[99–101] CdS,[102] ZnO,[103] CuIS$_2$/ZnS,[104] MoS$_2$,[105] and Cu$_2$O,[106] in composite systems with polystyrene, PMMA, PVK, polyvinylpyrrolidone (PVP), or polyimide, have been utilized in composites and have shown bistable switching properties (Scheme 2.5).

2.4 Mechanism Discussion of Organic Resistor Memory Devices

Resistive memory is based on the conductivity response of the active material to an applied electric field. Since conductivity is essentially a product of the carrier concentration and charge mobility, such a response can be induced by a change in the carrier concentration or/and a change in the charge mobility.[107] The conduction in polymers is normally explained by the intrinsic charge carrier generation and charge carrier injection from contacts in a high electric field. In resistive memory, charge injection from the electrodes to the polymers is thought to be more common. A number of conduction mechanisms, such as Schottky emission, thermionic emission, space charge-limited current, tunneling current, ionic conduction, hopping conduction, and impurity conduction, have been used to explain the conduction processes in polymers (See Table 2.1).[2,108,109] The question is: what is the mechanism that explains the transition of these conduction modes (switching behavior)? Until now, the conductivity transition has been explained by several mechanisms. Other properties, such as threshold voltage, switching polarity, and volatility, can also be explained by applying this mechanism. In this section, we will introduce several underlying mechanisms of the observed switching behavior by classification into four categories, filamentary conduction, space charge and traps, the charge transfer effect, and the conformational effect.

2.4.1 Filamentary Conduction

For decades, threshold and memory switching effects have been observed in insulating polymers, such as polystyrene,[110–113] polyethylene,[114] and poly(methyl methacrylate).[110,115] Electric switching in these materials is mostly due to the formation of filamentary conduction paths. As a general definition, when the high conductivity state current is highly localized to a small area of the memory cell, this phenomenon is called "filamentary" conduction. If a filament is formed in the material, the injected current will be insensitive to the memory cell area. This is due to the fact that the dimension of the filament is much smaller when compared to the device area.

Table 2.1 The basic conduction processes in insulators.[a]

Conduction mechanisms	J–V characteristics
Ohmic conduction: $J \propto V \exp\left(\dfrac{-\Delta E_{ae}}{kT}\right)$	Ionic conduction: $J \propto \dfrac{V}{T} \exp\left(\dfrac{-\Delta E_{ai}}{kT}\right)$
Hopping conduction: $J \propto V \exp\left(\dfrac{\varphi}{kT}\right)$	Space-charge-limited current: $J \propto \dfrac{9\varepsilon_i \mu V^2}{8d^3}$

Schottky emission:

$$J \propto T^2 \exp\left[\frac{-q\left(\varphi - \sqrt{qV/4d\pi\varepsilon}\right)}{kT}\right]$$

Thermionic emission:

$$J \propto T^2 \exp\left[\frac{-\left(\varphi - q\sqrt{qV/4d\pi\varepsilon}\right)}{kT}\right]$$

Frenkel–Poole emission:

$$J \propto V \exp\left[\frac{-q\left(\varphi - \sqrt{qV/4d\pi\varepsilon}\right)}{kT}\right]$$

Tunnel of field emission:

$$J \propto V^2 \exp\left[\frac{-4\sqrt{2m}\,(q\varphi)^{3/2}}{3qhV}\right]$$

Direct tunneling: $J \propto \dfrac{V}{d} \exp\left[\dfrac{\left(-2d\sqrt{2m\varphi}\right)}{h}\right]$

Fowler–Nordheim tunneling:

$$J \propto V^2 \exp\left[\frac{-4d\sqrt{2m}\,(\varphi)^{3/2}}{3qhV}\right]$$

[a] φ = barrier height, V = electric field, T = temperature, ε = insulator dynamic permittivity, m = effective mass, ΔE_{ae} = activation energy of electrons, ΔE_{ai} = activation energy of ions, d = insulator thickness, q = charge and μ = carrier mobility.

Two types of filamentary conduction have been widely reported so far. One is carbon-rich filament formation caused by local degradation of polymer films.[110–113,116] This is commonly observed in insulating polymers as listed above. The filaments can be formed by a two-stage process as follows. The initial breakdown of the film by the external electric field produces a carbon-rich area surrounding a central void. In the second step, further breakdown produces narrow and highly conductive carbon filaments linking the two electrodes. Therefore, switching is a consequence of the formation,

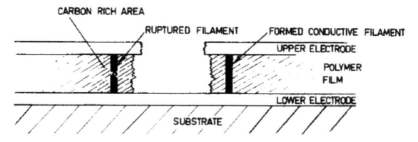

Figure 2.15 Proposed model for carbon rich filament conduction. Reprinted with permission from ref. 111. Copyright 2009, American Institute of Physics.

rupture, and reformation of these filaments. As the potential passes the range of threshold voltages, the current gradually increases as more and more filaments are reformed, or an interfilament triggering process may occur, producing a sudden increase in current. It can be said that the threshold voltage for a given filament will depend upon the length of its ruptured section. This means that if a filament is completely destroyed then the threshold voltage required to reform the filament will be a function of the film thickness. On the other hand, if only a small section is ruptured at a weak point within a filament, the threshold voltage will depend upon the length of this section and thus be independent of the film thickness (Figure 2.15).[111]

The other mechanism is metallic filament formation by local diffusion, migration, or sputtering of electrodes through polymer films.[113,116] The basic event happening during the switching process is the formation, rupture, and reformation of these filaments which are the same events happening in carbon-rich filament formation. For metallic filament formation, the functional polymers are required to possess π-conjugation and strongly coordinating atoms such as N and S, that can bind the metal ions regardless of the position in which they are introduced (backbone or side chain).[67] Therefore, the memory behavior of the device with the configuration Cu/P3HT/Al is driven by metallic filament formation.[68] For such a device configuration, high positive voltage is believed to play an important role in ionizing the copper electrode and injecting the ions into the polymer layer. Assisted by coordination to the heteroatom (S) of the P3HT, copper ions are uniformly distributed throughout the polymer layer. Copper ions are then metallized by the injected electrons to form the filament under a negative voltage bias (Figure 2.16).[68]

The filamentary conduction mechanism is also widely applied in order to understand the memory phenomena in polyimides. When an electric field is applied to the functional polyimide, the donor and acceptor sites are filled with holes and electrons, respectively. After the applied bias reaches a threshold, the trapped charges are able to move through the trapped donor or acceptor sites by a hopping process. This leads to the formation of a conductive filament, resulting in current flow between the two electrodes. Based on this mechanism, several phenomena, such as the film thickness, electrode,

Organic Resistor Memory Devices

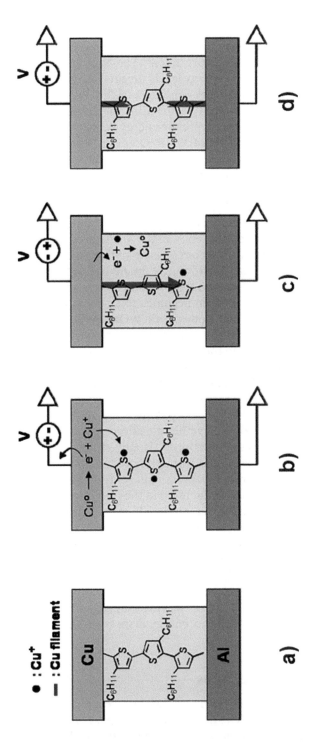

Figure 2.16 Schematic showing the formation of a metal filament within a polymer layer: (a) device structure, (b) ionization and drift processes of copper caused by positive voltage, (c) metal filament formation by the reduction of copper ions, and (d) the breakdown of the copper filament by Joule heating. Reprinted with permission from ref. 68. Copyright 2007, American Chemical Society.

current compliance, and the substituent of the polymer backbone effect, can be explained. As described in the Materials section, 6F-2TPA PI (Scheme 2.4) shows different types of memory behavior depending on the film thickness and voltage sweeping direction (Figure 2.14).[81] The highest occupied molecular orbital (HOMO) and LUMO energy levels for this polymer are −5.10 and −1.88 eV, respectively, while the work functions of the Au and Al electrodes are −5.1 and −4.2 eV, respectively. A 15 nm film thickness is too thin and the applied voltage can easily overcome the energy barrier between the work function of the electrodes and HOMO–LUMO energy levels. As a result, a short circuit current flows through the film without showing any switching characteristics. As the film thickness increases to 34 nm, the thickness will be large enough to generate an energy barrier between the work function of the Al electrode and the LUMO energy level of the polymer, preventing a short circuit in the positive flow. On the other hand, since there is a zero energy barrier between the work function of the Au electrode and the HOMO energy level of the polymer, the short circuit current flow can still be observed during the negative sweep. As a result, the 34 nm 6F-2TPA PI film shows unipolar switching non-volatile WORM-type memory behavior. By increasing the film thickness to 74 nm, the film becomes thick enough to prevent any short circuit, resulting in bipolar non-volatile WORM-type memory. By further increasing the thickness to 100 nm, the film becomes too thick to form a stable conductive filament and the device shows volatile DRAM-type behavior during the negative sweep. However, since the energy barrier between the Al electrode and the LUMO energy level is higher than that between the Au electrode and the HOMO energy level, a conductive filament cannot be formed during the positive voltage sweep. Finally, a 150 nm film thickness prevents any formation of conductive filaments resulting in non-switching behavior in both voltage sweep directions. Also, regardless of the direction, as the film thickness increases, the formation of the conductive filament becomes more difficult and requires more energy. Therefore, the threshold voltage increases together with the increase in the film thickness.

Although a significant number of memory effects are clearly explained by filamentary conduction and several results indirectly support the existence of conductive filaments, as the dimensions of the filaments are very small, it is difficult to obtain a direct image of the filamentary path connecting the top and bottom electrodes. Recently, the formation and rupture of Ag metallic bridges connecting the two electrodes have been directly observed in a conjugated polyelectrolyte (WPF-BT-FEO; Scheme 2.3) system using transmission electron microscopy (TEM) and energy-dispersive X-ray spectroscopy (EDX) (Figure 2.17).[62]

2.4.2 Space Charge and Traps

Electrical switching in several polymers has been explained by the space charge and trap mechanism. The definition of the space charge limited current is as follows. In the case of ohmic contact between the electrode and

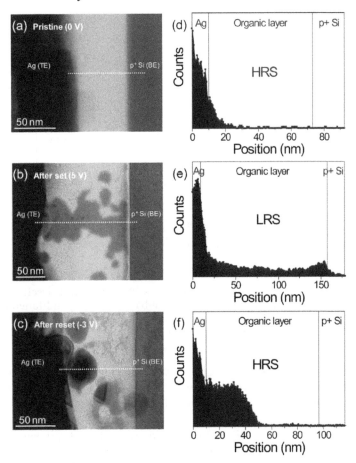

Figure 2.17 (a)–(c) Cross-sectional TEM images of memory cells in different resistance states. (d)–(f) EDX analysis of Ag element profiles along the dotted lines marked in (a)–(c). Reprinted with permission from ref. 62. Copyright 2011, Wiley.

memory material and the material being trap-free, a space charge is built up by the carriers accumulating near the electrode. Mutual repulsion of each charge generates a restriction on the total charge injection from the electrode to the memory material which limits the current. The space charge in materials arises from several sources, such as carrier injection from the electrode, ionized dopants in the interfacial depletion regions, and accumulation of mobile ions at the electrode interfaces.[117] On the other hand, traps are present in the bulk material and/or at interfaces where the presence of traps results in decreased carrier mobility. Also, traps at the interface will influence charge injection into the material.

For instance, the DRAM effect observed in PFOxPy, reported by Kang *et al.* in 2006, can be explained by the space charge and trap effect (Figure 2.18).[61]

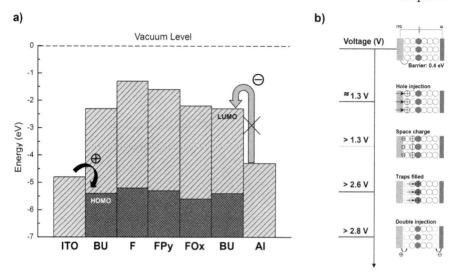

Figure 2.18 DFT molecular simulation results (B3LYP/6-31G(d) level): (a) LUMO and HOMO energy levels for the basic unit (BU) and functional segments of PFOxPy along with the work function of the electrodes. (b) Operational mechanism of the memory.

From the energy level diagram, it is obvious that the electron injection from the Al cathode into the LUMO energy level of PFOxPy is much more difficult than the hole injection from the ITO anode to the HOMO energy level. Hence, PFOxPy is a p-type material and holes dominate the conduction process. Based on molecular simulation of the basic unit of PFOxPy, it was observed that a positive electrostatic potential (ESP) is continuously spread along the conjugated backbone which indicates that holes can migrate through this open channel. Negative ESP regions located on the electron acceptor unit were also observed. These negative regions can serve as traps to block the mobility of the injected holes. Under a low negative bias, a Schottky barrier is formed near the ITO anode. When the field exceeds this barrier, hole injection near the anode occurs which generates a space charge. When near the turn-on voltage, the generated carriers fill some of the charge traps and the contact between PFOxPy and cathode also forms an electron-injecting contact. As a result, double injection takes place and a rapid increase in the carrier concentration (eventually leading to a current increase) switches the device to the ON state. Since the depths of the traps are relatively shallow, a reverse voltage pulse or turning off the power for a few minutes will allow the filled traps to be de-trapped, leading to volatile behavior.

The trapping site does not necessarily have to be composed of acceptor groups. For instance, volatile bistable electrical switching behavior was observed in devices with ITO/P3HT$_{52}$-b-P3PT$_{39}$ or P3HT$_{102}$-b-P3PT$_{37}$/Al structure.[6] This switching arises from the accumulation of space charges due to the existence of trapping sites in the P3PT domains. The poly(3-hexylthiophene)

(P3HT) device only shows semiconductor characteristics, while the P3HT$_{52}$-*b*-P3PT$_{39}$ and P3HT$_{102}$-*b*-P3PT$_{37}$ devices show unstable DRAM behavior because efficient P3HT interchain hopping percolates a conductive pathway without any available trapped P3PT domains in the active layer. In this case, the voltage-controlled multi-level states depend on charge hopping conduction between the P3HT blocks and the charge traps in the P3PT blocks. The P3HT$_{52}$-*b*-P3PT$_{39}$ device exhibits a higher threshold voltage to change from the OFF state to the ON state, as compared to the P3HT$_{102}$-*b*-P3PT$_{37}$ device, due to the relatively large trap-effective P3PT domains of P3HT$_{52}$-*b*-P3PT$_{39}$, thus requiring a higher external applied voltage to remove the trapped-charge.[6]

2.4.3 Charge Transfer Effect

Based on field induced CT theory, by applying an electrical bias to a material that is composed of D and A moieties, charges are transferred from the D moiety to the A moiety, forming a highly conductive CT state. Therefore, it can be said that the stability of the CT state upon removal of the external power supply determines the volatility of the resulting memory characteristics. If the CT state is unstable in the absence of electric power, the resulting memory property would be a volatile one and *vice versa*. In a material which shows volatile DRAM properties, the CT state immediately dissociates *via* back CT or a recombination process of the separated charges after removing the applied electric field due to the low stability of its CT state. This dissociation can be delayed or prevented by several factors, such as a conformational change, high LUMO energy level, high dipole moment, or high conjugation. For instance, based on DFT calculations, the dihedral angle changes between the D and A unit during the formation of the CT state of TP6F-PI,[75] P(BPPO)-PI,[7] and AZTA-PI[88] are 1.8°, 14.5°, and 17.1°, respectively (Figure 2.19), and the resulting memory characteristics are DRAM, SRAM, and WORM, respectively. A large conformational change generates a high energy barrier to prevent back CT resulting in a difference in volatility.

For the LUMO energy level effect, the comparison of metal complex polymers provides a clear example. A conjugated polymer with polyfluorene as the D unit and an Ir complex as the A unit (iamP1; Scheme 2.2) exhibits rewritable FLASH-type memory behavior,[52] while an Eu complex polymer with the same D unit (polyfluorene) shows non-erasable WORM-type memory characteristics.[49] The difference in the erasing ability can be interpreted as the ability to dissociate the field-induced CT state which clearly comes from the electron affinity difference between Ir and Eu,[5] in other words, the LUMO energy difference between the two polymers. The dipole effect is also a powerful tool to retain the CT state after removal of the external electric field. The C_{60} incorporated PVK polymer (C_{60} as A and PVK as D) shows non-volatile FLASH-type memory behavior which indicates that a stable CT state is induced in this polymer.[8] Despite the existence of C_{60} surrounding a positively charged carbazole unit, which indicates the dissociation of the CT state, the electron-withdrawing ability of C_{60} generates a dipole moment between

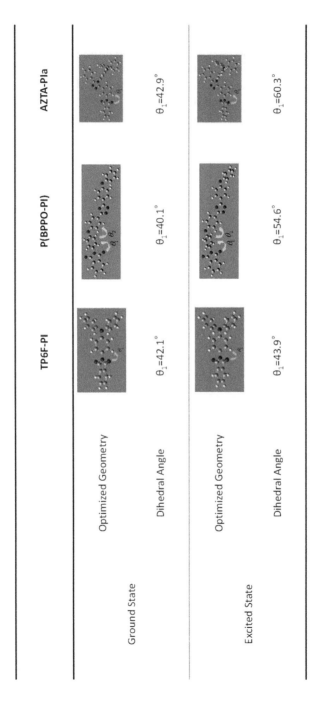

Figure 2.19 Optimized geometries, and dihedral angles (θ_1) between the phthalimide plane and the adjacent benzene ring in the ground and excited states of TP6F-PI, P(BPPO-PI), and AZTA-PIa.

the carbazole and C_{60} strong enough to create an internal electric field which allows the electrons to remain in the C_{60}. Recently, it has been revealed that the stability of the CT state is quite sensitive to the conjugation effect. For example, by embedding an extremely small amount of the perylene structure in the backbone of a polymer that shows DRAM properties,[75] the resulting memory characteristics drastically change to WORM type.[79] Introducing the pyrene moiety would result in the same behavior.[80] The stabilization of the radical anion and cation through the highly conjugated perylene and pyrene, respectively, in the CT state contributes to the formation of the stable CT state that induces non-volatile memory properties.[79,80] The stabilization of the radical anion and cation could also be achieved not only by incorporating a highly conjugated moiety into the backbone, but also by blending such a compound into the volatile memory system.[94]

2.4.4 The Conformation Effect and Other Effects

The conformation effect is usually considered when the active material is not composed of a D and A structure. For instance, the switching behavior of Rose Bengal embedded in a supramolecular structure can be explained by electro-reduction and conformation transition.[19] By applying an electric field, the Rose Bengal molecule switches between three conformational states, which are denoted as "Neutral State 1", "Anion State 2a, 2b", and "State 3" resulting in a switching behavior between the three conduction states (Figure 2.20). In the low voltage region, bias-induced electro-reduction of the Rose Bengal occurs and brings this molecule into the anionic state. By applying a higher voltage, the two parallel planes in the Rose Bengal are induced to change their conformation to perpendicular mode which further facilitates the conduction of the film, resulting in switching behavior.

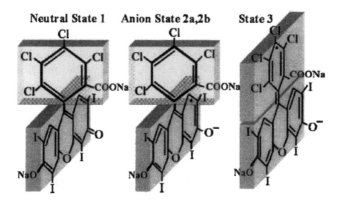

Figure 2.20 Chemical structure of Rose Bengal in its neutral state, State 1; reduced state without any conformational change, State 2a or State 2b; reduced state with conformational change, State 3. Reprinted with permission from ref. 19. Copyright 2004, American Institute of Physics.

The significant effect of the spacer embedded in the PVK system is also explained by the conformation effect.[37–39] Under an electric field, hole injection from ITO oxidizes the carbazole groups near the interface which then form positively charged species. Subsequently, nearby neutral carbazole groups undergo charge transfer or D–A interactions with positively charged carbazole groups to form a partial or full face-to-face conformation with the neighboring carbazole groups. The positive charge is then delocalized to the neighboring, ordered carbazole groups and eventually through the bulk polymer film. When the applied voltage exceeds the threshold value, a significant fraction of the carbazole groups undergo such a conformational change, resulting in enhanced charge transport through neighboring, aligned carbazole groups either on the same or neighboring polymer chains (intrachain or interchain hopping), producing the high conductivity state (ON state). Since the packing of carbazole in PVK is strong, PVK itself is initially in the high conductivity state and shows negligible switching characteristics.[34,38] By introducing a flexible spacer between the backbone and carbazole unit, the initial (before applying bias) orientation of the carbazole becomes random. After applying a voltage to these spacer-embedded PVK derivatives, a regioregular arrangement, which is similar to that of the face-to-face conformation of PVK, occurs resulting in the facilitation of carrier delocalization and transport. The effect of this spacer unit has been discussed based on X-ray diffraction (XRD) patterns, theoretical calculation of the initial conformation, and transmission electron microscopy (TEM) as shown in Figure 2.21.[37,38] While the XRD pattern of PVK shows a sharp peak at around $2\theta = 7.76°$, which corresponds to chain parallelism indicating the regioregularity of PVK, the XRD patterns of PMCz and PVBCz only show broad peaks around $2\theta = 18.73°$ and $19.77°$, respectively, corresponding to an amorphous halo (Figure 2.21a). This result indicates that the spacer interferes with the interaction of the carbazole unit in the amorphous polymer, eliminating the unusual degree of chain parallelism and destroying the regioregularity. This random orientation nature of PMCz and PVBCz is also suggested by the simulated 3D models (Figure 2.21b). After applying a voltage, the conformation ordering in these polymers was observed by TEM. As can be seen, ordered microdomains appear in the PMCz and PVBCz films after applying a bias, while PVK shows similar microdomains even before the voltage is applied (Figure 2.21c). The difference in the PMCz and PVBCz memory properties, which are determined to be WORM and volatile, respectively, comes from the greater conformational freedom of PVBCz which allows the induced conformation to relax to the initial state.[37]

2.5 Applications of Organic Resistor Memory Devices

With the development of high performance memory materials, memory device development for actual applications becomes important. These device developments include increasing the data storage capacity, providing

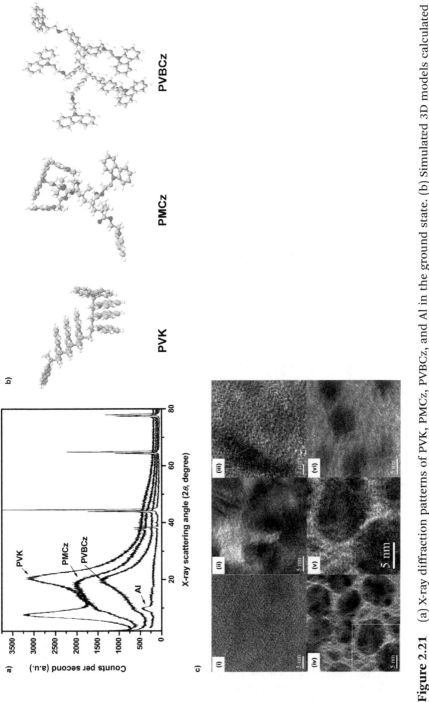

Figure 2.21 (a) X-ray diffraction patterns of PVK, PMCz, PVBCz, and Al in the ground state. (b) Simulated 3D models calculated using molecular mechanics showing the optimized geometry corresponding to the minimum energy states of PVK, PMCz, and PVBCz. (c) TEM images of (i) PMCz film without any bias applied (OFF state), (ii) PMCz film showing ordered microdomains after a voltage sweep from 0 to −4 V (ON state), (iii) PMCz film in the ON state with higher magnification, (iv) PVK film, (v) magnified region of the PVK film, and (vi) a PVBCz film after a voltage sweep from 0 to −4 V and conformation relaxation. Adapted with permission from ref. 37. Copyright 2007, American Chemical Society.

mechanical functionality, and improving the device fabrication process. In this section, taking the advantages of organic materials, such as solution processability and flexibility, the recent development of memory device processes for practical use will be introduced.

2.5.1 Three-Dimensional Devices

In 2010, Lee *et al.* demonstrated the fabrication of a three-dimensional stacked 8 × 8 crossbar three-layer resistive memory device with each memory cell being able to be independently controlled (Figure 2.22).[118] They utilized a composite of polyimide and PCBM as the memory element and a simple spin-coating process to deposit the active layer. The successful demonstration of 83% operation of the memory cells out of all three layers without any malfunction was achieved which could be further improved by a more careful fabrication process. The study showed the possibility of a highly integrated organic memory device with a significantly increased cell density from an architectural point of view. Although there are several problems when considering the practical use of such highly integrated memory devices, such as cell size,[119] sneak current problems,[120] *etc.*, numerous studies to overcome these problems are being simultaneously carried out.[121]

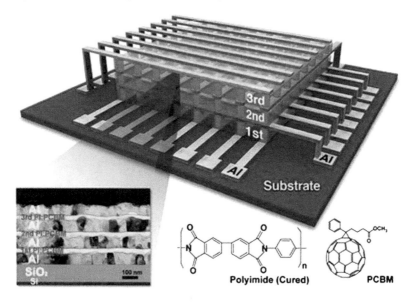

Figure 2.22 Schematic illustration of a 3D-stacked organic resistive memory assembly featuring an 8 × 8 crossbar structure. A total of 192 cells (coloured orange) was produced in three active layers sandwiched between pairs of Al electrode layers. Bottom left: a cross-sectional TEM image of the stacked device, highlighting the three layers of organic memory cells. Bottom right: the chemical structures of cured PI and PCBM used as the memory material. Adapted with permission from ref. 118. Copyright 2010, Wiley.

2.5.2 Increasing Density by Scale Down

In 2011, a junction of size 2 μm, and 100 nm memory devices were reported by Lee and coworkers.[122] The active material used in these devices was the same combination of polyimide and PCBM as that used in the three-dimensional device (Figure 2.23).[118] To avoid damage to the underlying organic active layer in such a small device size by the solution processes required in conventional optical lithography, a nonaqueous direct metal-transfer method was used to pattern the top electrode. The device with a 2 μm junction size (Figure 2.23a) exhibited stable switching characteristics with a high yield of operating memory cells (50 cells out of 64). On the other hand, the device with the 100 nm scale (Figure 2.23b) showed a notable reduction in the ON current level with a poor yield of the operating cell. This is attributed to the decrease in the device junction area and high sensitivity to the surface smoothness of each component (substrate, bottom electrode, active material, and top electrode). This issue is thought to be solvable by incorporation of a planarization layer.

Another approach to achieve a small scale high density memory device is to have each single molecule distributed at regular intervals and to individually read them out. This concept is similar to the hard-disk drive where the transition of magnetization of equally distributed ferromagnetic materials is read by the head. Gao and others have applied this concept to supermolecules such as rotaxane and have used the SPM technique to achieve reproducible nanometer scale recording.[16] It has been revealed that by applying a voltage to a rotaxane-based Langmuir–Blodgett film on a highly-oriented pyrolytic graphite substrate, WORM-type memory characteristics would result. Therefore, recording dots were fabricated written one by one by applying a positive voltage pulse from a scanning tunneling microscope (STM) tip (Figure 2.24). For polymers, Yoshida's group[123] and Zhuge's group[124] individually reported

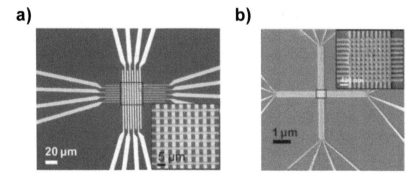

Figure 2.23 (a) An optical image of organic memory devices in a 2 μm scale 8 × 8 array structure. The inset image shows an overview of the entire area of the memory cells. (b) Scanning electron microscopy (SEM) image of 100 nm scale organic memory devices in an 8 × 8 array structure. The inset SEM image shows an overview of the entire area of the memory cells. Adapted with permission from ref. 122. Copyright 2011, Wiley.

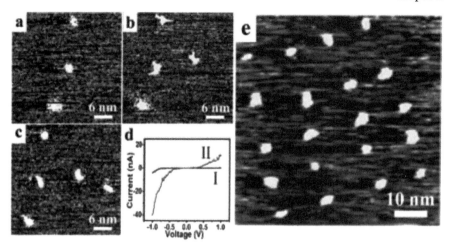

Figure 2.24 STM images of recording dots written on H1 thin films. (a)–(c) Recording dots written one by one through the application of voltage pulses from the STM tip. (d) Typical I–V characteristics measured using the original films (I) and using the induced recording dots (II). (e) STM image of a 5 × 4 recording dots array on the H1/HOPG sample. This film was left exposed to air for about 2 months after its preparation. STM was performed in constant current mode with set points of V_{bias} = 0.65 V and I_{ref} = 0.05 nA. Reprinted with permission from ref. 16. Copyright 2005, American Chemical Society.

high density memory cells by utilizing the SPM technique. For example, a dot pattern with a diameter and pitch of 30 and 50 nm, respectively, could be obtained by applying a pulse voltage using the SPM technique to a polyimide film.[123]

2.5.3 Flexible Memory Devices

One of the most notable features of electronic devices based on organic materials is their flexibility due to the mechanical properties of the functional materials. Flexibility is considered to be important for future electronics involving foldable and wearable devices. Similar to other flexible organic devices, flexible memory devices have also been developed.[51,63,80,125–128] The development of flexible resistive memory devices started in 2007. A 16 byte memory array based on polymer-stabilized gold nanoparticles was fabricated on which an indium zinc oxide coated polycarbonate sheet and a layer of cross-linked poly(vinylphenol) was used as the substrate.[127] It was confirmed that after connecting such a device to the current sensing circuit, the memory array can be correctly addressed and operated, while maintaining a low-power consumption. Kang and coworkers fabricated a flexible WORM memory device with a polypyrrole film/P6FBEu/Au configuration, in which the conducting polypyrrole film was simultaneously used as the substrate and electrode.[51] Conductance switching, with an ON/OFF current ratio up to 200,

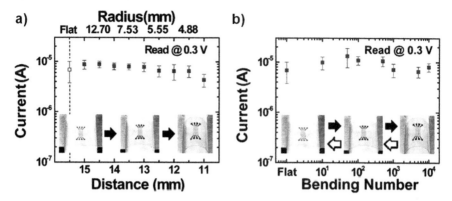

Figure 2.25 Statistical ON current distribution of flexible organic memory devices: (a) under different bending conditions and (b) for different bending cycle numbers. Adapted with permission from ref. 125. Copyright 2011, American Chemical Society.

was observed in this flexible memory device. By using polyimide, which is known to have a high mechanical strength, as the active material, Lee et al. reported the successful demonstration of a mechanically durable flexible memory device.[126] In fact, no significant drop in the ON/OFF ratio was observed even after bending this device down to a radius of 9 mm and also bending 140 times. Subsequently, the durability of this memory device was further enhanced to the point at which stable non-volatile behavior occurred, even after bending at a radius of curvature of 4.2 mm and for up to 10 000 bending cycles (Figure 2.25).[125] In that study, a patterned multilayer graphene electrode was utilized to enhance the mechanical strength of the device.

2.5.4 Transferrable Memory Devices

Very recently, Chen's group developed an efficient way to fabricate a WORM-type transferrable memory device, namely a graphene macromolecular memory label (GMML), using a mixture of poly(3-butylthiophene) (P3BT) and PMMA as the active material.[129,130] The important point in these reports is that the fabrication process of GMMLs is feasible from a practical point of view. First, the graphene was synthesized on copper foil using a typical chemical vapor deposition (CVD) technique. The as-prepared blended mixture of PMMA was then simply spin-coated on the graphene. Subsequently, the supporting copper foil was etched and finally the Al top electrode was deposited. The sticker-like GMML can be transferred onto various substrates, such as a glove, medical wristband, coin, p-type silicon, and organic diode without losing its function (Figure 2.26). This indicates that, as well as independently functioning, the GMML can be easily integrated with other circuit components which lowers the fabrication cost of future electronic modules. By changing the active layer to P3HT and PMMA and also the ratio of

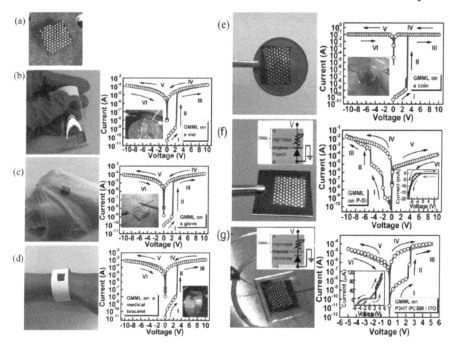

Figure 2.26 (a) Free-standing GMML after etching away copper foil. (b)–(g) Photographs with the corresponding I–V curves alongside for GMMLs transferred onto various non-conventional substrates, including the outside surface of (b) a cylindrical vial, (c) a glove (PVC), (d) a medical wristband, (e) a coin, (f) p-type silicon, and (g) an organic diode (the size of GMML is about 1 cm^2). Adapted with permission from ref. 129. Copyright 2013, Wiley.

semiconducting polymer to PMMA, a rewritable GMML was subsequently fabricated by the same group. Similar to the previous GMML, the rewritable one could also be easily labeled on arbitrary non-conventional substrates including rigid, rough, and non-planar surfaces without losing its original function. Together with the WORM-type GMML, this study provides a new opportunity for memory device applications in various areas.

2.6 Concluding Remarks

In this chapter, we have reviewed the basics and advances in organic resistive memory materials and devices. Due to the extensive activity by research groups worldwide, the fundamental effects of the material or/and device structures on the resulting memory characteristics have gradually been revealed and vast sums of experimental data offer a promising outlook toward the further development of high performance memory materials and devices. Although the structure and operating mechanism of the devices are quite simple, that is, a sandwich device structure utilizing the conductivity response, the mechanism underlying the switching behavior of the functional material is extremely complicated and elusive. So far, several theories have been adopted to explain such phenomena and have been working well to a significant extent. As has been clarified in this chapter, by careful design of the material structures based on this theory, control of the memory characteristics can certainly be achieved. For further advances in organic memory materials and devices, a more systematic investigation of the fundamental switching phenomena and observation of conclusive evidence of the key factors determining the switching behavior will be inevitable.

References

1. Q.-D. Ling, D.-J. Liaw, E. Y.-H. Teo, C. Zhu, D. S.-H. Chan, E.-T. Kang and K.-G. Neoh, *Polymer*, 2007, **48**, 5182.
2. Q. Ling, D. Liaw, C. Zhu, D. S. Chan, E. Kang and K. Neoh, *Prog. Polym. Sci.*, 2008, **33**, 917.
3. T. Kurosawa, T. Higashihara and M. Ueda, *Polym. Chem.*, 2013, **4**, 16.
4. Y. Chen, G. Liu, C. Wang, W. Zhang, R.-W. Li and L. Wang, *Mater. Horiz.*, 2014, **1**, 489.
5. W. P. Lin, S. J. Liu, T. Gong, Q. Zhao and W. Huang, *Adv. Mater.*, 2014, **26**, 570.
6. Y.-C. Lai, K. Ohshimizu, W.-Y. Lee, J.-C. Hsu, T. Higashihara, M. Ueda and W.-C. Chen, *J. Mater. Chem.*, 2011, **21**, 14502.
7. Y.-L. Liu, K.-L. Wang, G.-S. Huang, C.-X. Zhu, E.-S. Tok, K.-G. Neoh and E.-T. Kang, *Chem. Mater.*, 2009, **21**, 3391.
8. Q. Ling, S. Lim, Y. Song, C. Zhu and D. S. Chan, *Langmuir*, 2007, **23**, 312.
9. N.-H. You, C.-C. Chueh, C.-L. Liu, M. Ueda and W.-C. Chen, *Macromolecules*, 2009, **42**, 4456.
10. K.-L. Wang, Y.-L. Liu, J.-W. Lee, K.-G. Neoh and E.-T. Kang, *Macromolecules*, 2010, **43**, 7159.
11. A. Aviram and M. Ratner, *Chem. Phys. Lett.*, 1974, **29**, 277.
12. J. C. Scott and L. D. Bozano, *Adv. Mater.*, 2007, **19**, 1452.
13. C. P. Collier, E. W. Wong, M. Belohradsky and F. M. Raymo, *Science*, 1999, **285**, 391.
14. M. Asakawa, P. R. Ashton, V. Balzani, A. Credi, C. Hamers, G. Mattersteig, M. Montalti, A. N. Shipway, N. Spencer, J. F. Stoddart, M. S. Tolley, M. Venturi, A. J. P. White and D. J. Williams, *Angew. Chem. Int. Ed.*, 1998, **37**, 333.
15. M. Feng, L. Gao, Z. Deng, W. Ji, X. Guo, S. Du, D. Shi, D. Zhang, D. Zhu and H. Gao, *J. Am. Chem. Soc.*, 2007, **129**, 2204.
16. M. Feng, X. Guo, X. Lin, X. He, W. Ji, S. Du, D. Zhang, D. Zhu and H. Gao, *J. Am. Chem. Soc.*, 2005, **127**, 15338.
17. L. P. Ma, Y. L. Song, H. J. Gao, W. B. Zhao, H. Y. Chen, Z. Q. Xue and S. J. Pang, *Appl. Phys. Lett.*, 1996, **69**, 3752.
18. H. J. Gao, K. Sohlberg, Z. Q. Xue, H. Y. Chen, S. M. Hou, L. P. Ma, X. W. Fang, S. J. Pang and S. J. Pennycook, *Phys. Rev. Lett.*, 2008, **84**, 8.

19. A. Bandyopadhyay and A. J. Pal, *Appl. Phys. Lett.*, 2004, **84**, 999.
20. A. Bandhopadhyay and A. J. Pal, *J. Phys. Chem. B*, 2003, **107**, 2531.
21. J. Gong and Y. Osada, *Appl. Phys. Lett.*, 1992, **61**, 2787.
22. C. W. Chu, J. Ouyang, J.-H. Tseng and Y. Yang, *Adv. Mater.*, 2005, **17**, 1440.
23. S. Paul, *IEEE Trans. Nanotechnol.*, 2007, **6**, 191.
24. J. Lin, M. Zheng, J. Chen, X. Gao and D. Ma, *Inorg. Chem.*, 2007, **46**, 341.
25. C. Pearson, J. H. Ahn, M. F. Mabrook, D. A. Zeze, M. C. Petty, K. T. Kamtekar, C. Wang, M. R. Bryce, P. Dimitrakis and D. Tsoukalas, *Appl. Phys. Lett.*, 2007, **91**, 123506.
26. Y. Ma, X. Cao, G. Li, Y. Wen, Y. Yang, J. Wang, S. Du, L. Yang, H. Gao and Y. Song, *Adv. Funct. Mater.*, 2010, **20**, 803.
27. W. Y. Lee, T. Kurosawa, S. T. Lin, T. Higashihara, M. Ueda and W. C. Chen, *Chem. Mater.*, 2011, **23**, 4487.
28. H. Li, Q. Xu, N. Li, R. Sun, J. Ge, J. Lu, H. Gu and F. Yan, *J. Am. Chem. Soc.*, 2010, **132**, 5542.
29. S. Miao, H. Li, Q. Xu, N. Li, J. Zheng, R. Sun, J. Lu and C. M. Li, *J. Mater. Chem.*, 2012, **22**, 16582.
30. Y. Zhang, H. Zhuang, Y. Yang, X. Xu, Q. Bao, N. Li, H. Li, Q. Xu, J. Lu and L. Wang, *J. Phys. Chem. C*, 2012, **116**, 228932.
31. D. He, H. Zhuang, H. Liu, H. Liu, H. Li and J. Lu, *J. Mater. Chem. C*, 2013, **1**, 7883.
32. S. Miao, H. Li, Q. Xu, Y. Li, S. Ji, N. Li, L. Wang, J. Zheng and J. Lu, *Adv. Mater.*, 2012, **24**, 6210.
33. Z. Su, H. Zhuang, H. Liu, H. Li, Q. Xu, J. Lu and L. Wang, *J. Mater. Chem. C*, 2014, **2**, 5673.
34. Y. Sadaoka and Y. Sakai, *J. Chem. Soc., Faraday Trans. 2*, 1976, **72**, 1911.
35. D. M. Pai, *J. Chem. Phys.*, 1970, **52**, 2285.
36. G. Safoula, K. Napo, J. C. Bern, S. Touihri and K. Alimi, *Eur. Polym. J.*, 2001, **37**, 843.
37. S. Lim, Q. Ling, E. Teo and C. Zhu, *Chem. Mater.*, 2007, **19**, 5148.
38. E. Y. H. Teo, Q. D. Ling, Y. Song, Y. P. Tan, W. Wang, E. T. Kang, D. S. H. Chan and C. Zhu, *Org. Electron.*, 2006, **7**, 173.
39. B. Zhang, Y. Chen, Y. Zhang, X. Chen, Z. Chi, J. Yang, J. Ou, M. Q. Zhang, D. Li, D. Wang, M. Liu and J. Zhou, *Phys. Chem. Chem. Phys.*, 2012, **14**, 4640.
40. L. Xie, Q. Ling, X. Hou and W. Huang, *J. Am. Chem. Soc.*, 2008, **130**, 2120.
41. Y.-K. Fang, C.-L. Liu and W.-C. Chen, *J. Mater. Chem.*, 2011, **21**, 4778.
42. Y. Fang, C. Liu, G. Yang, P. Chen and W. Chen, *Macromolecules*, 2011, **44**, 2604.
43. I. Grabchev, I. Moneva, V. Bojinov and S. Guittonneau, *J. Mater. Chem.*, 2000, **10**, 1291.
44. S. Banerjee, E. B. Veale, C. M. Phelan, S. A. Murphy, G. M. Tocci, L. J. Gillespie, D. O. Frimannsson, J. M. Kelly and T. Gunnlaugsson, *Chem. Soc. Rev.*, 2013, **42**, 1601.

45. H. Li, N. Li, R. Sun, H. Gu, J. Ge, J. Lu, Q. Xu, X. Xia and L. Wang, *J. Phys. Chem. C*, 2011, **115**, 8288.
46. Z. Liu, A. A. Yasseri, J. S. Lindsey and D. F. Bocian, *Science*, 2003, **302**, 1543.
47. Q. Ling, Y. Song, S. J. Ding, C. Zhu, D. S. H. Chan, D.-L. Kwong, E.-T. Kang and K.-G. Neoh, *Adv. Mater.*, 2005, **17**, 455.
48. Q. Ling, W. Wang, Y. Song, C. Zhu and D. S. Chan, *J. Phys. Chem. B*, 2006, **110**, 23995.
49. Q.-D. Ling, Y. Song, E. Y. H. Teo, S.-L. Lim, C. Zhu, D. S. H. Chan, D.-L. Kwong, E.-T. Kang and K.-G. Neoh, *Electrochem. Solid-State Lett.*, 2006, **9**, G268.
50. Y. Song, Y. P. Tan, E. Y. H. Teo, C. Zhu, D. S. H. Chan, Q. D. Ling, K. G. Neoh and E. T. Kang, *J. Appl. Phys.*, 2006, **100**, 084508.
51. L. Li, Q.-D. Ling, S.-L. Lim, Y.-P. Tan, C. Zhu, D. S. H. Chan, E.-T. Kang and K.-G. Neoh, *Org. Electron.*, 2007, **8**, 401.
52. S.-J. Liu, Z.-H. Lin, Q. Zhao, Y. Ma, H.-F. Shi, M.-D. Yi, Q.-D. Ling, Q.-L. Fan, C.-X. Zhu, E.-T. Kang and W. Huang, *Adv. Funct. Mater.*, 2011, **21**, 979.
53. S.-J. Liu, W.-P. Lin, M.-D. Yi, W.-J. Xu, C. Tang, Q. Zhao, S.-H. Ye, X.-M. Liu and W. Huang, *J. Mater. Chem.*, 2012, **22**, 22964.
54. S.-J. Liu, P. Wang, Q. Zhao, H.-Y. Yang, J. Wong, H.-B. Sun, X.-C. Dong, W.-P. Lin and W. Huang, *Adv. Mater.*, 2012, **24**, 2901.
55. P. Wang, S.-J. Liu, Z.-H. Lin, X.-C. Dong, Q. Zhao, W.-P. Lin, M.-D. Yi, S.-H. Ye, C.-X. Zhu and W. Huang, *J. Mater. Chem.*, 2012, **22**, 9576.
56. W. Lin, H. Sun, S. Liu, H. Yang, S. Ye, W. Xu, Q. Zhao, X. Liu and W. Huang, *Macromol. Chem. Phys.*, 2012, **213**, 2472.
57. A. Bandyopadhyay, S. Sahu and M. Higuchi, *J. Am. Chem. Soc.*, 2011, **133**, 1168.
58. P. Ferrocene, T. Choi, K. Lee, W. Joo, S. Lee and T. Lee, *J. Am. Chem. Soc.*, 2007, **129**, 9842.
59. V. Cimrovb and D. Neher, *Synth. Met.*, 1996, **76**, 125.
60. S. K. Majee, H. S. Majumdar, A. Bolognesi and A. J. Pal, *Synth. Met.*, 2006, **156**, 828.
61. Q.-D. Ling, Y. Song, S.-L. Lim, E. Y.-H. Teo, Y.-P. Tan, C. Zhu, D. S. H. Chan, D.-L. Kwong, E.-T. Kang and K.-G. Neoh, *Angew. Chem., Int. Ed. Engl.*, 2006, **45**, 2947.
62. B. Cho, J.-M. Yun, S. Song, Y. Ji, D.-Y. Kim and T. Lee, *Adv. Funct. Mater.*, 2011, **21**, 3976.
63. H.-C. Wu, A.-D. Yu, W.-Y. Lee, C.-L. Liu and W.-C. Chen, *Chem. Commun.*, 2012, **48**, 9135.
64. T. Yamamoto, K. Sanechika and A. Yamamoto, *J. Polym. Sci., Polym. Lett. Ed.*, 1980, **18**, 9.
65. G. B. Street and T. C. Clarke, *IBM J. Res. Dev.*, 1981, **25**, 51.
66. H. S. Majumdar, A. Bandyopadhyay, A. Bolognesi and A. J. Pal, *J. Appl. Phys.*, 2002, **91**, 2433.

67. W.-J. Joo, T.-L. Choi, J. Lee, S. K. Lee, M.-S. Jung, N. Kim and J. M. Kim, *J. Phys. Chem. B*, 2006, **110**, 23812.
68. W.-J. Joo, T.-L. Choi, K.-H. Lee and Y. Chung, *J. Phys. Chem. B*, 2007, **111**, 7756.
69. M. C. Tester, B. Cho, S. Song, Y. Ji, H. Choi, H. C. Ko, J. Lee, G. Jung and T. Lee, *IEEE Electron Device Lett.*, 2013, **34**, 51.
70. J. Liu, Z. Yin, X. Cao, F. Zhao, A. Lin, L. Xie, Q. Fan, F. Boey, H. Zhang and W. Huang, *ACS Nano*, 2010, **4**, 3987.
71. C. O. N. Spectus, *Acc. Chem. Res.*, 2008, **41**, 1202.
72. Y.-K. Fang, C.-L. Liu, C. Li, C.-J. Lin, R. Mezzenga and W.-C. Chen, *Adv. Funct. Mater.*, 2010, **20**, 3012.
73. S. Baek, D. Lee, J. Kim, S.-H. Hong, O. Kim and M. Ree, *Adv. Funct. Mater.*, 2007, **17**, 2637.
74. X. Liu, Z. Ji, M. Liu, L. Shang, D. Li and Y. Dai, *Chin. Sci. Bull.*, 2011, **56**, 3178.
75. Q.-D. Ling, F.-C. Chang, Y. Song, C.-X. Zhu, D.-J. Liaw, D. S.-H. Chan, E.-T. Kang and K.-G. Neoh, *J. Am. Chem. Soc.*, 2006, **128**, 8732.
76. T. Kuorosawa, C.-C. Chueh, C.-L. Liu, T. Higashihara, M. Ueda and W.-C. Chen, *Macromolecules*, 2010, **43**, 1236.
77. C. Chen, H. Yen, W. Chen and G. Liou, *J. Polym. Sci., Part A: Polym. Chem.*, 2011, **49**, 3709.
78. C.-J. Chen, H.-J. Yen, W.-C. Chen and G.-S. Liou, *J. Mater. Chem.*, 2012, **22**, 14085.
79. T. Kurosawa, Y.-C. Lai, T. Higashihara, M. Ueda, C.-L. Liu and W.-C. Chen, *Macromolecules*, 2012, **45**, 4556.
80. A.-D. Yu, T. Kurosawa, Y.-C. Lai, T. Higashihara, M. Ueda, C.-L. Liu and W.-C. Chen, *J. Mater. Chem.*, 2012, **22**, 20754.
81. T. J. Lee, C.-W. Chang, S. G. Hahm, K. Kim, S. Park, D. M. Kim, J. Kim, W.-S. Kwon, G.-S. Liou and M. Ree, *Nanotechnology*, 2009, **20**, 135204.
82. D. M. Kim, S. Park, T. J. Lee, S. G. Hahm, K. Kim, J. C. Kim, W. Kwon and M. Ree, *Langmuir*, 2009, **25**, 11713.
83. D. M. Kim, Y.-G. Ko, J. K. Choi, K. Kim, W. Kwon, J. Jung, T.-H. Yoon and M. Ree, *Polymer*, 2012, **53**, 1703.
84. C. Liu, T. Kurosawa, A. Yu, T. Higashihara, M. Ueda and W. Chen, *J. Phys. Chem. C*, 2011, **115**, 5930.
85. G. Tian, D. Wu, S. Qi, Z. Wu and X. Wang, *Macromol. Rapid Commun.*, 2011, **32**, 384.
86. Y. Liu, Y. Zhang, Q. Lan, S. Liu, Z. Qin, L. Chen, C. Zhao, Z. Chi, J. Xu and J. Economy, *Chem. Mater.*, 2012, **24**, 1212.
87. Y.-L. Liu, Q.-D. Ling, E.-T. Kang, K.-G. Neoh, D.-J. Liaw, K.-L. Wang, W.-T. Liou, C.-X. Zhu and D. S.-H. Chan, *J. Appl. Phys.*, 2009, **105**, 044501.
88. K.-L. Wang, Y.-L. Liu, I.-H. Shih, K.-G. Neoh and E.-T. Kang, *J. Polym. Sci., Part A: Polym. Chem.*, 2010, **48**, 5790.
89. S. G. Hahm, S. Choi, S.-H. Hong, T. J. Lee, S. Park, D. M. Kim, W.-S. Kwon, K. Kim, O. Kim and M. Ree, *Adv. Funct. Mater.*, 2008, **18**, 3276.

90. S. G. Hahm, S. Choi, S.-H. Hong, T. J. Lee, S. Park, D. M. Kim, J. C. Kim, W. Kwon, K. Kim, M.-J. Kim, O. Kim and M. Ree, *J. Mater. Chem.*, 2009, **19**, 2207.
91. K. Kim, S. Park, S. G. Hahm, T. J. Lee, D. M. Kim, J. C. Kim, W. Kwon, Y.-G. Ko and M. Ree, *J. Phys. Chem. B*, 2009, **113**, 9143.
92. S. Park, K. Kim, D. M. Kim, W. Kwon, J. Choi and M. Ree, *ACS Appl. Mater. Interfaces*, 2011, **3**, 765.
93. Q. Liu, K. Jiang, L. Wang, Y. Wen, J. Wang, Y. Ma and Y. Song, *Appl. Phys. Lett.*, 2010, **96**, 213305.
94. A. D. Yu, T. Kurosawa, Y. H. Chou, K. Aoyagi, Y. Shoji, T. Higashihara, M. Ueda, C. L. Liu and W. C. Chen, *ACS Appl. Mater. Interfaces*, 2013, **5**, 4921.
95. J. Ouyang, C.-W. Chu, C. R. Szmanda, L. Ma and Y. Yang, *Nat. Mater.*, 2004, **3**, 918.
96. J. Ouyang, C. Chu, R. J. Tseng and A. Prakash, *Proc. IEEE*, 2005, **93**, 1287.
97. A. Prakash, J. Ouyang, J.-L. Lin and Y. Yang, *J. Appl. Phys.*, 2006, **100**, 054309.
98. Y. Song, Q. D. Ling, S. L. Lim, E. Y. H. Teo, Y. P. Tan, L. Li, E. T. Kang, D. S. H. Chan and C. Zhu, *IEEE Electron Device Lett.*, 2007, **28**, 107.
99. S. I. White, P. M. Vora, J. M. Kikkawa and K. I. Winey, *Adv. Funct. Mater.*, 2011, **21**, 233.
100. W. T. Kim, J. H. Jung, T. W. Kim and D. I. Son, *Appl. Phys. Lett.*, 2010, **96**, 253301.
101. G. Tian, D. Wu, L. Shi, S. Qi and Z. Wu, *RSC Adv.*, 2012, **2**, 9846.
102. B. C. Das and A. J. Pal, *ACS Nano*, 2008, **2**, 1930.
103. K. H. Park, J. H. Jung, F. Li, D. I. Son and T. W. Kim, *Appl. Phys. Lett.*, 2008, **93**, 132104.
104. J. H. Shim, J. H. Jung, M. H. Lee, T. W. Kim, D. I. Son, A. N. Han and S. W. Kim, *Org. Electron.*, 2011, **12**, 1566.
105. J. Liu, Z. Zeng, X. Cao, G. Lu, L.-H. Wang, Q.-L. Fan, W. Huang and H. Zhang, *Small*, 2012, **8**, 3517.
106. J. H. Jung, J.-H. Kim, T. W. Kim, M. S. Song, Y.-H. Kim and S. Jin, *Appl. Phys. Lett.*, 2006, **89**, 122110.
107. D. Adler, *CRC Crit. Rev. Solid State Sci.*, 1971, **2**, 317.
108. M. Popo and C. Swenberg, *Electronic Processes in Organic Crystals and Polymers*, Oxford University Press, New York, 2nd edn, 1999.
109. S. Sze and K. Ng, *Physics of Semiconductor Devices*, Wiley-InterScience, 3rd edn, 2007.
110. H. K. Henisch, *Appl. Phys. Lett.*, 1974, **24**, 589.
111. L. F. Pender and R. J. Fleming, *J. Appl. Phys.*, 1975, **46**, 3426.
112. Y. Segui, *J. Appl. Phys.*, 1976, **47**, 140.
113. P. Sliva, G. Dir and C. Griffiths, *J. Non-Cryst. Solids*, 1970, **2**, 316.
114. J. Gazso, *Thin Solid Films*, 2000, **21**, 43.
115. Y. Sakai, Y. Sadaoka and G. Okada, *J. Mater. Sci.*, 1984, **19**, 1333.
116. H. Carchano, *Appl. Phys. Lett.*, 1971, **19**, 414.

117. D. M. Taylor and D. Street, *IEEE Trans. Dielectr. Electr. Insul.*, 2006, **13**, 1063.
118. S. Song, B. Cho, T.-W. Kim, Y. Ji, M. Jo, G. Wang, M. Choe, Y. H. Kahng, H. Hwang and T. Lee, *Adv. Mater.*, 2010, **22**, 5048.
119. N. Mosfets, A. Asenov, A. R. Brown, J. H. Davies, S. Kaya and G. Slavcheva, *IEEE Trans. Electron Devices*, 2003, **50**, 1837.
120. M. M. Zieglert and M. R. Stan, *IEEE-NANO 2002. Proc. 2002 2nd IEEE Conf. Nanotechnology, 2002*, 2002, 323.
121. J. Y. Seok, S. J. Song, J. H. Yoon, K. J. Yoon, T. H. Park, D. E. Kwon, H. Lim, G. H. Kim, D. S. Jeong and C. S. Hwang, *Adv. Funct. Mater.*, 2014, **24**, 5316.
122. J. J. Kim, B. Cho, K. S. Kim, T. Lee and G. Y. Jung, *Adv. Mater.*, 2011, **23**, 2104.
123. S. Yoshida, T. Ono and M. Esashi, *Nanotechnology*, 2011, **22**, 335302.
124. B. Hu, F. Zhuge, X. Zhu, S. Peng, X. Chen, L. Pan, Q. Yan and R.-W. Li, *J. Mater. Chem.*, 2012, **22**, 520.
125. M. G. Electrodes, Y. Ji, S. Lee, B. Cho, S. Song and T. Lee, *ACS Nano*, 2011, **5**, 5995.
126. Y. Ji, B. Cho, S. Song, T.-W. Kim, M. Choe, Y. H. Kahng and T. Lee, *Adv. Mater.*, 2010, **22**, 3071.
127. H. Lin, Z. Pei, J. Chen, C. Kung, Y. Lin, C. Tseng and Y. Chan, *Electron Device Meet. 2007, IEDM 2007. IEEE Int.*, 2007, 233.
128. D. Son, J. Kim, D. Park, W. K. Choi, F. Li and J. H. Ham, *Nanotechnology*, 2008, **19**, 055204.
129. Y.-C. Lai, F.-C. Hsu, J.-Y. Chen, J.-H. He, T.-C. Chang, Y.-P. Hsieh, T.-Y. Lin, Y.-J. Yang and Y.-F. Chen, *Adv. Mater.*, 2013, **25**, 2733.
130. Y.-C. Lai, Y.-X. Wang, Y.-C. Huang, T.-Y. Lin, Y.-P. Hsieh, Y.-J. Yang and Y.-F. Chen, *Adv. Funct. Mater.*, 2014, **24**, 1430.

CHAPTER 3

Donor–Acceptor Organic Molecule Resistor Switching Memory Devices

JIANMEI LU*[a], HUA LI[a], AND QING-FENG XU[a]

[a]College of Chemistry, Chemical Engineering and Materials Science, Collaborative Innovation Center of Suzhou Nano Science and Technology, Soochow University, Suzhou 215123, China
*E-mail: lujm@suda.edu.cn

3.1 Introduction

Organic molecules have been proposed as an active part of a large variety of nonvolatile memory devices, including resistors, diodes and transistors.[1–5] Organic memory devices can be classified into: volatile and nonvolatile according to the retention of their conductivity state; 'WORM', 'FLASH', 'DRAM' and 'SRAM' according to their different 'read–write–erase' behavior; 'binary', 'ternary' and 'multilevel-bit' according to the memory bits.[6–10]

Materials with two distinct stable conductivity states under an electric field are called binary electric memory materials.[11] Compared to their polymer and inorganic counterparts, organic molecules have distinct advantages such as defined structures and low production cost, and their tunable molecular energy levels were inclined to obtain the exact structure–performance relationship, and low production cost. Since a highly conducting complex, tetrathioful-valene–tetracyanoquinodimethane (TTF–TCNQ), was studied in the 1970s,[12,13] charge transfer molecules have been one of the most attractive candidates for high density electrical data storage.[14-22] Among the reported organic memory materials, organic molecules can be divided into two types: organic molecules without a metal and organometallic complexes.

In this chapter, we review those organic molecules with the D–A structure and their interesting memory performances.

3.2 Organic Molecules as Active Material in Memory Devices

Initially, organic films for data recording based on charge transfer were mainly metal–organic complexes represented by Ag-TCNQ and Cu-TCNQ.[23-25] Compared with their organometallic counterparts, pure organic films without a metal have good and uniform physico-chemical properties. Therefore, organic donor compounds have been proposed to replace metal donors, and a series of organic electron donor and electron acceptor materials have been explored to prepare organic composite thin films for electrical recording.[26-28] However, recent molecules based on the donor–acceptor (D–A) model have attracted considerable attention owing to their significant advantages: (i) the formation of charge transfer (CT) complexes under an external electric field, which can realize transformation of the memory device from the OFF state to the ON state; (ii) a decrease of the carrier injection barrier (the simultaneous increase of the LUMO level and decrease of the HOMO level); (iii) potential application in single-molecule based devices for ultrahigh density data storage.

Molecule design is predominant in molecule based memory devices due to the lack of a 'molecular library' for memory materials. According to numerous studies, a suitable combination of electron donor and acceptor moieties is necessary for improving memory performance. In this chapter, we list most of the reported molecules with different donor and acceptor groups and their memory performances. In particular, the molecules were simply classified according to their pendant donor or acceptor groups.

3.2.1 The Common Electron Donor and Acceptor Groups Used in Molecular Based Memory Devices

3.2.1.1 Molecules with the TPA Group

The TPA (triphenylamine) unit is a classical electron donor group in organic electronics. One of the earliest studies on memory performance by Song and coworkers used TPA as backbone or terminal group.[29,30] They employed molecules containing the electron donor TPA for memory device applications. They demonstrated that different structures of cyanide-substituted triphenylamine molecules such as BDOYM and TABM affected the ON/OFF ratio and the reversibility of their memory devices significantly. For example, when the BDOYM molecule was deposited on Au(111), charge transfer was observed in

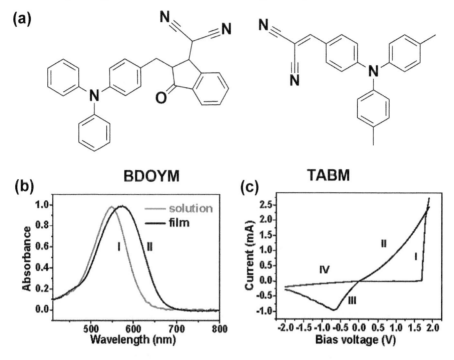

Figure 3.1 (a) Structures of BDOYM (left) and TABM (right); (b) UV-vis spectra of BDOYM (I, in cyclohexane solution; II, film on ITO substrate); (c) macroscopic I–V characteristics of the film. Reproduced with permission from ref. 30. Copyright 2007 American Chemical Society.

UV-vis spectra (Figure 3.1b). In addition, by using an STM tip, the molecule was switched from 'OFF' to 'ON' to fulfill a 'writing' process when bias was applied.

Recently, Song and coworkers extended the TPA containing D–A structure to form D–π–A–π–D and A–π–D–π–A molecules[31] (Figure 3.2) for better memory performance (Figure 3.3). The results show that the devices prepared from the D–π–A–π–D molecule (TPDBCN) had FLASH (write-read-erase) characteristics with an ON/OFF ratio of up to 10^6. In contrast, the A–π–D–π–A molecules (TPDYCN1 and TPDYCN2) only showed 'WORM' performance with a relatively low ON/OFF ratio. This study indicates that the arrangement of the donor and acceptor moieties affects the electrical switching characteristics.

3.2.1.2 Molecules with the Azo Group

Azobenzene compounds are usually selected as electrical and optical materials due to their good charge transfer ability.[32]

Song and coworkers once prepared an azo molecule with the D–A structure (CDHAB, Figure 3.4).[33] Its crystalline thin film was drop-cast by molecular

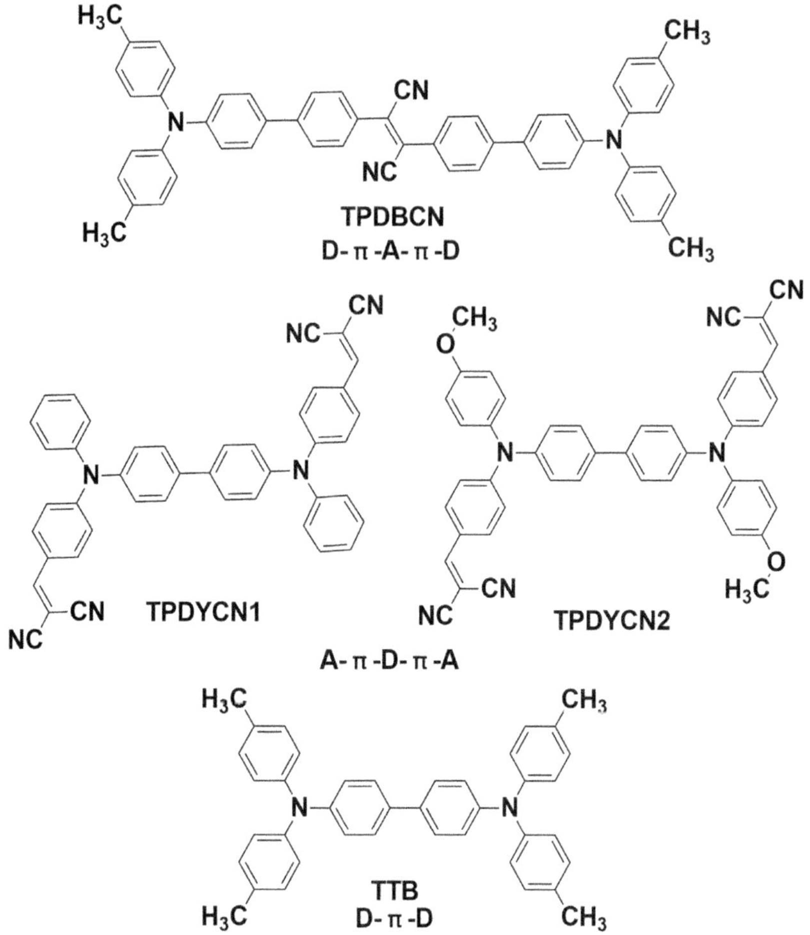

Figure 3.2 Structures of molecules bearing TPA moiety. Reproduced with permission from ref. 31. Copyright 2010 WILEY-VCH Verlag GmbH & Co. KGaA, Weinheim.

self-assembly on a HOPG substrate. Reversible data storage properties were realized on the thin film by an STM tip (surface unit is 1.8 × 0.8 nm^2) when applying pulsed voltages. A mechanism was suggested whereby formation of the recording dots occurred due to reversible intermolecular charge transfer induced by the electric field.

This result also inspired our recent research, in particular on multilevel memory performance mentioned later. We have synthesized a dozen azobenzene derivatives and have fabricated them into sandwich memory devices. For example, we synthesized three azo compounds in which the end electron donors of different alkyl chains were adjusted for better film morphology (Figure 3.5).[34] The film of Azo3 is more planar and the switching threshold voltage of the fabricated device is much lower as the alkyl chain of Azo3 is

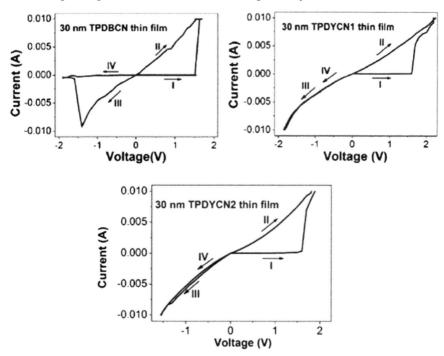

Figure 3.3 Macroscopic I–V characteristics of ITO/TPDBCN (30 nm thick)/Al, ITO/TPDYCN1 (30 nm thick)/Al and ITO/TPDYCN2 (30 nm thick)/Al. Reproduced with permission from ref. 31. Copyright 2010 WILEY-VCH Verlag GmbH & Co. KGaA, Weinheim.

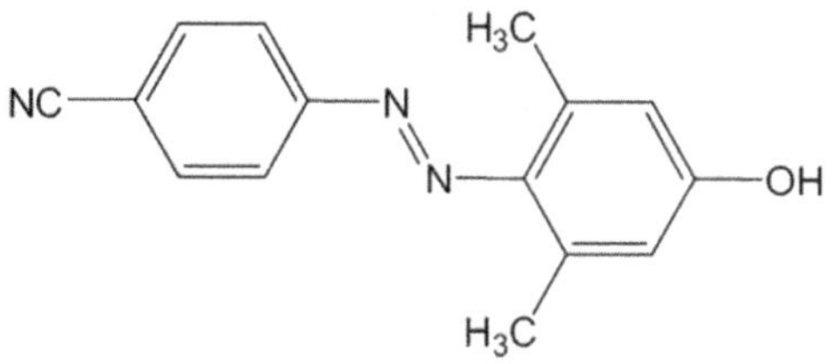

Figure 3.4 Molecular structure of 4′-cyano-2,6-dimethyl-4-hydroxy azobenzene (CDHAB). Reproduced with permission from ref. 33. Copyright 2006 WILEY-VCH Verlag GmbH & Co. KGaA, Weinheim.

the longest. This indicates that a suitable alkyl chain length designed in the target molecule can potentially achieve better performance.

3.2.1.3 Molecules with the Naphthalimide Group

The naphthalimide group is considered to be an electron acceptor which shows unique optical and electronic properties.[35,36] We used the naphthalimide group as backbone to construct D–A molecules with higher memory

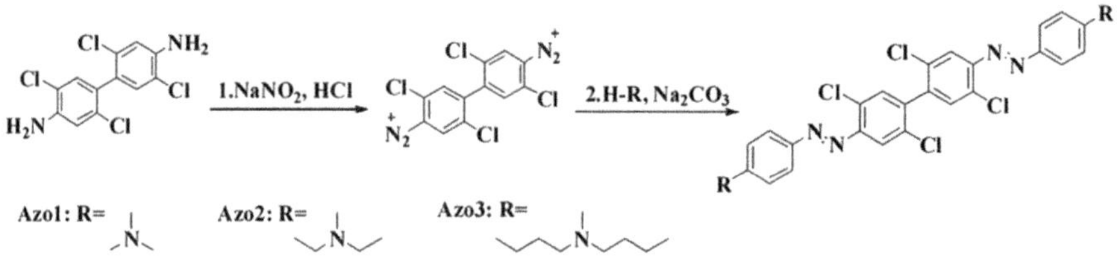

Figure 3.5 Synthetic route for Azo1–3 (Azo1: 4,4′-(1E,1′E)-(2,2′,5,5′-tetrachlorobiphenyl-4,4′-diyl)-bis-(diazene-2,1-diyl)-bis(N,N-dimethylaniline), Azo2: 4′-(1E,1′E)-(2,2′,5,5′-tetrachlorobiphenyl-4,4′-diyl)bis-(diazene-2,1-diyl)-bis(N,N-diethylaniline), Azo3: 4,4′-(1E,1′E)-(2,2′,5,5′-tetrachlorobiphenyl-4,4′-diyl)bis-(diazene-2,1-diyl)bis(N,N-dibutylaniline)). Reproduced with permission from ref. 34.

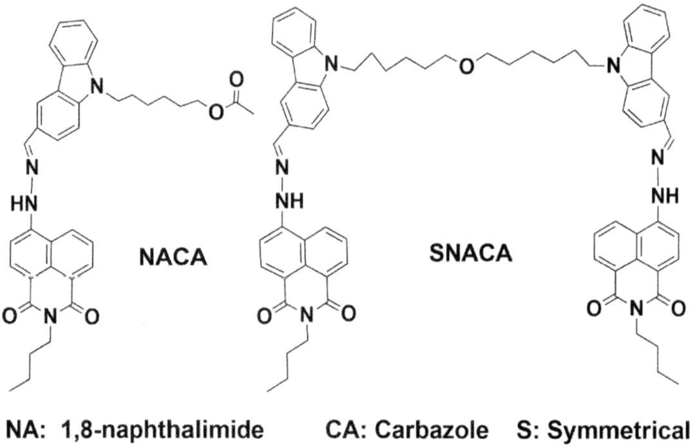

Figure 3.6 Molecular structures of NACA and SNACA. Reproduced with permission from ref. 37.

performance. Compound NACA (3-(N-butyl-4-carbaldehyde-1,8-naphthalimide)-9-hexylacetate-9H-carbazole) comprised naphthalimide (NA) as the electron acceptor and carbazole (CA) as electron donor linked by a hydrazone bridge.[37] Compound SNACA was composed of two NA-CA moieties with a long ether chain as shown in Figure 3.6. SNACA has better solubility and film-formation properties than the NACA molecule due to the longer chain. When they are fabricated into sandwich devices by spin coating, the SNACA molecule based memory device shows volatile memory behavior (DRAM) with a high ON/OFF current ratio (10^9), while the NACA molecule does not show any memory properties. This indicates that the longer alkyl chain improves the film quality and memory performance. A detailed study of the memory mechanism revealed that charge trapping–detrapping occurred under an electric field. Moreover, the SNACA molecule based memory device exhibited electro-optical switching behavior (Figure 3.7b), which may have potential in optical sensors.

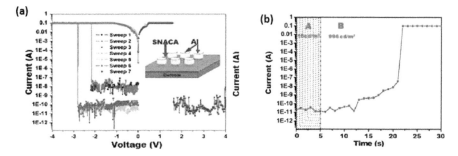

Figure 3.7 (a) Current–voltage (*I*–*V*) characteristics of SNACA based device with ITO as bottom electrode. Sweeps 1–3 are measured under ambient conditions. (b) Electro-optical switch performance of the SNACA based device (area A indicates a light intensity of 15 cd m^{-2}, area B indicates a light intensity of 996 cd m^{-2}). Reproduced with permission from ref. 37.

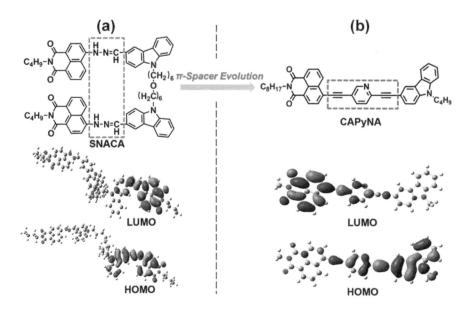

Figure 3.8 Molecular structures, HOMO and LUMO energy levels of SNACA (a) and CAPyNA (b). Reproduced with permission from ref. 38.

Based on the above results, a π-spacer, pyridyl acetylene, was introduced into the conjugated molecular skeleton to replace the original hydrazone bond between the carbazole unit and the naphthalimide moiety (CAPyNA, see Figure 3.8).[38] The incorporation of a rigid group could alter the distribution of electron density throughout the molecular backbone and tune the memory performance of the device. As shown in Figure 3.8b, the theoretical simulation results showed that the electron density observably shifted from the carbazole group to the naphthalimide moiety during the HOMO to

LUMO excitation, indicating that the new π-spacer affects the electron density distribution of the molecule. In addition, the pyridyl acetylene linkage is expected to facilitate molecular assembly into one-dimensional molecular tapes *via* hydrogen bonding to afford well-ordered structures due to the complete planarity and highly polarized properties, leading to intermolecular interactions, such as heteroatom contact or π–π stacking. The CAPyNA-based sandwich memory device obtained from the solution process exhibited WORM bistable memory behavior.

3.2.1.4 Molecules with the 4,4′-(Hexafluoroisopropylidene)-diphthalic Anhydride Group

The 4,4′-(hexafluoroisopropylidene)-diphthalic anhydride group (6FDA) is often used in copolymers as an electron acceptor. Along these lines, Chen and coworkers introduced two small molecules with 6FDA as electron acceptor and TPA or APC as electron donor (Figure 3.9). Devices based on D–A–D oligoimides (ATPA-6FDA) revealed reversible nonvolatile negative-differential-resistance (NDR) characteristics.[39] Without applying voltage stress, the ON and OFF states of the devices showed no obvious degradation over an

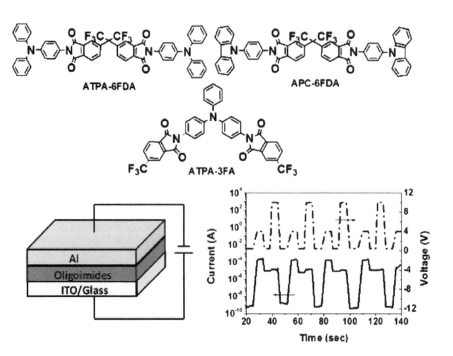

Figure 3.9 Molecular structures of ATPA-6FDA, APC-6FDA, and ATPA-3FA, their memory device and their retention time. Reproduced with permission from ref. 39. Copyright 2011 American Chemical Society.

operation time of 10 s and 10^8 read pulses. However, devices prepared from the A–D–A (APC-6FDA) oligoimide showed only insulating properties.

3.2.1.5 Molecules with the Carbazole Group

The carbazole group is a well-known candidate for hole transport materials in organic photo-electronic devices due to its good hole transport capability.[40] It has been frequently used in functional polymers. Our group once synthesized several bicarbazole derivatives to investigate their memory behavior (Figure 3.10).[41] In this work, bicarbazole served as the electron donor, and benzothiazole (BCZ-BT), nitryl (BCZ-NO$_2$) and 1,10-dicyanovinyl (BCZ-CN) were used as electron acceptors to form a D–A structure with a variable electron-delocalization extent and electron-withdrawing strength. The BCZ-BT based memory device exhibited volatile static random access memory (SRAM) switching behavior, while devices based on BCZ-NO$_2$ were found to exhibit stable nonvolatile WORM characteristics, and the BCZ-CN device

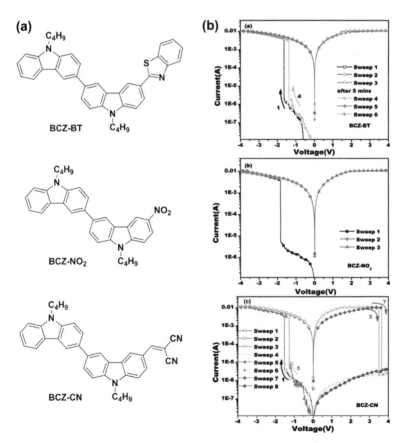

Figure 3.10 Molecular structures of BCZ-BT, BCZ-NO$_2$ and BCZ-CZ (a) and their I–V curves (b). Reproduced with permission from ref. 41.

acted as rewritable FLASH memory with an ON/OFF current ratio of about 10^4. Therefore, from a structural point of view, the retention of the conductivity state of the devices is tunable by adjusting the electron withdrawing strength of the terminal acceptor groups.

3.3 Organometallic Complexes as Active Materials in Memory Devices

Organometallic complex based memory behavior was studied from the 1970s. Generation of an ionic coordination complex with a potential organic bistable memory application was a challenge.

In the early stages, the use of metal–organic molecules as active materials was because metal–organic molecules fulfill the requirements of good (cat)ionic conductors. Cu(TCNQ) (7,7,8,8-tetracyano-*p*-quinodimethane)[42] was proposed as a solid ionic conductor to provide mobile Cu(I) to form conductive (metallic) filaments by the electrochemical reduction of Cu(I)/Cu(0). This study showed that the reduction occurs at the negative electrode, and the filaments grow in a (porous) oxide layer (for example native aluminum oxide) at that electrode. This proposed role of CuTCNQ also was probed with a scanning tunneling microscope tip, where metallic filaments grew from the tip and bridged the nanometer gap between the material and the tip.[43]

However, recent studies showed that the stable metal–N coordination would induce reversible charge transfer through the applied electric field. There are several studies which show that metal complexes can act as charge transfer complexes which have an important effect on the 'read–write–erase' memory performance.

Ma's group reported memory effects in single-layer organic light-emitting devices (OLEDs) based on lanthanide complexes (M = Sm^{3+}, Gd^{3+}, and Eu^{3+}) (Figure 3.11).[44] The device structure used in this study was ITO/PEDOT/PVK:RE complex/LiF/Ca/Ag (PEDOT is 3,4-poly(ethylenedioxythiophene)-poly(styrenesulfonate)). These devices can be switched electrically between two states with an ON/OFF ratio of about 2 orders of magnitude, and the switches are nonvolatile. Over 10^6 write–read–erase cycles were realized, and no obvious current degradation was observed. The results provided a simple approach of doping RE organic complexes into a conducting polymer as active medium in memory devices. For the case of lanthanide complexes doped with PVK, it is reasonable to attribute the electronic transition to an electrical-field-induced charge transfer between the polymer and metal complexes. The high-conductance state of the device is due to the higher hole mobility of PVK.

Subsequently, an ionic ruthenium dye with different charge transfer ligands (bis(2,2′-bipyridyl)(triazolopyridyl)ruthenium(II) complex) was synthesized and used as a key element in organic bistable devices (Figure 3.12).[45] The complex was cast by spin coating on ITO glass and Al as top electrode was deposited by thermal evaporation. This device is much simpler than the OLED device.

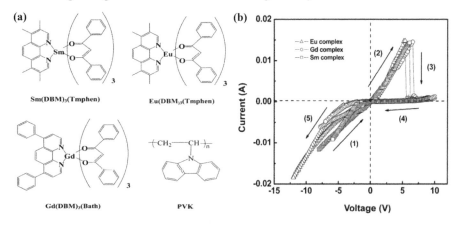

Figure 3.11 (a) Molecular structures of Sm(DBM)$_3$(Tmphen), Eu(DBM)$_3$-(Tmphen), Gd(DBM)$_3$(Bath), and PVK. (b) Current–voltage characteristics of the devices (ITO/PEDOT/PVK: M(DBM)$_3$(Tmphen)/LiF/Ca/Ag, M = Sm, Eu and Gd). Reproduced with permission from ref. 44. Copyright 2006 American Chemical Society.

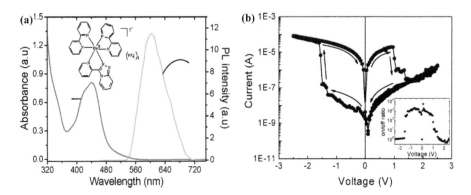

Figure 3.12 (a) Absorbance and photoluminance spectra of ruthenium complex in acetonitrile solution. Inset shows the molecular structure of the ruthenium complex. (b) Current–voltage characteristics of a device based on a spin-cast film of the ruthenium complex for two sweep directions in log scale. The ratio between the ON- and OFF-state current as a function of applied voltage for the same sweeps is represented in the inset figure. Reproduced with permission from ref. 45. Copyright 2008 American Chemical Society.

High conductance switching of the complex with an ON/OFF ratio of 10^3 and rewritability with associated memory effects was observed in these devices.

The basic concept behind the porphyrin-based memory behavior is outlined in Figure 3.13.[46] Porphyrins were chosen due to the characteristics of their redox properties, which provide the basis for writing/reading the memory cell. The important redox characteristics of porphyrins include: (1) π-cation radicals that are stable under ambient conditions, facilitating practical

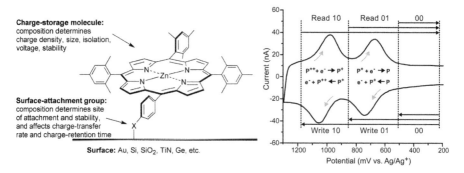

Figure 3.13 Porphyrin-based memory element (left panel). Redox-based read/write process; P = porphyrin (right panel). Reproduced with permission from ref. 46. Copyright 2011 American Chemical Society.

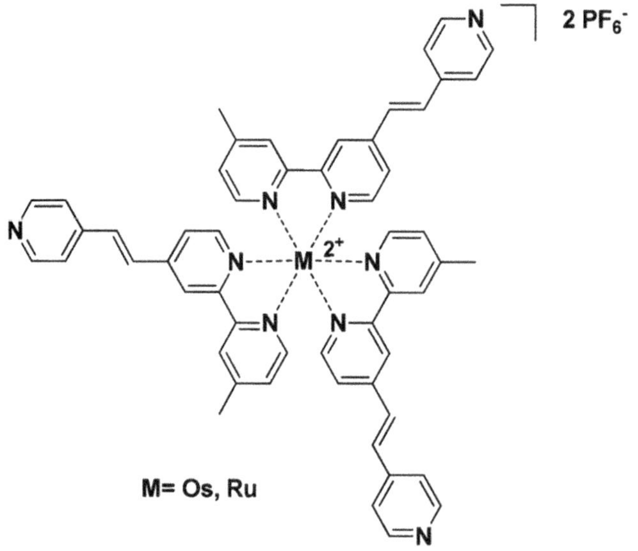

Figure 3.14 Typical polypyridyl complexes used for active molecules in multilevel memory. Reproduced with permission from ref. 47. Copyright 2011 American Chemical Society.

applications; (2) multiple cationic states that are accessible at low potentials, affording multibit information storage with low power consumption; (3) capability of storing charge for extended periods (up to several minutes) in the absence of an applied potential, further diminishing power consumption and significantly attenuating the refresh rates required in a memory device.

Moreover, metal polypyridyl complexes with electrochromic properties also play important roles in molecule based memory devices.[47] A self-propagating molecular-based assembly (SPMA) was used by Ruiter and coworkers to demonstrate binary and ternary memory as shown Figure 3.14.[47,48]

In conclusion, metal complexes have been considered as solid ion conductors or charge transfer complexes for switching the conductivity states of memory devices. However, the number of coordinated ligand types is still small. More attention should be paid to molecule design and mechanism investigation.

3.4 Organic Molecules for Multilevel Resistive Memory Devices

It is well known that applying multivalued digits in data storage could increase memory capacity. Multilevel memory devices mean that three or more stable conductivity states exist in the same bit. Updating from classical binary memory, 2^n, to multi-state memory, M^n (M is the number of available states for each bit and n is the number of digits), multilevel memory leads to a significant increase in memory capacity for ultrahigh density data storage. Undoubtedly, molecule based multilevel memory devices are talented because of their tunable electronic properties, and molecule-scale miniaturization in device design. To obtain molecules with these electronic properties is becoming of particular interest. For instance, electroactive molecules such as porphyrin metal complexes and polypyridyl complexes with two or more reversible oxidation waves are promising candidates. Their redox mechanism is also clearly illustrated by electrical chemistry. There are several reviews mentioned above which contribute to this topic.

Another approach is the recent use of pure organic molecules with chemically tunable "traps" (electron acceptors) which is raising great hopes for simplification and miniaturization of multilevel devices.[49] These molecule based devices form three or more stable electric conductivity states under an electric field. Their particular structure, distinct memory behavior and structure–performance relationship are discussed in the following section.

3.4.1 Azo Benzene Derivative Based Ternary Memory Devices

A series of azo molecules with two kinds of different electron acceptors (acting as charge traps) have been shown to exhibit stable ternary nonvolatile memory performance (Figure 3.15). The first reported molecule Azo1 has a symmetrical structure with a dimethyl amino (donor) azobenzene linked by a sulfone group. There are two electron acceptors: sulfone (acceptor 1) and azo (acceptor 2) groups with different electron accepting properties, which means that there are two kinds of charge traps, deep and shallow charge traps, respectively.[50] The I–V characteristics of the ITO/Azo1/Al device show that two

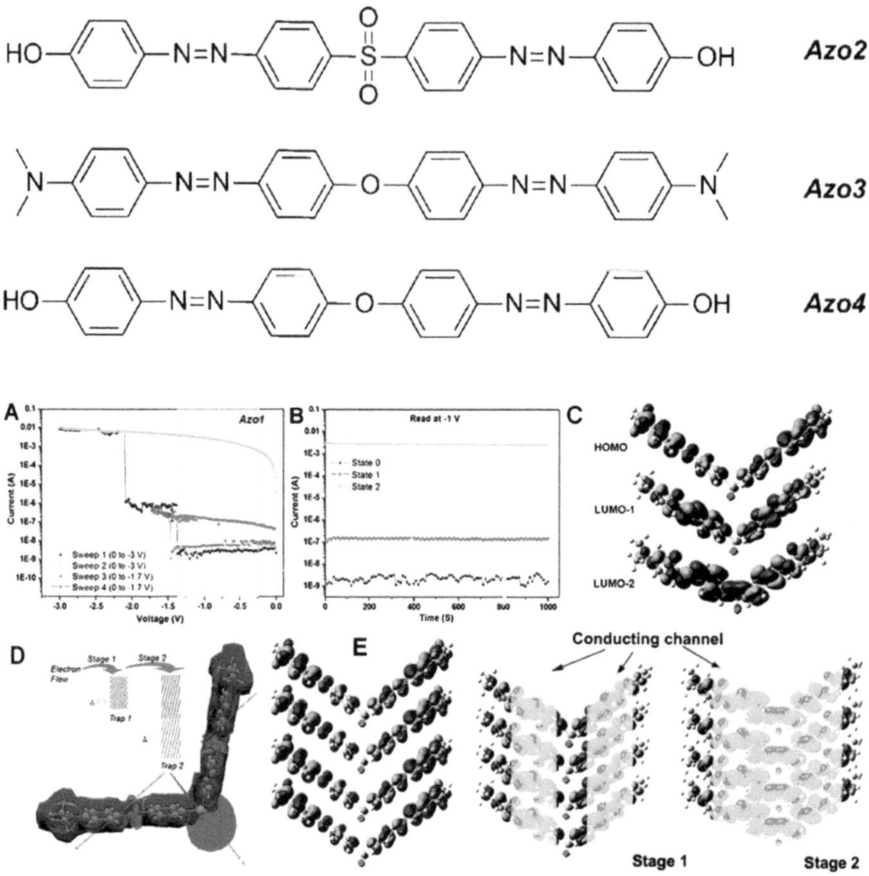

Figure 3.15 Molecular structures of Azo1–4; (A) I–V characteristics of the memory device fabricated using Azo1. (B) Stability of the device in three states under a constant "read" voltage of −1 V. (C) DFT molecular simulation results (B3LYP/6-311G* level): LUMO−1, LUMO−2, HOMO. (D) Molecular simulation results for ESP surfaces of Azo1 and the proposed electron-flow mechanism at the molecular level. (E) Molecular stacking model and illustrations of the conducting channels of the stacking molecules. Reproduced with permission from ref. 50. Copyright 2010 American Chemical Society.

abrupt current increases occurred at switching voltages of −1.37 (from OFF to ON1) and −2.09 V (from ON1 to ON2) in the first sweep from 0 to −3.0 V; the current ratio of the "0", "1", and "2" states is $1:10^2:10^6$. This device exhibits ternary WORM memory behavior which is the first case of WORM behavior found using a pure organic molecule.[51] For further rational molecular design and better understanding of the mechanism, several derivatives were reported thereafter and their ternary memory performances were also studied. For instance, two molecules DPAPIT and DPAPPD[52] were synthesized to study the

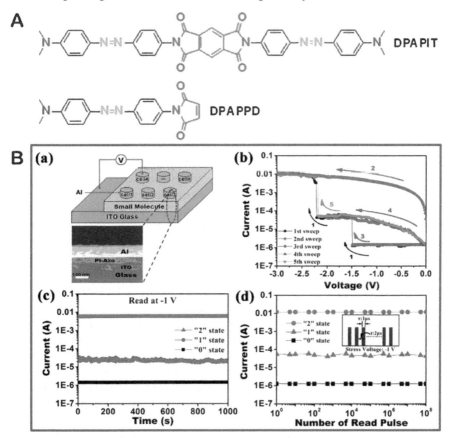

Figure 3.16 (A) Chemical structure of DPAPIT and DPAPPD; (B) (a) (top) illustration of sandwich device; (bottom) SEM image of a cross-section view of the device; (b) current–voltage (I–V) characteristics of the memory device fabricated using a DPAPIT film annealed at 80 °C; (c) stability of the device in three states under a constant "read" voltage of −1 V; (d) effect of read cycles on three states under a stress of voltage of −1 V. Reproduced with permission from ref. 52.

effect of molecular conjugate length (Figure 3.16). Both molecules have an electron-donating dimethylamino azobenzene block, however DPAPIT has an electron accepting phthalimide core unit which is bridged by another dimethylamino azobenzene block while DPAPPD is only ended by a small pyrrole group.

The I–V characteristics of the memory devices show that the DPAPIT-based device has a nonvolatile ternary memory effect with a current ratio of $1:10^{1.7}:10^4$ for the "0", "1" and "2" states and with both of the switching threshold voltages lower than −3 V.

In contrast, the shorter molecule DPAPPD showed an amorphous microstructure verified by XRD. Its memory devices had no obvious conductive switching behavior observed when a bias was applied. This illustrates that higher conjugation of the molecule is favorable for formation of a crystalline film and memory performance.

Subsequent research showed that molecule planarity is also important for high performance ternary memory behavior. Two compounds, DPKAZO 1 and FKAZO 2 (Figure 3.17), similar to the above mentioned AZO compounds, were studied. The DPKAZO 1 based device showed switching threshold voltages (STVs) of −1.50 and −2.61 V with the current ratio of the "0", "1", and "2" states being $1:10^2:10^6$, while the FKAZO 2 based device showed STVs of −1.05 and −1.81 V with the current ratio of "0", "1", and "2" states being $1:1.2 \times 10^2:4 \times 10^4$ (Figure 3.18). The lower threshold voltage of the FKAZO 2 based device was attributed to the more planar structure of FKAZO 2. According to theoretical calculations,[53] because of the twisted conformation of DPKAZO 1, its π–π stacking interactions are weaker than those of FKAZO 2, resulting in an increased interlayer packing distance and weaker molecular packing. The density of DPKAZO 1 is 1.15 g cm^{-3} and the interlayer packing distance is 5.1 Å. FKAZO 2 shows good planarity, and is almost parallel to the ITO substrate without intralayer π–π interactions. The π–π stacking between the layers in FKAZO 2 is stronger than that in DPKAZO 1, which has a density of 1.21 g cm^{-3} and an interlayer packing distance of 3.8 Å. The calculated results are in agreement with the XRD, AFM, and SAED results, and indicate that FKAZO 2 is more planar and packs more efficiently than DPKAZO 1. Efficient molecular packing and an orderly arrangement in the nano-film can facilitate the formation of free carrier transport pathways and reduce the charge carrier injection and

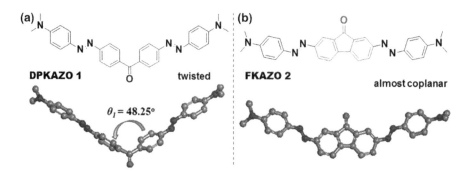

Figure 3.17 Chemical structures, optimized geometries, and dihedral angle (θ_1) between the planes of two benzene rings adjacent to the central electron-deficient unit: (a) DPKAZO 1 and (b) FKAZO 2. Reproduced with permission from ref. 53. Copyright 2012 WILEY-VCH Verlag GmbH & Co. KGaA, Weinheim.

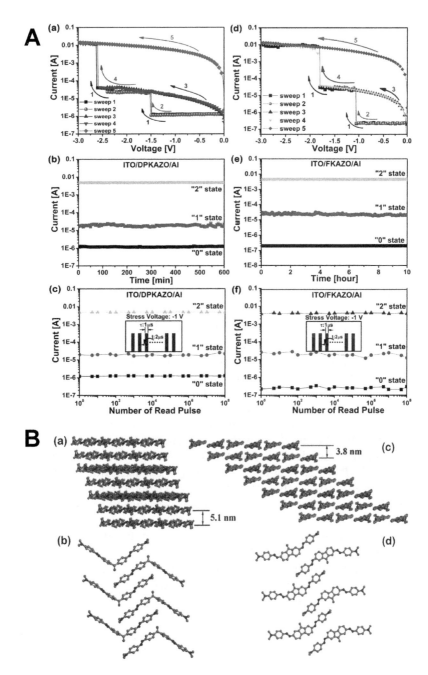

Figure 3.18 (A) Current–voltage (*I–V*) characteristics, effect of the operation time (at −1 V) on the device current in the "0", "1", and "2" states and effect of a read pulse of −1 V on the three states of an ITO/DPKAZO 1/Al device (a)–(c) and an ITO/FKAZO 2/Al device (d)–(f). (B) (a) The stacking layers of DPKAZO 1. (b) One layer of DPKAZO 1 including multiple DPKAZO 1 molecules in twisted conformations. (c) The stacking layers of FKAZO 2. (d) One layer of FKAZO 2 including multiple FKAZO 2 molecules in planar conformation. Due to the twisted conformation of DPKAZO 1, the π–π stacking interaction in DPKAZO 1 is weaker than in FKAZO 2. Reproduced with permission from ref. 53. Copyright 2012 WILEY-VCH Verlag GmbH & Co. KGaA, Weinheim.

conduction energy barriers. Compared with DPKAZO 1 devices, the V_{th} of FKAZO 2 based memory devices is markedly lowered.

In addition to symmetric molecules, a series of asymmetric conjugated donor–acceptor (D–A) molecules with triphenylamine (TPA) as donor and an azobenzene chromophore and/or cyano group as acceptor were synthesized to explore their ternary memory behavior.

Devices based on a D–π–A1–π–A2 molecule with two electron-withdrawing groups exhibited excellent ternary memory behavior, while those based on a D–π–A1 (TPAVC 2) or D–π–A2 (TPAAH 3) molecule containing only one acceptor showed binary memory characteristics (Figure 3.19).[54] The molecule TPAVH 4 without a D–A structure shows no memory behavior. This indicates that two kinds of electron acceptors in a single molecule are important for the molecular design of ternary materials.

In addition, the bistable memory effects of TPAVC 2 and TPAAH 3 based data storage devices varied from WORM to FLASH when the acceptor was changed from a cyano group to an azobenzene chromophore.

Based on this study, three other triphenylamine (TPA) based molecules with different terminal acceptors (*i.e.* nitro, acetyl and bromine, the electron accepting strength decreasing in turn) were examined.[55] The influence of the terminal electron acceptor strength on the film morphology and the device storage performance was investigated. Nonvolatile ternary ("0", "1" and "2" states) memory devices for high density data storage could be achieved with a simple ITO/D–A molecule/Al sandwich configuration for TPA-NAP and TPA-AAP (Figure 3.20). It is noteworthy that the memory device based on TPA-AAP exhibited better reproducibility and stability with lower operation voltage than that of TPA-NAP, promising low power consumption data storage. The results demonstrate that changing the electron accepting strength can alter the film morphology and the device performance of organic electronic devices.

Three twisted anthraquinone/azo D–A molecules were synthesized, and the effects of conjugated backbone length, alkyl chain length, and the thermal annealing temperature on the memory characteristics were explored (Figure 3.21). The device based on Azo-02,[56] which has a longer conjugated backbone and alkyl chains, exhibited stable nonvolatile ternary memory behavior as the thermal annealing temperature increased, with basically the same switching threshold voltages and current ratios of $\sim 1:10^{3.43}:10^{5.50}$. For the "counterpart" molecule (Azo-01 and Azo-03) films, defects appeared, or disordered crystal packing occurred as the thermal annealing temperature increased, and the corresponding devices exhibited no obvious conductance switching behaviors. The mechanism related to the electrical switching properties and WORM (write-once-read-many-times) behavior was elucidated through molecular simulation. The results demonstrated that long-term thermally stable high density data storage devices can be fabricated by adjusting the structures of twisted molecules.

Donor–Acceptor Organic Molecule Resistor Switching Memory Devices 119

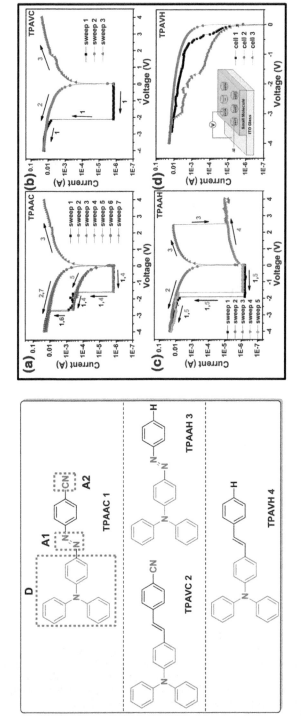

Figure 3.19 Chemical structures of four conjugated small molecules: TPAAC 1, TPAVC 2, TPAAH 3, and TPAVH 4 (left panel). *I–V* characteristics of the four small molecule based memory devices: (a) TPAAC 1, (b) TPAVC 2, (c) TPAAH 3, and (d) TPAVH 4; the inset of (d) shows a schematic diagram of the ITO/small molecule/Al sandwich-structure memory device (right panel). Reproduced with permission from ref. 54.

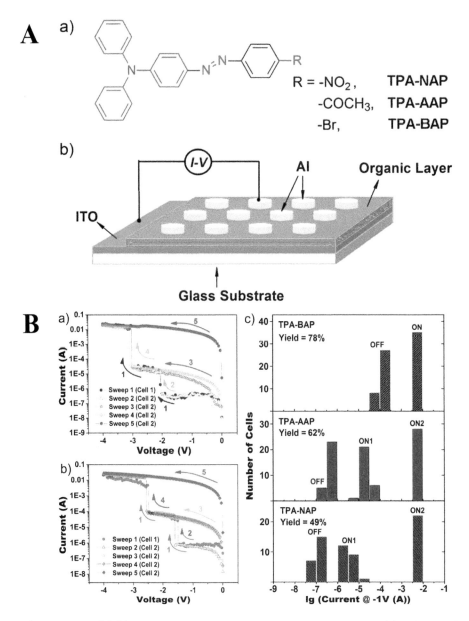

Figure 3.20 (A) (a) Molecular structures of the D–A molecules and (b) scheme of the ITO/D–A molecule/Al memory device. (B) Typical I–V curves of the (a) ITO/TPA-NAP or (b) TPA-AAP/Al memory device: the first sweep was carried out from 0 to −4 V on storage cell 1; the other voltage sweeps were carried out on storage cell 2; (c) the statistical data of reproducibility and current distribution (read at −1 V; 45 cells for each kind of device). Reproduced with permission from ref. 55.

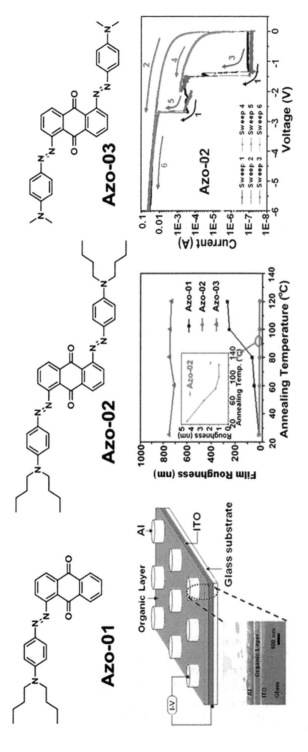

Figure 3.21 (a) Molecular structures of the Azo-01, Azo-02, and Azo-03 molecules; (b) schematic diagram of device consisting of an organic layer sandwiched between an ITO bottom electrode and an Al top electrode; SEM image of one storage cell (cross-section view); (c) roughness of films at different thermal annealing temperatures; (d) current–voltage (I–V) characteristics of the ITO/Azo-02 (25 °C)/Al device. Reproduced with permission from ref. 56. Copyright 2012 American Chemical Society.

3.4.2 Other Functional Molecules with Ternary Memory

In addition to azo derivatives, some other N rich organic molecules have also been studied for ternary memory performance. The large stable heteroacene, CDPzN, contains two different types of heteroatoms (N and O atoms) in its backbone and has nine linearly fused rings. Larger heteroacenes are good candidates for organic electronics because: (1) more heteroatoms might provide multilevel stable oxidation states, which is very important to realize 3^n or larger data storage; (2) synthetic work would be more challenging; (3) there will be greater chance of tuning the stability and properties of as-designed molecules.[57,58] Sandwich-structure memory devices based on CDPzN exhibited excellent ternary memory behavior with high ON2/ON1/OFF current ratios of $10^{6.3} : 10^{4.3} : 1$ and good stability for the three states (Figure 3.22). In this case, we first observed that there are two onset oxidation potentials for CDPzN according to the cyclic voltammogram (CV) curve of the CDPzN film on an indium-tin oxide (ITO) glass substrate. This is helpful for clarifying the conduction mechanism of CDPzN based devices. The device shows multilevel memory characteristics due to the two charge traps with the different electron-withdrawing abilities of cyano and pyrazine.

Based on this topic, we synthesized another derivative, a novel larger oxacalix-[4]arene(4,6,25,27-tetranitro-2,8,23,29-tetraoxacalix[4]-36,37-bis-(decyloxy)phenazine), abbreviated as 4N4OPz, which has two different types of electron-withdrawing groups (nitro and pyrazine). The introduction of O atoms could make 4N4OPz more stable both in the ground state and oxidation state. As expected, memory devices based on 4N4OPz exhibited ternary memory behavior with ON2/ON1/OFF current ratios of $10^{8.7} : 10^{4.2} : 1$, low switching threshold voltage of −1.80 V/−2.87 V, and good stability for these three states. The memory performance was confirmed by ITO/4N4OPz/LiF/Al and ITO/4N4OPz/Pt devices (Figure 3.23). The conduction mechanism of the ITO/4N4OPz/Al device shows multilevel memory characteristics due to the two charge traps with different electron-withdrawing abilities of the tetranitro and pyrazine groups.

3.4.3 Molecules with Ternary Memory Performance Through Photoelectric Effects

Song and coworkers used a DBA meta-conjugated molecule (Figure 3.24) in which triphenylamine (TPA) acts as an electron donor (D) and 2-dicyanomethylen-3-cyano-4,5,5-trimethyl-2,5-dihydrofuran (TCF) acts as an electron acceptor (A) for developing a photoinduced multilevel memory device.[59] The DBA compound shows bistable switching behavior between "0" (OFF) and "2" (HC-ON) states in the dark, with a large ON/OFF ratio of 10^6. Under UV light, a new "1" (LC-ON) state is addressable. This "1" state is "written" from the "0" state, by a combination of UV light and a

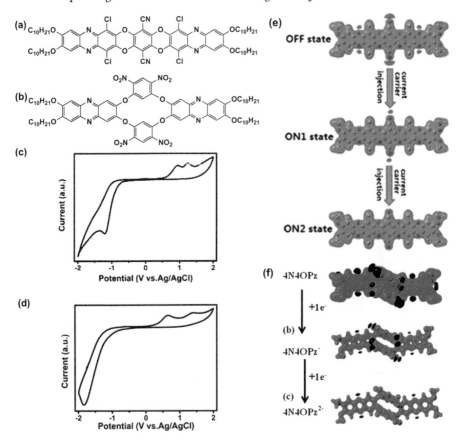

Figure 3.22 (a and b) Chemical structures of the two molecules; (c and d) voltammogram curves of thin films on ITO glass in a 0.1 mol L^{-1} solution of TBAPF6 in acetonitrile solution. Scan rate: 100 mV s^{-1}; (e and f) DFT molecular simulation results. (a, c and e) CDPzN; (b, d and f) 4N4OPz. Reproduced with permission from ref. 57 and 58. Copyright 2013 American Chemical Society & 2014 The Royal Society of Chemistry.

positive voltage of +1.0 V, or from the "2" (HC-ON) state, by a combination of UV light and a negative voltage of −3.4 V (Figure 3.25). Based on CV measurements, charge carrier transport modeling, quantum chemical calculations, and absorption spectra analysis, the mechanism of the DBA photoinduced multilevel memory is attributed to substep charge transfer processes through the cooperation of UV light and the electrical field. The write–read–erase switching cycles among the ternary conductivity (0, 1, 2) states reveal that the capability of tailoring photoelectrical properties provides a new approach to multilevel memory and encrypted storage. The photoelectric effect will open up new opportunities for designing multifunctional devices.

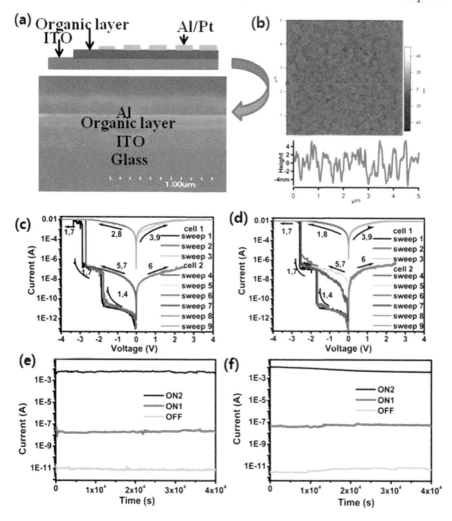

Figure 3.23 Memory device characteristics of 4N4OPz. (a) Scheme of the sandwich device and SEM image of a cross section of the device. (b) Tapping-mode (5 μm × 5 μm) AFM topography and typical cross-section profile of AFM topographic image of 4N4OPz film on ITO substrate. (c and d) I–V characteristics of the memory device (ITO/4N4OPz/Al and ITO/4N4OPz/LiF/Al, respectively) fabricated with 4N4OPz. (e and f) Stability of the memory device (ITO/4N4OPz/Al and ITO/4N4OPz/LiF/Al, respectively) in three states under a constant "read" voltage of −1 V. Reproduced with permission from ref. 58.

3.5 Structural Effects on Memory Performance

Molecular structures not only change the memory bit to realize high density data storage, but also affect the practical application of these molecule based memory devices such as the switching voltage, ON/OFF ratio, and the stability of conductivity states.

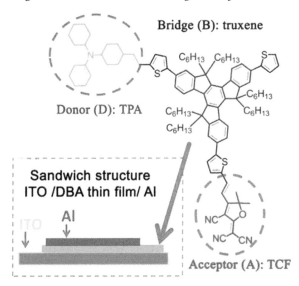

Figure 3.24 Molecular structure of the donor–bridge–acceptor compound (DBA). Inset: schematic diagram of an ITO/DBA/Al sandwich memory device. Reproduced with permission from ref. 59. Copyright 2012 American Chemical Society.

3.5.1 Switching Voltage and ON/OFF Ratio

Lower switching voltage means that the 'writing' process is energy saving. Ways if reducing the V_{th} value are attracting much attention.

For example, we have successfully synthesized two D–A molecules (NI–Cz, NI–TPA, Figure 3.26) and have demonstrated the effects of the coplanarity of the donor moiety on the reproducibility of switching phenomena of memory devices.[60] The results showed that both devices show FLASH memory behavior due to the formation and dissociation of a charge transfer complex and their ON/OFF ratios exceed 10^3 with a long retention time of about 10^4 s. However, the switching voltage of the device based on NI–Cz was much lower than that based on NI–TPA due to the rigid carbazole moiety which is favorable for improving the surface morphology. Some other studies also verified that increasing the molecular planarity would reduce the switching voltage effectively.[53]

A larger ON/OFF ratio means a lower misreading rate. This is necessary in practical applications. Our group once used TPA as electron donor and linked different electron acceptors to form a series of D–π–A molecules with the alkyl chain varying from a cyclic alkyl chain to methyl (Figure 3.27).[61] The results showed that the planarity of the molecules was obviously improved to generate an ordered molecule arrangement in the film which decreased the hole injection energy barrier between the active layer and the electrode, leading to lower threshold voltages and higher ON/OFF current ratios.

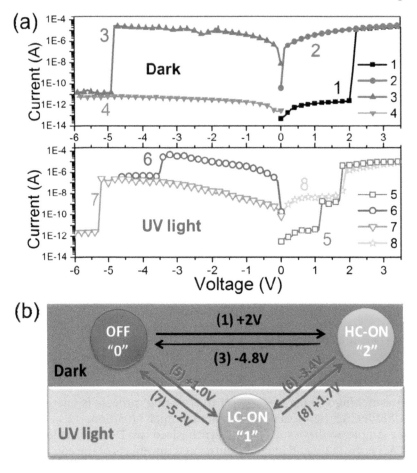

Figure 3.25 (a) *I–V* characteristics of memory device. Sweeps 1–4 were scanned in the dark. Sweeps 5–8 were scanned under UV light of 1.01 mW cm^{-2}. (b) Schematic graph showing the possible transitions and threshold voltage among three states: "0" (OFF state); "1" (low-conductivity state, LC-ON); "2" (high-conductivity state, HC-ON). In cooperation with the UV light, a new LC-ON state was addressable. A photoinduced multilevel memory device has been achieved. The arrows among the three states show the switching thresholds and the corresponding *I–V* sweeps in part (a). Reproduced with permission from ref. 59. Copyright 2012 American Chemical Society.

The following case examines the arrangement of donor and acceptor in the molecule. Two D–A molecules NACANA and CANACA, based on a carbazole (CA) donor and a naphthalimide (NA) acceptor, with different D–A arrangements (A–D–A and D–A–D) were synthesized (Figure 3.28).[62] Although the devices ITO/NACANA or CANACA layer/Al both exhibited a volatile nature after shutting off the external electric field, the ON-state retention time (*ca.* 12 min) of the NACANA based device is longer than that of the CANACA

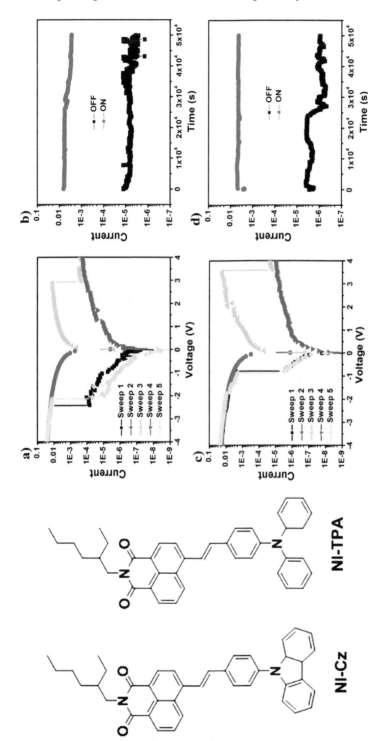

Figure 3.26 Molecular structures of NI-Cz and NI-TPA (left panel); current density–voltage (*I–V*) characteristics and effect of the operation time (at 1 V) on the device current density in the OFF and ON states of the device (right panel (a for NI-Cz and c for NI-TPA), b for NI-Cz and d for NI-TPA). Reproduced with permission from ref. 60. Copyright 2014 Elsevier.

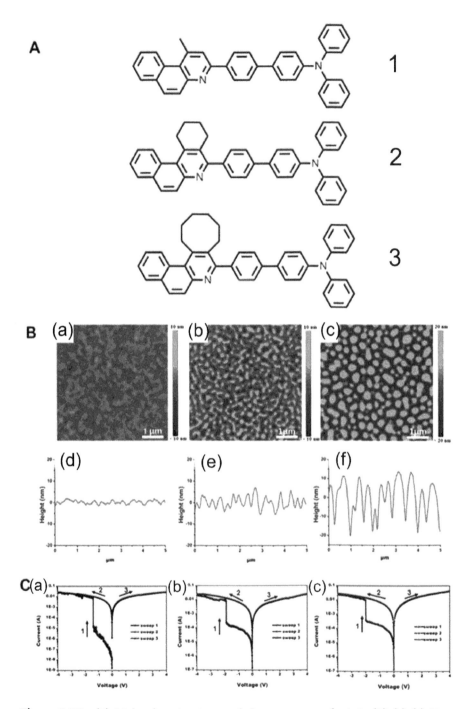

Figure 3.27 (A) Molecular structures of three compounds 1–3; (B) (a)–(c) Tapping-mode AFM height images of the thin films of compounds 1–3 respectively vacuum-deposited onto an ITO substrate; (d)–(f) typical cross section profiles of AFM topographic images for compounds 1–3 respectively. Scale bars are equal to 1 mm. (C) (a)–(c) Current–voltage (I–V) characteristics of the three compounds 1–3 respectively. Reproduced with permission from ref. 61. Copyright 2014 Elsevier.

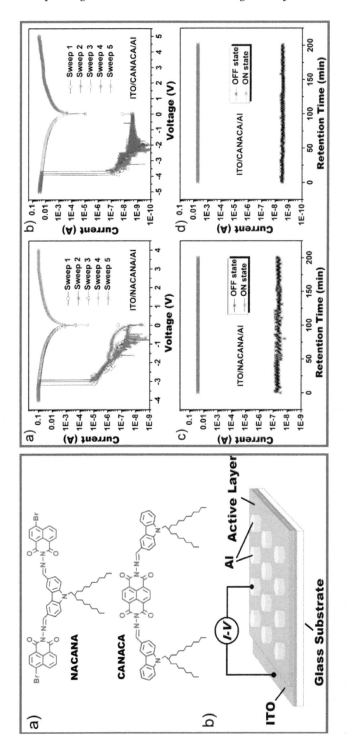

Figure 3.28 (Left panel) (a) molecular structure of NACANA and CANACA and (b) schematic of the prototype sandwich device. (Right panel) current–voltage (*I–V*) characteristics of (a) ITO/NACANA/Al and (b) ITO/CANACA/Al memory devices; retention time measurements for the ON and OFF states of the (c) ITO/NACANA/Al and (d) ITO/CANACA/Al devices under a continuous readout voltage of −1 V. Reproduced with permission from ref. 62. Copyright 2014 American Chemical Society.

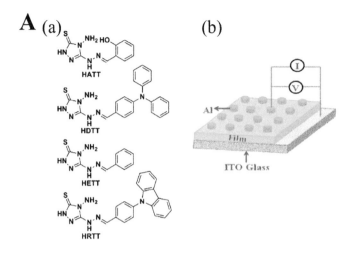
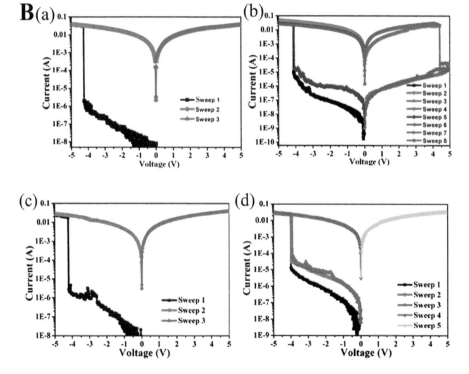

Figure 3.29 (A) (a) Chemical structure of HATT, HDTT, HETT and HRTT; (b) schematic of the sandwich devices. (B) Typical current–voltage (*I*–*V*) characteristics of memory devices in the ON and OFF states fabricated with: (a) HATT; (b) HDTT; (c) HETT and (d) HRTT. Reproduced with permission from ref. 63.

device (*ca.* 6 min). The difference in retention ability of the programmed states could be attributed to the difference in the D–A arrangement.

3.5.2 Memory Types

The reported memory types include 'WORM', 'DRAM', and 'FLASH' depending on the stability of the conductivity state. Molecular structures also play an important role in this aspect.

We once reported four donor–acceptor organic molecules (HATT, HDTT, HETT and HRTT) which have the same electron acceptor (triazole) but are terminated by different electron donors (phenol, triphenylamine, benzene and carbazole) which were used as the active layer in NVM (nonvolatile memory) devices (Figure 3.29).[63] *I–V* measurements showed that the ITO/HATT/Al and ITO/HETT/Al devices presented WORM characteristics, ITO/HDTT/Al exhibited a stable FLASH-type effect and the ITO/HRTT/Al device showed volatile DRAM switching behavior. These performances were preserved even if the electrodes were changed to Pt. Mechanism analysis indicated that the differences were because of the different charge delocalization abilities of the four donors due to the varying effects of electron donating ability and the consequent differences in charge separation states.

Chen and coworkers reported a D-A-A molecule TPA–NI–DCN consisting of triphenylamine (TPA) donors and naphthalimide (NI)/dicyanovinylene (DCN) acceptors, similar to NI–TPA (Figure 3.30).[64] The additional DCN attached as end group can result in stable intramolecular charge transfer and a charge separated state to maintain the ON state current. The TPA–NI–DCN based device was demonstrated to exhibit the WORM switching characteristics of organic nonvolatile memory due to the strong polarity of the TPA–NI–DCN moiety.

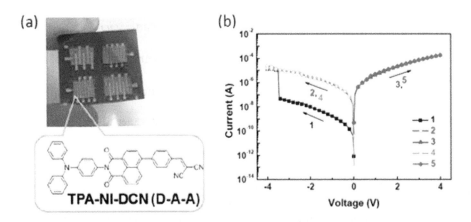

Figure 3.30 (a) Molecular structure of TPA–NI–DCN; (b) *I–V* characteristics of TPA–NI–DCN based devices. Reproduced with permission from ref. 64. Copyright 2014 WILEY-VCH Verlag GmbH & Co. KGaA, Weinheim.

3.6 Conclusion

In conclusion, we have outlined the recent progress made in the field of organic molecule based memory devices from a structural point of view. The D–A structure of molecules plays a vital role in fundamental research on memory devices. Through structural design, the memory performance can be improved in many aspects including high density data storage, lower energy cost and higher 'writing and reading' retention time. It should be mentioned that use of D–A structured molecules to realize multilevel memory effects is very promising in high density data storage. However, the performance of memory devices is also significantly affected by lots of other effects such as the interface between the metal electrode and molecular layer, local conditions and the purity of organics. However, construction of a 'molecule' library for organic memory is necessary for future applications. Furthermore, despite considerable advances in fundamental research on molecule based memory devices, it is still at the beginning and obstacles persist against commercially viable devices; in particular, the underlying switching and conduction mechanism have yet to be clearly understood. In future research, illustration of their conduction mechanism, fabrication of memory cells at nanometer size and extension of their application from data storage to logic gates and sensors are promising new topics.

References

1. Q. D. Ling, F. C. Chang, Y. Song, C. X. Zhu, D. J. Liaw, D. S. H. Chan, E. T. Kang and K. G. Neoh, *J. Am. Chem. Soc.*, 2006, **128**, 8732.
2. C. W. Tang and S. A. Vanslyke, *Appl. Phys. Lett.*, 1987, **51**, 913.
3. R. H. Friend, R. W. Gymer, A. B. Holmes, J. H. Burroughes, R. N. Marks, C. Taliani, D. D. C. Bradley, D. A. Dos Santos, J. L. Bredas, M. Logdlund and W. R. Salaneck, *Nature*, 1999, **397**, 121.
4. N. S. Sariciftci, L. Smilowitz, A. J. Heeger and F. Wudl, *Science*, 1992, **258**, 1474.
5. D. J. Gundlach, Y. Y. Lin, T. N. Jackson, S. F. Nelson and D. G. Schlom, *IEEE Electron Device Lett.*, 1997, **18**, 87.
6. L. D. Bozano, B. W. Kean, V. R. Deline, J. R. Salem and J. C. Scott, *Appl. Phys. Lett.*, 2004, **84**, 607.
7. R. J. Tseng, J. Ouyang, C.-W. Chu, J. Huang and Y. Yang, *Appl. Phys. Lett.*, 2006, **88**, 123506.
8. Y. Chen, G. Liu, C. Wang, W. Zhang, R.-W. Li and L. Wang, *Mater. Horiz.*, 2014, **1**, 489.
9. C. Li, W. Fan, B. Lei, D. Zhang, S. Han, T. Tang, X. Liu, Z. Liu, S. Asano, M. Meyyappan, J. Han and C. Zhou, *Appl. Phys. Lett.*, 2004, **84**, 1949.
10. D. Tondelier, K. Lmimouni, D. Vuillaume, C. Fery and G. Haas, *Appl. Phys. Lett.*, 2004, **85**, 5763.
11. A.-D. Yu, T. Kurosawa, Y.-H. Chou, K. Aoyagi, Y. Shoji, T. Higashihara, M. Ueda, C.-L. Liu and W.-C. Chen, *ACS Appl. Mater. Interfaces*, 2013, **5**, 4921.

12. L. B. Coleman, *Solid State Commun.*, 1973, **12**, 1125.
13. J. Ferraris, D. O. Cowan, V. J. Walatka and J. H. Peristein, *J. Am. Chem. Soc.*, 1973, **95**, 948.
14. R. S. Potember, R. C. Hoffman and T. O. Poehler, *APL Tech. Dig.*, 1986, **7**, 129.
15. J. Gong and Y. Osada, *Appl. Phys. Lett.*, 1992, **61**, 2787.
16. Z. Q. Xue, M. Ouyang, K. Z. Wang, H. X. Zhang and C. H. Huang, *Thin Solid Films*, 1996, **288**, 296.
17. S. Yamaguchi, C. A. Vianda and R. S. Potember, *J. Vac. Sci. Technol., B: Microelectron. Nanometer Struct.*, 1991, **9**, 1129.
18. R. S. Potember, T. O. Poehler and D. O. Cowan, *Appl. Phys. Lett.*, 1979, **34**, 405.
19. L. P. Ma, S. Pyo, J. Y. Ouyang, Q. F. Xu and Y. Yang, *Appl. Phys. Lett.*, 2003, **82**, 1419.
20. L. D. Bozano, B. W. Kean, V. R. Deline, J. R. Salem and J. C. Scott, *Appl. Phys. Lett.*, 2004, **84**, 607.
21. J. Y. Ouyang, C. W. Chu, C. R. Szmanda, L. P. Ma and Y. Yang, *Nat. Mater.*, 2004, **3**, 918.
22. J. C. Scott and L. D. Bozano, *Adv. Mater.*, 2007, **19**, 1452.
23. Z. Zhang, H. Zhao, M. M. Matsushita, K. Awaga and K. R. Dunbar, *J. Mater. Chem. C*, 2014, **2**, 399.
24. B. Mukherjee, M. Mukherjee, J.-e. Park and S. Pyo, *J. Phys. Chem. C*, 2009, **114**, 567.
25. J. Xiao, Z. Yin, H. Li, Q. Zhang, F. Boey, H. Zhang and Q. Zhang, *J. Am. Chem. Soc.*, 2010, **132**, 6926.
26. C. W. Chu, J. Y. Ouyang, J. H. Tseng and Y. Yang, *Adv. Mater.*, 2005, **17**, 1440.
27. L. P. Ma, Y. L. Song, H. J. Gao, W. B. Zhao, H. Y. Chen, Z. Q. Xue and S. J. Pang, *Appl. Phys. Lett.*, 1996, **69**, 3752.
28. J. C. Li, Z. Q. Xue, W. M. Liu, S. M. Hou, X. L. Li and X. Y. Zhao, *Phys. Lett. A*, 2000, **266**, 441.
29. Y. Zhao, C. Ye, Y. Qiao, W. Xu, Y. Song and D. Zhu, *Tetrahedron*, 2012, **68**, 1547.
30. Y. Shang, Y. Wen, S. Li, S. Du, X. He, L. Cai, Y. Li, L. Yang, H. Gao and Y. Song, *J. Am. Chem. Soc.*, 2007, **129**, 11674.
31. Y. Ma, X. Cao, G. Li, Y. Wen, Y. Yang, J. Wang, S. Du, L. Yang, H. Gao and Y. Song, *Adv. Funct. Mater.*, 2010, **20**, 803.
32. S. L. Lim, N. J. Li, J. M. Lu, Q. D. Ling, C. X. Zhu, E. T. Kang and K. G. Neoh, *ACS Appl. Mater. Interfaces*, 2009, **1**, 60.
33. Y. Q. Wen, J. X. Wang, J. P. Hu, L. Jiang, H. J. Gao, Y. L. Song and D. B. Zhu, *Adv. Mater.*, 2006, **18**, 1983.
34. W. Ren, Y. Zhu, J. Ge, X. Xu, R. Sun, N. Li, H. Li, Q. Xu, J. Zheng and J. Lu, *Phys. Chem. Chem. Phys.*, 2013, **15**, 9212.
35. D. M. Kim, S. Park, T. J. Lee, S. G. Hahm, K. Kim, J. C. Kim, W. Kwon and M. Ree, *Langmuir*, 2009, **25**, 11713.
36. Y. Liu, Y. Zhang, Q. Lan, S. Liu, Z. Qin, L. Chen, C. Zhao, Z. Chi, J. Xu and J. Economy, *Chem. Mater.*, 2012, **24**, 1212.

37. H. Li, Z. Jin, N. Li, Q. Xu, H. Gu, J. Lu, X. Xia and L. Wang, *J. Mater. Chem.*, 2011, **21**, 5860.
38. G. Wang, S. Miao, Q. Zhang, H. Liu, H. Li, N. Li, Q. Xu, J. Lu and L. Wang, *Chem. Commun.*, 2013, **49**, 9470.
39. W.-Y. Lee, T. Kurosawa, S.-T. Lin, T. Higashihara, M. Ueda and W.-C. Chen, *Chem. Mater.*, 2011, **23**, 4487.
40. H. Liu, R. Bo, H. Liu, N. Li, Q. Xu, H. Li, J. Lu and L. Wang, *J. Mater. Chem. C*, 2014, **2**, 5709.
41. H. Liu, H. Zhuang, H. Li, J. Lu and L. Wang, *Phys. Chem. Chem. Phys.*, 2014, **16**, 17125.
42. P. Heremans, G. H. Gelinck, R. Müller, K.-J. Baeg, D.-Y. Kim and Y.-Y. Noh, *Chem. Mater.*, 2011, **23**, 341.
43. A. R. Harris, A. K. Neufeld, A. P. O'Mullane and A. M. Bond, *J. Mater. Chem.*, 2006, **16**, 4397.
44. J. F. Fang, H. You, J. S. Chen, J. Lin and D. G. Ma, *Inorg. Chem.*, 2006, **45**, 3701.
45. B. Pradhan and S. Das, *Chem. Mater.*, 2008, **20**, 1209.
46. J. S. Lindsey and D. F. Bocian, *Acc. Chem. Res.*, 2011, **44**, 638.
47. G. D. Ruiter and M. E. V. D. Boom, *Acc. Chem. Res.*, 2011, **44**, 563.
48. G. Ruiter, L. Motiei, J. Choudhury, N. Oded and M. E. van der Boom, *Angew. Chem., Int. Ed.*, 2010, **49**, 4780.
49. M. Mas-Torrent, C. Rovira and J. Veciana, *Adv. Mater.*, 2013, **25**, 462.
50. H. Li, Q. Xu, N. Li, R. Sun, J. Ge, J. Lu, H. Gu and F. Yan, *J. Am. Chem. Soc.*, 2010, **132**, 5542.
51. S. Rovner, Data storage goes organic, *Chem. Eng. News*, 2010, **4**, 9.
52. S. Miao, H. Li, Q. Xu, N. Li, J. Zheng, R. Sun, J. Lu and C. M. Li, *J. Mater. Chem.*, 2012, **22**, 16582.
53. S. Miao, H. Li, Q. Xu, Y. Li, S. Ji, N. Li, L. Wang, J. Zheng and J. Lu, *Adv. Mater.*, 2012, **24**, 6210.
54. S. Miao, Y. Zhu, H. Zhuang, X. Xu, H. Li, R. Sun, N. Li, S. Ji and J. Lu, *J. Mater. Chem. C*, 2013, **1**, 2320.
55. H. Zhuang, Q. Zhang, Y. Zhu, X. Xu, H. Liu, N. Li, Q. Xu, H. Li, J. Lu and L. Wang, *J. Mater. Chem. C*, 2013, **1**, 3816.
56. Y. Zhang, H. Zhuang, Y. Yang, X. Xu, Q. Bao, N. Li, H. Li, Q. Xu, J. Lu and L. Wang, *J. Phys. Chem. C*, 2012, **116**, 22832.
57. P. Y. Gu, F. Zhou, J. Gao, G. Li, C. Wang, Q. F. Xu, Q. Zhang and J. M. Lu, *J. Am. Chem. Soc.*, 2013, **135**, 14086.
58. P.-Y. Gu, J. Gao, C.-J. Lu, W. Chen, C. Wang, G. Li, F. Zhou, Q.-F. Xu, J.-M. Lu and Q. Zhang, *Mater. Horiz.*, 2014, **1**, 446.
59. C. Ye, Q. Peng, M. Li, J. Luo, Z. Tang, J. Pei, J. Chen, Z. Shuai, L. Jiang and Y. Song, *J. Am. Chem. Soc.*, 2012, **134**, 20053.
60. W. Ren, H. Zhuang, Q. Bao, S. Miao, H. Li, J. Lu and L. Wang, *Dyes Pigm.*, 2014, **100**, 127.
61. R. Bo, H. Li, H. Liu, H. Zhuang, N. Li, Q. Xu, J. Lu and L. Wang, *Dyes Pigm.*, 2014, **109**, 155.

62. H. Zhuang, Q. Zhou, Y. Li, Q. Zhang, H. Li, Q. Xu, N. Li, J. Lu and L. Wang, *ACS Appl. Mater. Interfaces*, 2014, **6**, 94.
63. F. Zhou, J.-H. He, Q. Liu, P.-Y. Gu, H. Li, G.-Q. Xu, Q.-F. Xu and J.-M. Lu, *J. Mater. Chem. C*, 2014, **2**, 7674.
64. L. Dong, G. Li, A.-D. Yu, Z. Bo, C.-L. Liu and W.-C. Chen, *Chem.–Asian J.*, 2014, **9**, 3403.

CHAPTER 4

High Performance Polyimides for Resistive Switching Memory Devices

HUNG-JU YEN[a], JIA-HAO WU[b], AND GUEY-SHENG LIOU*[b]

[a]Physical Chemistry and Applied Spectroscopy (C-PCS), Chemistry Division, Los Alamos National Laboratory, Los Alamos, New Mexico 87545, United States; [b]Functional Polymeric Materials Laboratory, Institute of Polymer Science and Engineering, National Taiwan University, 1 Roosevelt Road, 4th Scc., Taipei 10617, Taiwan
*E-mail: gsliou@ntu.edu.tw

4.1 Introduction to Polyimide-Based Electrical Memory Devices

Aromatic polyimides (PIs) have been considered as one of the most important classes of high-performance polymers with a combination of exceptional thermal, mechanical, optical, and electrical properties accompanied by chemical and solvent resistance.[1–3] The excellent combination of properties makes them suitable for various applications, from engineering plastics in the aerospace industry to membranes and optoelectronic devices. However, PIs exhibit low solubility in common organic solvents due to their high chemical resistance. Therefore, thermal imidization is usually carried out by heating the poly(amic acid) (PAA) precursor film to obtain a PI film. Accordingly, intermolecular interaction is promoted *via* thermal imidization and the packing behavior is highly dependent on the film-formation method as well. Both phenomena might significantly affect the electrical properties of the resultant polymer.[4] In addition, the facile functionalization of PIs could offer an opportunity to fine-tune their optoelectronic properties to give organo-solubility. Therefore, PIs exhibit promising applications in a wide range of optoelectronic devices, including polymer resistive memory devices.[5] A PI memory device distinguishes itself from other polymer devices by its superior chemical and thermal properties, while still possessing good scalability and processability. In PIs, the phthalimide groups serve as strong electron acceptors (As). By introducing an electron donor (D; triphenylamine or carbazole groups) into a PI to form a D–A structure, an electric field-induced charge transfer (CT) state can be formed, which is the primary mechanism responsible for the memory characteristics. In addition, trap-limited space charge limited conduction and local filament formation have also been used as mechanisms to explain the memory behavior. Thus, memory behavior which can be tuned through the introduction of various electron Ds/As and the structural design of a soluble PI, while maintaining high thermal stability, mechanical strength, and tunable memory properties, is highly required.

4.2 Mechanism

Polymer resistive memory devices are differentiated by the changes in the conductivities of the active polymers under different electric fields. Plenty of research has been dedicated to understanding the electric switching phenomena within memory devices. Although this research field is still controversial, researchers have proposed several well established switching mechanisms based on theoretical simulations, experimental results, and advanced analytical techniques.[6-14] In this chapter, we summarize the most widely used mechanisms in polyimide resistive memory devices, charge transfer, space charge traps, and filamentary conduction.

4.2.1 Charge Transfer

Charge transfer (CT) can be classified as a process of partial transfer of electronic charge from the donor to the acceptor moiety in an electron D–A system by applying an suitable voltage, which can result in a sharp increase in conductivity.[15] In order to obtain a better understanding of the switching mechanism, several study methods, such as density functional theory (DFT) calculations, UV/vis absorption spectroscopy, *in situ* fluorescence spectroscopy, and transmission electron microscope (TEM) techniques, could be used to investigate and explain the CT phenomenon.[16-19] CT is anticipated to occur most frequently in polymers with the D–A structure.[20-22] Memory behavior based on a D–A polymer can be tuned adequately through modification of the polymer structure. By tuning the electron-donating or -accepting capability of D–A polymers, different memory behavior can be achieved.[23] A strong dipole moment in a polymer is also beneficial to maintain the conductive CT state, usually leading to non-volatile behavior. Otherwise, if the dipole moment is not strong enough, the conductive CT state is not stable after removal of the electric field and a volatile memory device will be observed.

4.2.1.1 Conformational Change

Conformational change is also a widely used mechanism in polymer resistive memory devices. In order to investigate the linkage effect between the TPA donor and the 6FDA acceptor, Ueda and Chen *et al.*[24] reported two PIs, **P1** and **P2**, consisting of electron-donating 4-amino-4′-(*p*-aminophenoxy)-triphenylamine (AAPT) or 4,4′-bis(*p*-aminophenoxy)-triphenylamine (APT) (Scheme 4.1), and their corresponding memory devices exhibited dynamic random access memory (DRAM) and static random access memory (SRAM) behaviors, respectively. The difference in linkage conformation plays an important role in the DRAM and SRAM properties of **P1** and **P2**. The dual-mediated

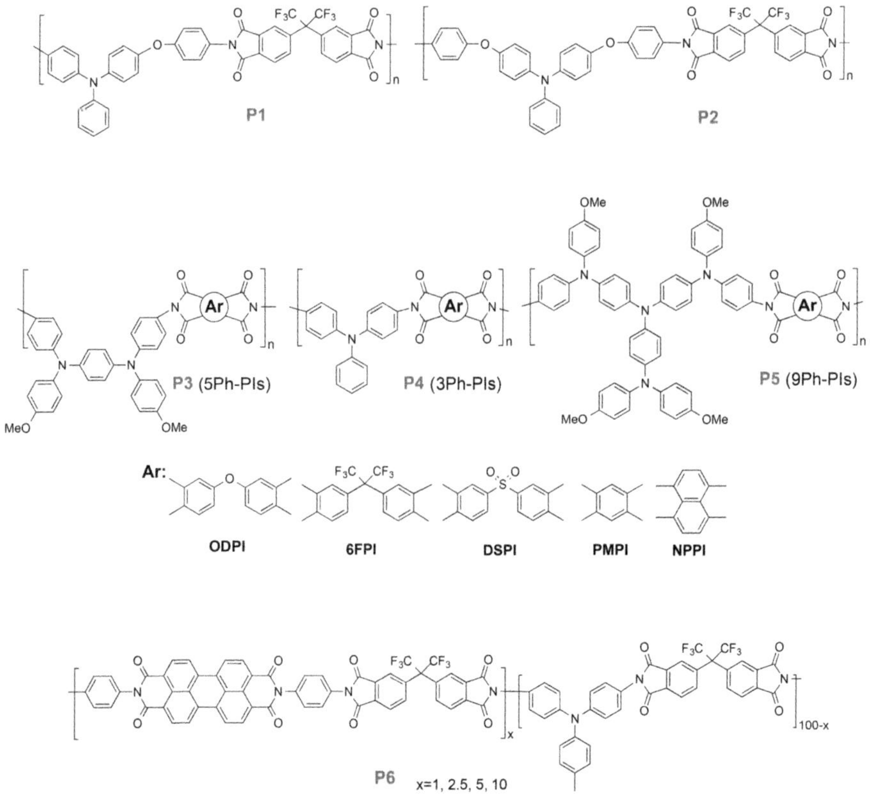

Scheme 4.1 Chemical structures of polyimide resistive memory devices demonstrating charge transfer mechanism.

phenoxy linkage of **P2** leads to a more twisted conformation as opposed to the mono-substituted **P1** based on theoretical calculations. Thus, a potential barrier will be produced to delay the back CT process by the electric field, resulting in SRAM characteristics.

4.2.1.2   LUMO Energy Level

The LUMO energy also plays a major role in stabilizing the CT state. Since the CT state is a meta-stable state, a lower LUMO energy level provides a more stable CT state. Our group reported a series of functional PIs, **P3**[21] with different dianhydrides such as oxydiphthalic dianhydride (ODPA), hexafluoroisopropyl bis(phthalic dianhydride) (6FDA), 3,3′,4,4′-diphenylsulfonetetracarboxylic dianhydride (DSDA), pyromellitic dianhydride (PMDA), and 1,4,5,8-naphthalenetetracarboxylic dianhydride (NPDA). With the electron-withdrawing capability of the PIs increasing (lowered LUMO energy level), the retention time of the corresponding memory device increases, making the resulting memory properties vary from DRAM to SRAM to WORM.

4.2.1.3   HOMO Energy Level

The HOMO energy level has been found to play another important role in the CT state. In our recent study,[25] 3Ph-PIs **P4**, 5Ph-PIs **P3**, and 9Ph-PIs **P5** based on different electron-donating ability were discussed. With the electron-donating ability increasing from 3Ph-PIs (**P4**) to 5Ph-PIs (**P3**) to 9Ph-PIs (**P5**), the retention time of the memory device exhibits a systematic increase. For example, the PIs with the 6FDA dianhydride possess only DRAM properties for **P4-6FPI** and **P3-6FPI**, while they show SRAM behavior for **P5-6FPI**. Therefore, by facile design of the chemical structure, tunable memory behavior with different retention times could be readily achieved.

4.2.1.4   Dipole Moment

According to our work,[21] an interesting exception was also reported. Although **P3-DSPI** has a higher LUMO energy level than **P3-PMPI**, it revealed only non-volatile WORM behavior instead of volatile properties. This mismatch in LUMO energy and memory characteristics could be explained from the viewpoint of the dipole moment. The main difference between these two PIs is the higher dipole moment of **P3-DSPI** (5.45 D) compared to **P3-PMPI** (3.70 D). Accordingly, a higher dipole moment led to non-volatile memory properties as it resulted in a more stable CT state.

4.2.1.5   High Conjugation

Introducing a highly conjugated or high electron affinity unit is a simple and effective method for the development of PIs with non-volatile memory characteristics.[26] For example, the introduction of a small amount of perylene bisimide (PBI) (*i.e.*, 5 mol%) dramatically changes the memory properties from volatile DRAM to non-volatile WORM type (the case of **P6**), which is attributed to the separated charge, *i.e.*, the radical anion, being stabilized by the high conjugation and also being trapped by the deep LUMO energy level arising from the high electron affinity PBI.

4.2.2   Space Charge Traps

If the interface between the electrode and polymer is ohmic and the polymer is trap-free, the carriers near the electrode would be accumulated and build up a space charge. Mutual repulsion between individual charges restricts the total charge injected into the polymer, and the resulting current is defined as a space charge-limited current (SCLC). Space charges in materials may occur from several sources, such as (1) electrode injection of electrons and/or holes, (2) ionized dopants in interfacial depletion regions, and (3) accumulation of mobile ions at electrode interfaces. Traps may be present in the bulk material or at interfaces in which they will reduce carrier mobility. When present at interfaces, they may also affect charge injection into the material. The electrical switching behaviors of some polymer materials have been reported to be associated with space charges and traps.[17]

4.2.3   Filament Conduction

When the ON state current is highly localized to a small area of the memory device, the phenomenon is termed "filamentary" conduction. It has been suggested that filament conduction is confined to that arising from device physical damage in resistive random access memories. Two types of filament conduction have been widely reported in polymer resistive memory devices, and the formed filaments could be observed under an optical microscope or scanning electron microscope.[27,28] One type is related to the carbon-rich filaments formed by the local degradation of polymer films.[28,29] The other is associated with the metallic filaments that result from the migration of electrodes through polymer films.[30,31] For filamentary conduction, polymers with both coordinating atoms and π-conjugation that can bind to metal ions, regardless of the position of the binding sites (side chain or main chain), are essential for the production of metal filaments.[6,32] Therefore, the filamentary conduction mechanism has often been suggested to explain switching phenomena observed in a variety of polymer memory devices.

4.3   Polyimides for Volatile Memory Devices

For volatile memory effects, a device cannot be maintained in the ON state steadily and will relax to the OFF state after the power is turned off for a period of time. Nevertheless, the ON state can be maintained by refreshing the voltage pulse. According to the retention time of the ON state after the removal of the applied voltage, the volatile memory effects can be divided into dynamic random access memory (DRAM) and static random access memory (SRAM). For the DRAM effect (Figure 4.1),[25] the ON state can be retained for a short period after the removal of the applied voltage (less than 1 min). For SRAM behavior (Figure 4.2),[25] the device can maintain the ON state after turning off the power for a longer period of time than that observed in a DRAM device. Even with the longer retention time in the ON state for SRAM memory devices, they are also volatile and the ON state can relax to the OFF state without an erasing process.

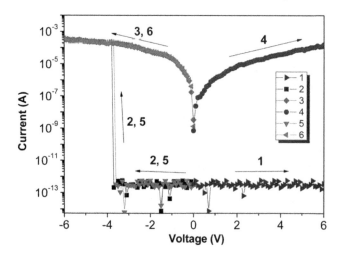

Figure 4.1 Current–voltage (*I–V*) characteristics of ITO/**P5-ODPI** (~50 nm)/Al memory device. Reproduced from ref. 25.

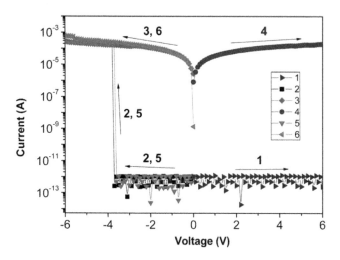

Figure 4.2 Current–voltage (*I–V*) characteristics of ITO/**P5-6FPI** (~50 nm)/Al memory device. (The fifth sweep was conducted after turning off the power for 3 min.) Reproduced from ref. 25.

4.3.1 DRAM Properties

Based on the CT mechanism, Kang *et al.*[16] first reported the DRAM memory characteristics of a triphenylamine (TPA)-functionalized PI **P7** (Scheme 4.2) in the device Al/**P7**/ITO. From the *I–V* curve, the memory device can be switched from a low-conductivity (OFF) state to a high-conductivity (ON) state, and remained in this high-conductivity state during the subsequent positive scan with an ON/OFF current ratio up to 10^5. The ON state can be switched back to the OFF state during the subsequent negative sweep and the device remained in the OFF state during the subsequent negative sweep, revealing an erasable memory property. In addition, this memory device was rewritable. After turning off the power for 1 minute, the ON state relaxed to the steady OFF state without an erasing process, indicating the volatile DRAM memory behavior of the device. Furthermore, the memory device could be reprogrammed to the ON state. Moreover, the ON state could be electrically sustained by a refreshing voltage pulse of 1 V every 5 s. According to theoretical calculations, the formation of unstable CT states is responsible for the observed DRAM behavior.

Compared with **P7**, PI **P8** reported by Kang *et al.*[33] with TPA-substituted diphenylpyridine moieties exhibits superior DRAM behavior with lower switch ON/OFF voltages and a more volatile (less stable) ON state.

P9 containing electron acceptor 6FDA and electron donor carbazole was synthesized by Qi *et al.*[34] Electrical switching results for the sandwiched PI memory device (ITO/**P9**/Au) indicate that the PI possesses electrical bistability with two accessible conductivity states, which can be reversibly switched from the OFF state to the ON state with an ON/OFF current ratio of about 10^4. The ON state of the device was lost immediately after removing the applied voltage, while the ON state can be electrically sustained by applying a constant bias (~3 V). The roles of the D and A components in the PI main chain were elucidated through molecular simulation.

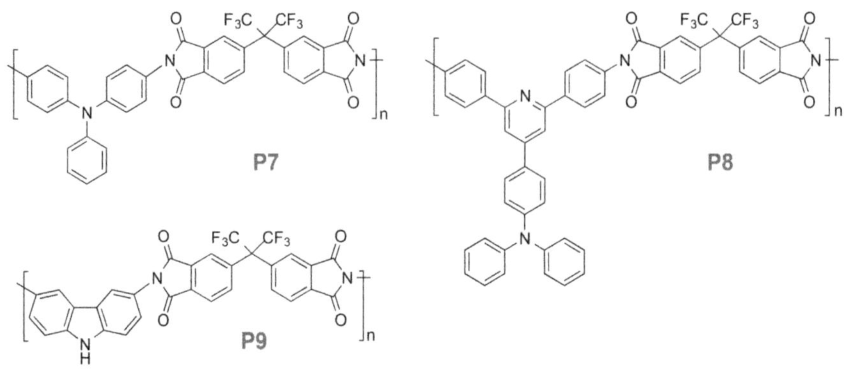

Scheme 4.2 Chemical structures of typical polyimides reported for volatile memory devices with DRAM properties.

4.3.2 SRAM Properties

Oxadiazole-based polymers are promising electron-transporting materials in light-emitting devices. However, when bonded to stronger electron acceptors, such as itaconimide or maleimide, the oxadiazole groups can also act as electron donors. Thus, a solution-processable functional PI **P10** (Scheme 4.3), containing oxadiazole and phthalimide moieties, was synthesized and reported by Kang et al.[35] During electrical sweeps, the device based on the ITO/**P10**/Al sandwich structure exhibits volatile bistable electrical switching characteristic of SRAM. As elucidated from DFT simulations, the oxadiazole unit can act as the electron donor in the presence of a phthalimide moiety of higher electron affinity. Electric field-induced CT from the oxadiazole donor to the phthalimide acceptor gives rise to a conductive CT state, which accounts for the transition of the **P10** device from the OFF state to the ON state. The conformational change induced by the electric field gives rise to a potential barrier to the back transfer of electrons, leading to the "remanence" (temporary retention) of the ON state. As the small dipole moment of **P10** in the ON state probably cannot maintain the conformation-coupled CT state, the ON state is volatile in nature and is eventually lost after removing the applied voltage. The conformation-coupled CT mechanism was further elucidated through molecular simulations of **P10** in the ground and excited states, and the changes in the photoelectronic spectra of **P10** film in the ON and OFF states.

Zhang and Xu's group produced a series of PIs (**P11a–d**) by the polymerization of diamine with a triphenylethylene rigid nonplanar conjugated moiety and four dianhydrides.[36,37] The PIs exhibited special fluorescence and resistive switching (ON/OFF) characteristics, a low dielectric constant, exceptional thermal stability, and excellent mechanical properties, mainly due to their rigid nonplanar conjugated structure. Memory devices with the configuration ITO/PI/Al exhibited distinct volatile memory characteristics of SRAM. Theoretical analysis showed that the chemical and steric structure of the dianhydride may affect the distribution of the electron cloud of the molecular orbitals, and the CT mechanism could be used to explain the memory characteristics of the aromatic rigid nonplanar conjugated structure to build high performance functional PIs.

A functional PI **P12** with a carbazole-tethered triphenylamine unit as the electron-donating moiety and 6FDA as the electron-accepting species was firstly synthesized by Liou et al.[38] The memory device fabricated by Qi's group[39] as a simple ITO/PI/Au sandwich structure exhibits two bi-directionally accessible conductivity states during both the positive and negative voltage sweeping processes with the achievement of an ON/OFF current ratio as high as about 10^5, and displays no polarity except for a trivial variation in the switching voltage. Moreover, both the ON and OFF states were stable under a constant voltage stress of −1 V and survived up to 10^8 read cycles with a −1 V periodic read pulse. The I–V characteristics of the device in the OFF state agree fairly well with the space charge-limited current (SCLC) model,

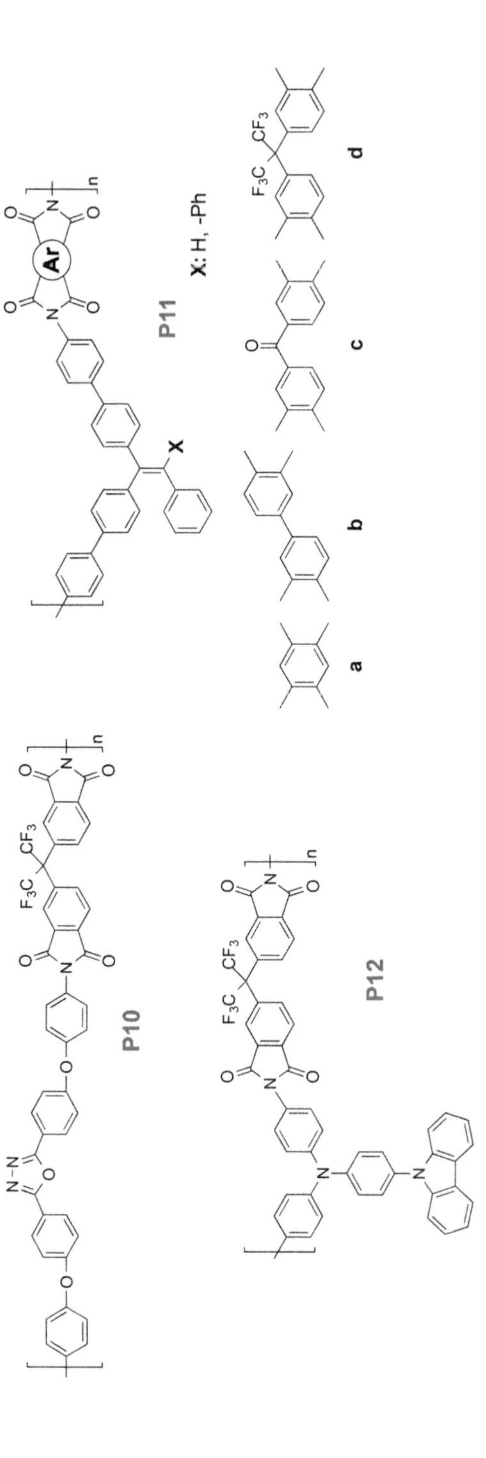

Scheme 4.3 Chemical structures of typical polyimides reported for volatile memory devices with SRAM properties.

while the conduction mechanism changes to the ohmic conduction model in the ON state. The roles of the donor and acceptor units in the memory mechanism of the electroactive polyimide were elucidated *via* molecular simulation. The strong electron-donating ability of the carbazole-tethered triphenylamine moieties and the stoichiometric excess of charge-trapping sites as compared to the 6FDA units are suggested to play an important role in realizing the temporarily inerasable, but eventually volatile SRAM memory properties observed in the current system.

4.4 Polyimides for Non-Volatile Memory Devices

A non-volatile memory device can stay in the ON state steadily without an applied voltage bias. Non-volatile memory effects can be divided into two classes, namely write-once read-many times (WORM) memory and rewritable (flash) memory, depending on whether a suitable voltage can switch the ON state to the OFF state or not. If the ON state can be switched back to the OFF state by applying a suitable voltage, which is an erasing process, the memory effect is called rewritable memory (Figure 4.3).[40] WORM memory, however, is capable of maintaining the ON state (holding data) permanently even after applying a voltage and being read repeatedly (Figure 4.4).[25]

4.4.1 WORM Properties

Ree *et al.* reported a PI derivative **P13** (Scheme 4.4) with a pendant hydroxyl group on the TPA group.[41] The device with the structure ITO/**P13**/Al exhibited WORM memory behavior with a high ON/OFF current ratio of up to 10^6,

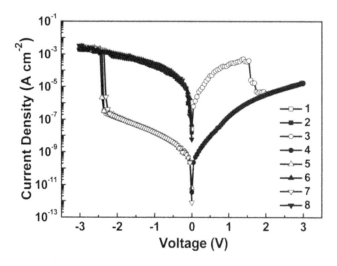

Figure 4.3 Current–voltage (*I*–*V*) characteristics of the ITO/**P50** (~50 nm)/Al memory device. Reproduced from ref. 40.

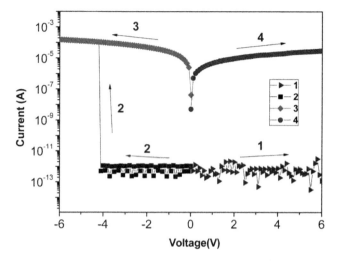

Figure 4.4 Current–voltage (*I*–*V*) characteristics of the ITO/**P5-DSPI** (~50 nm)/Al memory device. Reproduced from ref. 25.

a long retention time and low threshold voltage. In particular, WORM characteristics were found to persist even at high temperatures up to 150 °C, demonstrating the advantage of excellent thermal stability of PI-based memory devices. For the memory mechanism, a trap-limited SCLC mechanism is dominant when the device is in the OFF state according to the fitted *I*–*V* data. Moreover, the ON state current level of the device was found to be independent of the device cell size, suggesting that electrical transition in the device is due to filament formation inside the active polymer layer. Once filaments have formed, they act as channels through which the carriers flow by means of a hopping process, leading to the ON state.

A thermally stable PI **P14** with the pendant spiropyran moiety was designed as a functional material and fabricated into a memory device, and its optoelectrical dual-mode memory was studied by Song *et al.*[42] In an Al/**P14**/Al device, the memory exhibits WORM memory behavior with an ON/OFF current ratio of about 10^4. These properties open up the possibility of potential application of the multimode data storage memory devices with the thermally stable polyimide as the functioning medium.

Wang *et al.*[43] synthesized a series of functional PIs **P15a–e** with different phthalimide electron acceptors, showing excellent thermal stability and high glass transition temperatures (T_g). Memory devices based on **P15a** thin films exhibited symmetric bistable WORM memory behavior with switching threshold voltages of ±1.8 V. The ON state of the devices is non-volatile and can withstand a constant voltage stress of −1 V for 6 h and 10^8 pulse read cycles under ambient conditions. Such electrical switching behavior can be attributed to electric field-induced electron transfer from the TPA donor to the phthalimide acceptor through the oxadiazole spacer. The excellent

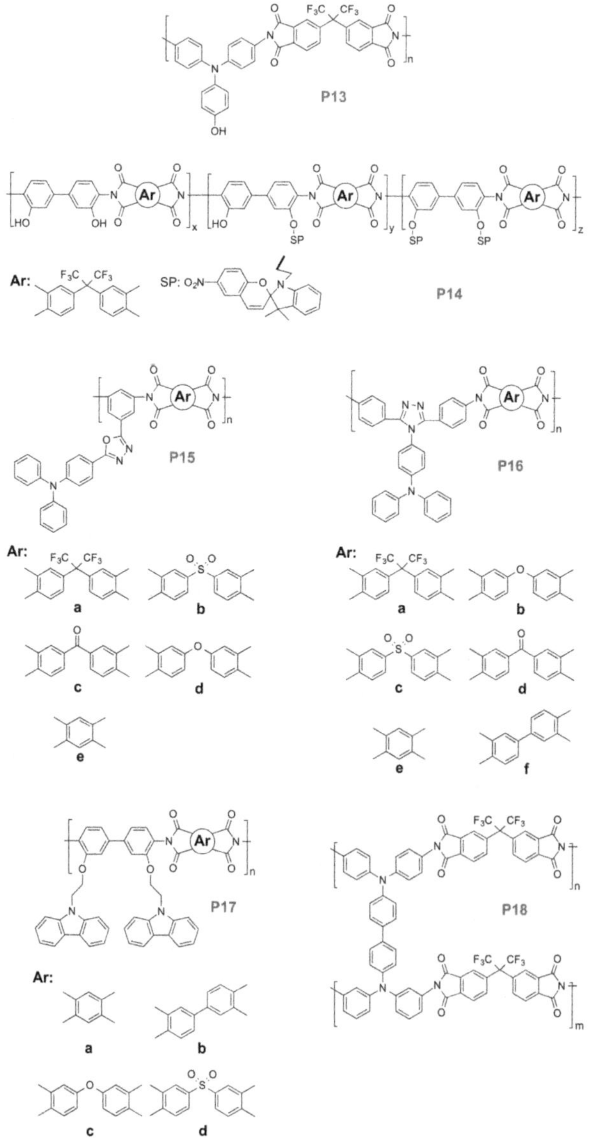

Scheme 4.4 Chemical structures of typical polyimides reported for non-volatile memory devices with WORM properties.

memory performance further demonstrates the advantages of PI derivatives for memory device applications.

Wang et al. further developed a series of PIs **P16a–f** containing TPA-substituted triazole moieties.[44] A memory device based on polymer **P16a** exhibited the WORM memory effect with excellent device performance, including a high ON/OFF current ratio (10^5) and good stability. According to theoretical calculations, the memory mechanism can be attributed to electric field-induced "conformation-coupled CT". The TPA moieties act as electron donors and the phthalimide moieties act as electron acceptors, and are linked by a single bond. Thus, a conformation-coupled CT process can probably occur under an electric field. The twisted molecular chain in the conformation-coupled CT state can give rise to a large potential energy barrier for the dissociation of the CT complexes. As a result, the conductive CT state is stabilized and the high-conductivity (ON) state is retained. Once a high-conductivity state is formed, carriers can move in both directions. Thus, a reverse bias of the same magnitude will not return the device to the low-conductivity state.

Another series of PIs **P17a–d** containing carbazole as electron donor and different phthalimides as electron acceptor was synthesized by Ree *et al.*[45] All of the polymers exhibited high thermal and dimensional stability. In particular, **P17a** and **P17b** revealed excellent chemical resistance toward organic solvents, which is very promising for the fabrication of high performance memory devices in three-dimensionally multi-stacked structures. Nanoscale thin films of these PIs can be fabricated *via* solution spin-coating of soluble poly(amic acid) precursors and a subsequent thermal imidization process. The PI devices exhibit excellent unipolar WORM memory behavior with an ON/OFF current ratio as high as 10^9–10^{10} in both positive and negative voltage sweeps and turn-on threshold voltages of 2.3–3.3 V, depending on the chemical structures of the polymers. According to the fitted *I–V* data, the trap-limited SCLC mechanism is dominant in the OFF states of the devices at high voltages, and ohmic conduction is dominant both in the OFF state at lower voltages and in the ON state.

Qi *et al.* reported an aromatic hyperbranched PI **P18** in which TPA moieties function as electron-donating groups and 6FDA serves as the electron-accepting species.[46] The fabricated memory device in a simple ITO/**P18**/Au sandwich structure exhibits two distinct but irreversible conductivity states, and displays a threshold transition voltage at around 2.0 V with an ON/OFF current ratio of about 300 and good operational stability. The non-volatile and irreversible electrical WORM memory effect observed in the current system was ascribed to electric field-induced charge transfer interactions between the 6FDA and TPA groups, and the strong electron affinity of the 6FDA units which supposedly possess the ability to retain the charge-separation state of the generated CT complexes.

4.4.2 Flash Properties

By changing the hole-transporting and electron-donating moieties, four PI derivatives containing diphenylcarbamyloxy (**P19**),[47] diphenylaminobenzylidenylimine (**P20**)[48] and carbazole (**P21**[49] and **P22**[50]) on the side chains have also been reported for application in excellent rewritable memory devices (Scheme 4.5). For example, **P19** exhibits a high ON/OFF current ratio up to 10^9, low power consumption, and long retention time both in the ON and OFF states.[47] The observed rewritable memory behavior for **P20** is due

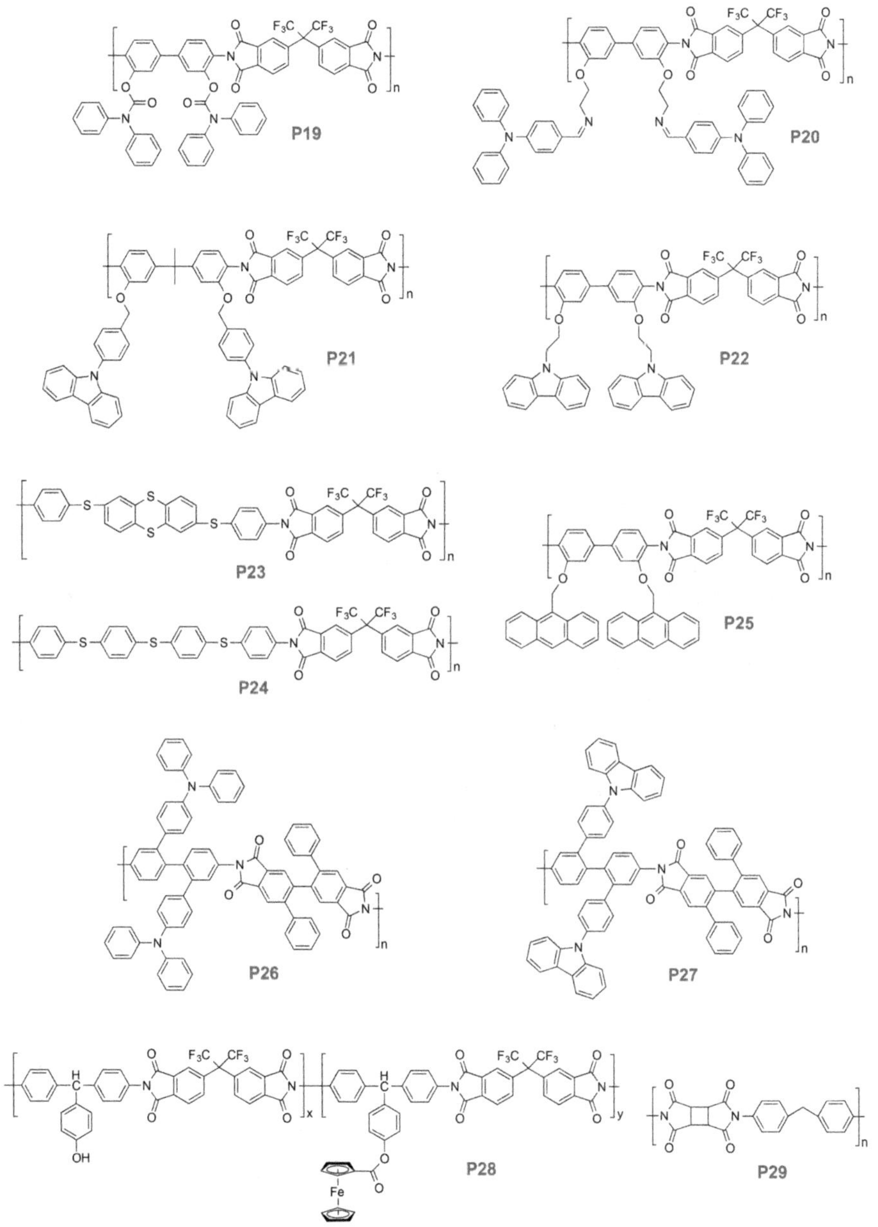

Scheme 4.5 Chemical structures of typical polyimides reported for non-volatile memory devices with flash properties.

to Schottky emission and local filament formation.[48] For polymer **P21**, a filamentary mechanism was demonstrated by *in situ* conductive atomic force microscopy measurements.[49] A memory device based on **P22** exhibits excellent unipolar ON and OFF switching behavior with very low power consumption and a high ON/OFF current ratio up to 10^{11}.[50]

In Ueda and Chen's study,[7] hole-transporting and electron-donating moieties, such as sulfur-containing, TPA and phenyl groups, could also be introduced into the main chain of PI derivatives. Electron-rich sulfur-containing groups generally have high dipole moments as well as good electron-donating and charge-transporting characteristics. Two sulfur-containing PIs **P23** and **P24** with 2,7-bis(phenylenesulfanyl)thianthrene (APTT) and 4,4′-thiobis(*p*-phenylenesulfanyl) (SDA) moieties in the main chains of the polymers were synthesized for memory device applications. Memory devices with the configuration ITO/PI/Al showed non-volatile flash memory characteristics with low turn-on threshold voltages of 1.5 V (**P23**) and 2.5 V (**P24**). The ON/OFF current ratios of the devices were all around 10^4 in ambient atmosphere. According to theoretical analysis, a field-induced CT mechanism between the APTT/3SDA and 6FDA moieties could be used to explain the memory characteristics of these polymers. The high dipole moments of **P23** and **P24** result in a stable CT complex even after the driving power is turned off; thus the ON state of these devices could still be retained. However, a reverse bias applied to **P23** or **P24** could dissociate the CT complex and return them to the OFF state, leading to flash memory behavior.

Ree *et al.*[51] reported PI **P25** by introducing anthracene as hole-transporting and electron-donating moiety on the side chain. This polymer is soluble in common solvents and high quality nanoscale thin films with excellent thermal and dimensional stability could be easily fabricated by means of spin-coating and subsequent drying. A memory device with the structure Al/**P25**/Al exhibits excellent non-volatile and rewritable bipolar and unipolar switching behaviors over a very small voltage range of less than ±2 V with a high ON/OFF ratio of up to 10^7. This device could be written, read, and erased repeatedly with long-term reliability under ambient air conditions as well as at high temperatures up to 200 °C, which can be produced in three-dimensional arrays for very high density storage. By fitting the *I–V* data for the OFF and ON states, the memory mechanism of the device was attributed to trap-limited SCLC and local filament formation. Furthermore, a polymer similar to **P25** without the anthracene unit exhibited no electrically bistable characteristics. Hence, both the anthracene moieties acting as hole trapping sites and the phthalimide moieties as electron trapping sites in polymer **P25** play a key role in the electrical switching properties. In addition, a device with an Au bottom electrode (Au/PI/Al device) was also found to exhibit switching behavior similar to that of the device with an Al bottom electrode.

In addition, Li *et al.*[52] designed and synthesized two aromatic PIs **P26** and **P27** with pendant carbazole and TPA-substituted groups as donors, respectively, for the fabrication of memory devices. Both the PIs showed good solubility, fine film-forming capability, and excellent thermal stability. A device based on a **P26** film exhibited bipolar rewritable flash memory characteristics with a turn-on voltage at around +2.7 V and a turn-off voltage at −1.5 V. On the other hand, a device based on **P27** film demonstrated unipolar rewritable flash memory behavior with a turn-on voltage at around ±2.2 V and a turn-off voltage around ±1.0 V. The ON/OFF current ratios of these devices were about 10^4. The current of the OFF state in Al/**P26**/ITO is fitted to the SCLC model, while Al/**P27**/ITO is fitted to the Schottky emission model. Both the ON state currents are well fitted by the ohmic model.

Ferrocene is also a typical electron donor. A polyimide (**P28**) grafted with ferrocene was synthesized by Wu et al.,[53] in which phthalimide can act as an electron acceptor. The resulting polymer **P28** exhibits electrical bistability and non-volatile memory behavior with an ON/OFF current ratio of about 10^3. The CT mechanism was reasonably fitted using thermionic emission and the space charge-limited current model.

Most recently, Tsai et al.[54] reported a new PI **P29** containing an electron donor and electron acceptor, which exhibited the non-volatile memory effect. PI **P29** exhibits asymmetric bipolar switching behavior with a high ON/OFF current ratio of >10^5. The switching mechanism of the **P29** layer could be attributed to the formation and dissociation of the CT complex. As positive voltage is applied, the electron donor and electron acceptor interact through electron transfer to form a conductive CT complex. In contrast, as opposite voltage is applied, the conductive CT complex dissociates, and the device returns to the initial low-conductivity state. Furthermore, the switching properties of **P29** are related to the thickness of the **P29** layer. With increasing thickness of the **P29** layer, the ON/OFF current ratio increases. In addition, the active area closely affects the stability of a device. That is, a small-sized **P29** device exhibits better stability than a large-sized device. Moreover, it was found that the molecular chain lengths of PI dominate the electron conduction capability in the low- and high-conductivity states to affect the switching properties. Therefore, molecular chain length is an important issue for PI memory research.

4.5 Molecular Design of the Volatility

For resistive memory devices, the writing operation is performed by applying a voltage bias or pulse to the device, resulting in switching between the high-resistance (OFF) and low-resistance (ON) states. Memory devices reveal non-volatile and volatile memory behavior depending on whether electrical power is required to maintain the given state (ON state) or not. The memory properties can be tuned through the following: changes of electron donor or acceptor for functional donor–acceptor type polymers, use of different linkages, conformation changes, and change of film thickness of the active polymer layer within the devices.[21,55-57]

4.5.1 Donor Effect

Shen et al. synthesized polymers **P30**/**P31** (Scheme 4.6) consisting of alternating electron-donating carbazole/TPA moieties and electron-accepting phthalimide moieties.[18] These polymers are thermally stable with 5% weight loss over 500 °C and T_gs were higher than 290 °C. The memory device with the configuration Al/**P30**/ITO showed flash memory behavior with low turn-on

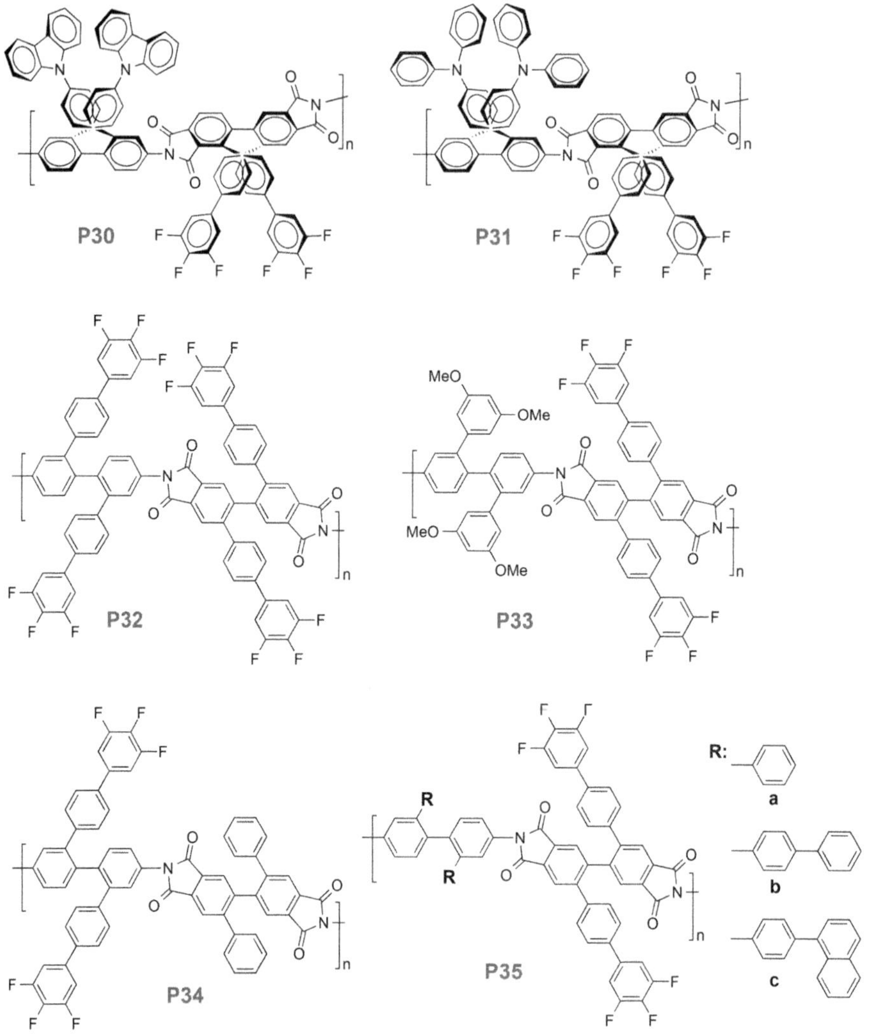

Scheme 4.6 Molecular design of the volatility of 2,2′-biphenyl-based polyimides through donor effects.

and turn-off threshold voltages at −1.6 and +2.2 V, respectively. The memory device constructed of Al/**P31**/ITO exhibited excellent WORM memory performance, including a low turn-on threshold voltage around ±2.7 V. Both devices exhibited high ON/OFF current ratios of around 10^6 and long retention times of 10^4 s even under ambient atmosphere. According to DFT theoretical calculations, switching mechanisms of CT between the electron donor and acceptor moieties, as well as electric field-induced conformational ordering of the polymer side chains/backbones are proposed to explain the memory behavior.

According to Shen and Ding's research, PI **P32**[58] containing 3′,4′,5′-trifluorobiphenyl and phenyl moieties as electron donors was synthesized, and the fabricated Al/**P32**/ITO devices exhibited WORM type memory behavior with a low turn-on threshold voltage of −1.7 V and an ON/OFF current ratio of 10^4. The *I–V* curves in the OFF and ON states were fitted according to various theoretical models. It was found that the charge transport in the OFF state was governed by the Poole–Frenkel model and that in the ON state was governed by an ohmic model. Similar to polymer **P32**, methoxy-containing PI **P33**[59] with excellent solubility, and high thermal and dimensional stability was also synthesized to investigate its memory behavior. The memory device with the configuration Al/**P33**/ITO exhibited bipolar WORM memory behavior with a low switch-on voltage less than 3 V. According to theoretical calculations, the relatively strong dipole moments of the methoxy-containing PI **P33** produced a stable CT complex for the WORM memory device as opposed to the **P32** device.

Shen and Ding also prepared another PI **P34**[58] with the introduction of 3′,4′,5′-trifluorobiphenyl and phenyl moieties as electron donors. The fabricated Al/**P34**/ITO device was demonstrated to reveal flash type memory behavior with "write" and "erase" voltages of about −1.3 V and 4.0 V, respectively. A series of PIs **P35a–c**[59] with the same main chain but different electron-donating groups on the 2,2′-position of the biphenyl moieties also exhibited bipolar flash type memory characteristics in Al/**P35**/ITO devices with low switch-on and switch-off voltages less than 3 V. According to theoretical calculations, the CT mechanism could be used to explain the memory characteristics of these PIs.

Two thermally, dimensionally stable aromatic PIs of TPA derivatives bearing an electron-accepting bis(trifluoromethyl)phenyl group (**P36**) and an electron-donating bithiophene group (**P37**) were synthesized (Scheme 4.7).[60] In Al/PI/Al devices, all of the PIs initially exhibited a high resistance (OFF state). Under positive and negative voltage sweeps, the PIs demonstrated various memory behaviors (volatile DRAM and non-volatile WORM memory) depending on the substituents of the TPA units. The electron-donating bithiophene substituents increased the nucleophilicity and aromatic resonance power (*i.e.*, π-conjugation length) of the TPA unit, improving the instabilities in the memory behavior. In contrast, the bis(trifluoromethyl)phenyl substituents significantly reduced the nucleophilic power of the TPA units, altering the memory behavior in a negative way. All of the PI films revealed high ON/OFF current ratios (10^5–10^8) and moreover high stabilities in both the OFF and ON states, even under ambient air conditions. All of the memory behaviors were governed by a mechanism involving field-enhanced thermal emission of trapped charges and filament formation. Overall, this study has demonstrated that thermally, dimensionally stable **P36** and **P37** are highly suitable active materials for the low-cost mass production of high performance, polarity-free programmable memory devices that can be operated with very low power consumption, high ON/OFF current ratios, and high thermal and dimensional stability.

Three functional PIs, **P38**, **P39**, and **P40**, bearing conjugated bis(triphenylamine) (2TPA) derivatives with electron-donating and accepting substituents

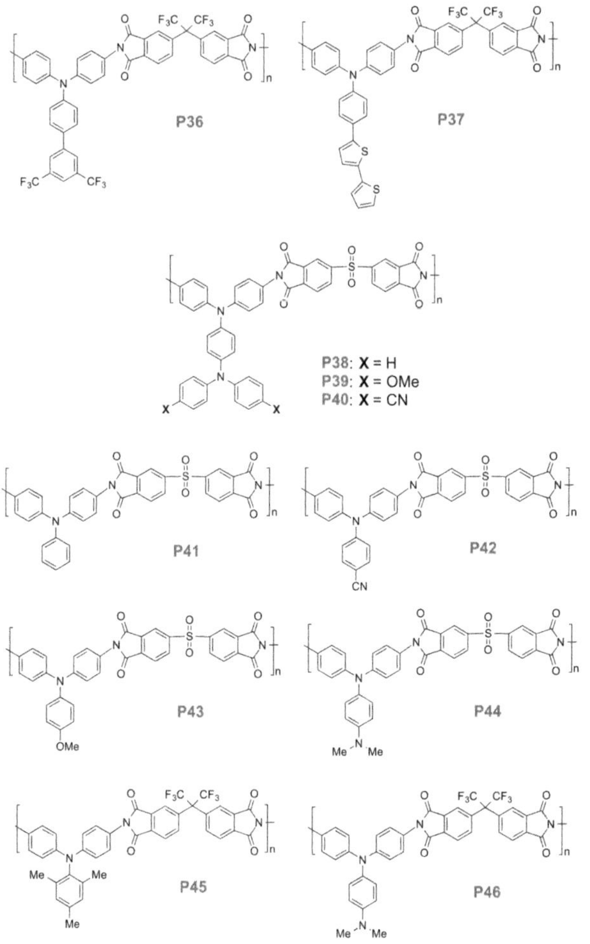

Scheme 4.7 Molecular design of the volatility of TPA-based polyimides through donor effects.

were synthesized.[61] All of the PIs initially revealed a high resistance (OFF state) in Al/PI/Al devices. However, under positive and negative voltage sweeps, the PIs demonstrated volatile or non-volatile digital memory behavior depending on the substituents of the 2TPA unit. The 2TPA-based PI, as well as PI **P39** bearing 2TPA with electron-donating methoxy substituents showed unipolar write-once-read-many-times (WORM) memory behavior, whereas PI **P40** containing electron-accepting cyano groups exhibited unipolar dynamic random access memory (DRAM) behavior. All of the PI films revealed excellent retention abilities in both the OFF and ON states, even under ambient air conditions. Moreover, they all revealed high ON/OFF current ratios (10^6–10^{10}). All of the memory behaviors were found to be governed by a mechanism involving trap-limited space charge limited conduction and local filament formation. Such memory behaviors were further investigated in detail taking into consideration the chemical nature and molecular orbital levels of the PIs, possible trapping sites, substituent effects, and the work function of the metal electrodes. Overall, this study demonstrated that thermally, dimensionally stable PIs are highly suitable for the low-cost mass production of high performance digital memory devices that can be operated with very low power consumption. Moreover, the memory mode can be tuned by changing the substituent in the 2TPA unit.

To further investigate the function of the modified TPA of PIs for memory device performance, a series of DSDA (3,3′,4,4′-diphenylsulfonyltetracarboximide)-based PIs **P41–P44**[62] bearing three different TPA derivatives was synthesized, which showed high thermal and dimensional stabilities compared with conventional PIs. In Al/PI/Al devices, all of the PIs initially exhibited a high resistance. Under positive and negative voltage sweeps, polymers **P41–P44** exhibited unipolar WORM memory behavior. Interestingly, for **P42**, only ~80% of the device cells showed unipolar WORM memory behavior, and the other device cells showed unipolar DRAM behavior, which can be attributed to the cooperative role played by the cyano substituents as charging sites and electron-accepting groups. On the other hand, polymer **P44** revealed rewritable memory behavior. The original low-conductivity state (OFF state) can be switched to the high-conductivity state (ON state) when a positive or negative voltage is applied to the device. In the ON state, the device was switched off by applying a voltage with a current compliance of 0.1 A, which was higher than that (0.01 A) chosen in the switching-ON process. All of the PIs switched on at approximately ±2 V. They all showed excellent retention abilities in both the OFF and ON states, even under ambient air conditions, and demonstrated high ON/OFF current ratios (10^6–10^8). The memory behavior was found to be governed by mechanisms involving trap-limited SCLC conduction and local filament formation.

Liou et al.[57] reported two PIs (**P45** and **P46**) by modifying the TPA with different substituents to optimize the polymer structure. The polymers displayed different non-volatile memory behaviors in Al/PI/Al devices depending on the substituents of the TPA unit. Polymer **P45** showed unipolar WORM memory behavior, whereas a polymer **P46** film revealed unipolar and bipolar switching memory behavior. Both of the polymers showed excellent stability both in the OFF and ON states even under ambient conditions with an ON/OFF current ratio up to 10^8–10^9. The memory behaviors of these two polymers can be attributed to the mechanism of trap-limited space charge limited conduction and local filament formation. The unipolar WORM memory characteristics of polymer **P45** can be attributed to the enhanced electron-donating ability of the TPA unit provided by three methyl substituents in addition to the phthalimide units, which acted as effective charge-trapping sites. In comparison, the unipolar and bipolar switching memory characteristics of polymer **P46** originated from the dimethylamine substituents acting as charge-trapping sites and enhancing the electron-donating ability of the TPA unit. These results demonstrated that the memory modes of PIs can be tuned by changing the substituents on the TPA unit.

Subsequently, Liou designed and prepared another aromatic PI **P5**[25] by introducing a highly electron-donating starburst triarylamine group. The introduction of the starburst triarylamine group onto the polymer main chain not only stabilizes its radical cations but also enhances the solubility and film-forming capability of the resulted PIs. In addition, the obtained polymers possessed high T_gs and good thermal stability, which play an important role in resistive memory devices. In order to investigate the effects of donor moieties in PIs on their memory behavior, the corresponding **P4** (3Ph-PIs) and **P3** (5Ph-PIs) were also prepared and discussed. With the intensity of electron donation increasing from **P4** to **P3** and then to **P5**, the retention time of the memory device reveals a systematic increase.

4.5.2 Acceptor Effect

A new series of aromatic PIs **P47a–P47c**[63] was synthesized *via* a two-step procedure from newly synthesized tetracarboxylic dianhydride monomers reacting with 2,2'-bis[4'-(3″,4″,5″-trifluorophenyl)phenyl]-4,4'-biphenyl diamine (Scheme 4.8). The resulting polymers exhibited excellent organo-solubility and thermal properties. In this study, each repeat unit of the polymers contained different aryl pendants on two imide rings which could have electric polarization characteristics. The aryl pendants are electron donors and thus probably act as nucleophilic sites, whereas the imide rings are electron acceptors and probably act as electrophilic sites. Thus, all of these groups are likely to act as charge-trapping sites, depending on their association. The fabricated sandwich structured memory devices of Al/**P47-a**/ITO were determined to present flash-type memory behavior, while Al/**P47-b**/ITO and Al/**P47-c**/ITO exhibited WORM memory capability with different threshold voltages. In addition, the Al/PI/ITO devices showed high stability under a constant stress or continuous read pulse voltage of −1.0 V. All of the devices exhibited a long retention time under ambient air conditions and an ON/OFF current ratio of about 10^3. This series of PIs exhibited a good combination of properties required for high performance materials and demonstrated promising potential for future organic memory applications.

Recently, Chen and Liou *et al.* reported a series of PIs **P3**[21] with different phthalimide moieties and found that the differences between the HOMO and LUMO energy levels, and the dipole moment can significantly influence the memory behavior. PIs **P3-DSPI** and **P3-NTPI** exhibit non-volatile WORM memory behavior in ITO/polymer/Al sandwich devices with threshold voltages of −4.8 V and −4.9 V, and high ON/OFF current ratios of 10^9 and 10^{10}, respectively. According to theoretical analysis, the low LUMO energy level and the strong CT effect of **P3-NTPI** produce the WORM behavior. Interestingly, although **P3-DSPI** has a higher LUMO energy level, it also showed WORM behavior because of the high dipole moment of 5.45 D. By reducing the electron-withdrawing intensity of the phthalimide moieties, the retention time of the corresponding memory devices also decreased, with polymer **P3-6FPI** and **P3-PMPI** exhibiting DRAM and SRAM memory behaviors,

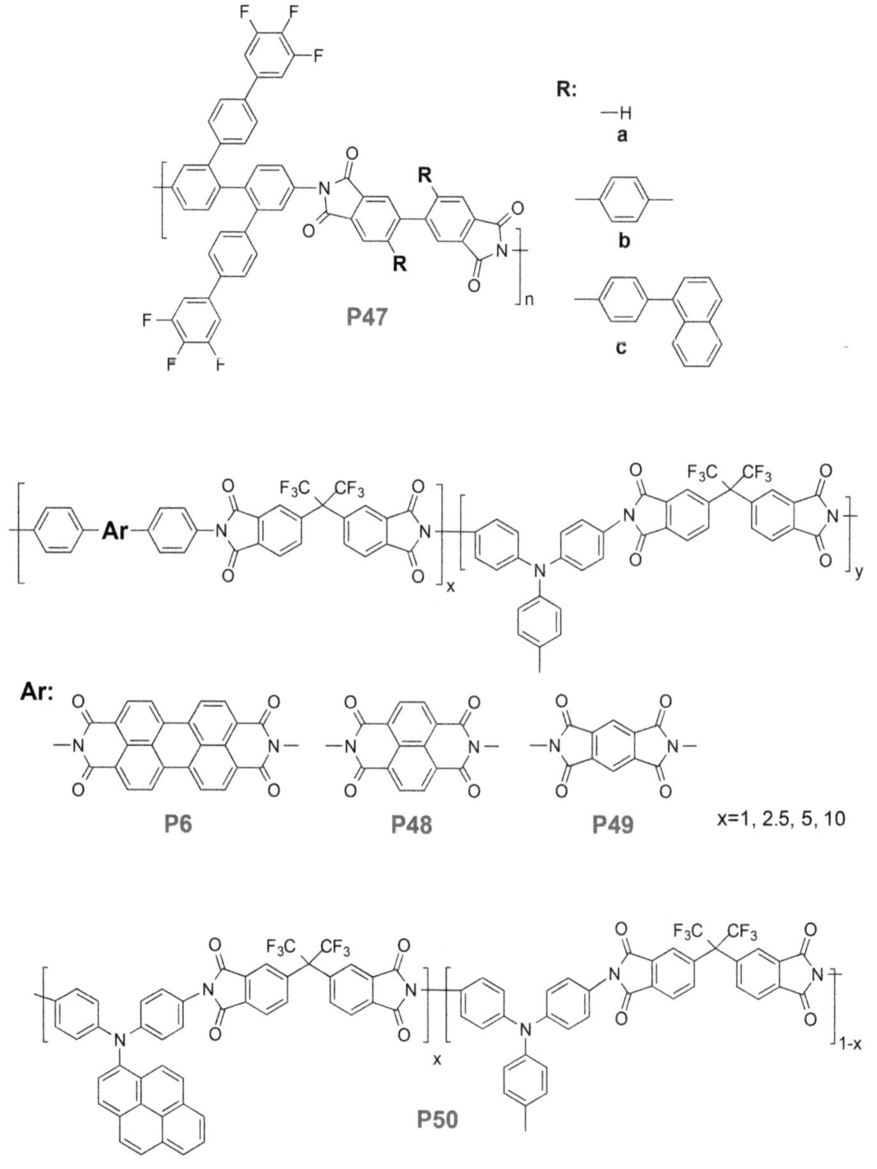

Scheme 4.8 Molecular design of the volatility through acceptor effects.

respectively. These results demonstrated that the energy levels and polarity of PIs can be tuned by changing the electron acceptor, which can influence the memory performance significantly.

Ueda and Chen et al.[26,64] reported a series of random coPIs, **P6**, **P48**, and **P49**, by varying the feeding ratio of acceptor units, perylenetetracarboxydiimide (PTI), naphthalenetetracarboxydiimide (NTI), and benzenetetracarboxydiimide (BTI), respectively. The memory devices prepared from these coPIs sandwiched between ITO and Al electrodes exhibited tunable electrical bistability from volatile to non-volatile WORM memory characteristics with the increase in PBI, NTI, or BTI composition. By summarizing the results, the effects of the conjugation length of the acceptor moiety (PTI > NTI > BTI) on the resulted memory characteristics were fully studied. A stable CT complex could be obtained through stabilization of the radical anion by means of the long conjugation and high electron affinity of the NTI and PTI moieties, leading to non-volatile WORM memory characteristics. On the other hand, due to the low HOMO energy level of the BTI unit, a large energy barrier was generated which prevents hole injection from the Al electrode to the polymer, resulting in non-switching behavior at a high BTI composition. The study also demonstrated that the composition of the acceptor moiety with different conjugation lengths significantly affected the memory characteristics of the donor–acceptor coPIs.

New coPIs **P50**[40] with main chain TPA/pyrene groups randomly copolymerized with 6FDA acceptors by chemical imidization were prepared by Ueda and Chen *et al.* and employed as the active layer for flexible resistance memory devices in a cross-point PEN/Al/**P50**/Al platform. The memory behaviors vary with the APAP molar content from volatile to non-volatile flash, flash/WORM and WORM memory for **P50-0, P50-5, P50-10** and **P50-100**, respectively. Electrical bistability was ascribed to CT as suggested by theoretical simulation, and the highly conjugated pyrene moiety is beneficial for a stable CT complex with excellent non-volatile electrical properties. The resistance switching characteristics were successfully analyzed using an appropriate theoretical *J–V* dependence model. The fabricated coPIs exhibited superior memory performance and a high level of mechanical durability when operated under various bending conditions. A possible switching mechanism based on CT interaction was proposed through molecular simulation and fitted with physical conduction models in the OFF and ON states. The manipulation of the memory volatility by tailoring the molecular design and fabrication of plastic electronic devices demonstrated promising applications of donor–acceptor PIs in integrated memory devices.

4.5.3 Linkage Effect

Liou *et al.* reported two TPA-based aromatic PIs (**P51** and **P52**)[65] containing pendent anthraquinone that was directly attached to or incorporated *via* ether linkages into the backbone as electron acceptor and the PIs were used for the comparison of memory behaviors (Scheme 4.9). All PIs exhibited excellent thermal stability and high glass transition temperature. The memory devices with the configuration ITO/**P51**/Al exhibited distinct volatile DRAM characteristics while the ITO/**P52**/Al device showed volatile SRAM properties. Thus, the results indicated that the isolated D–A system could effectively extend the retention time in memory device application. The ON/OFF current ratios of these memory devices are up to 10^9. The theoretical analysis results suggest

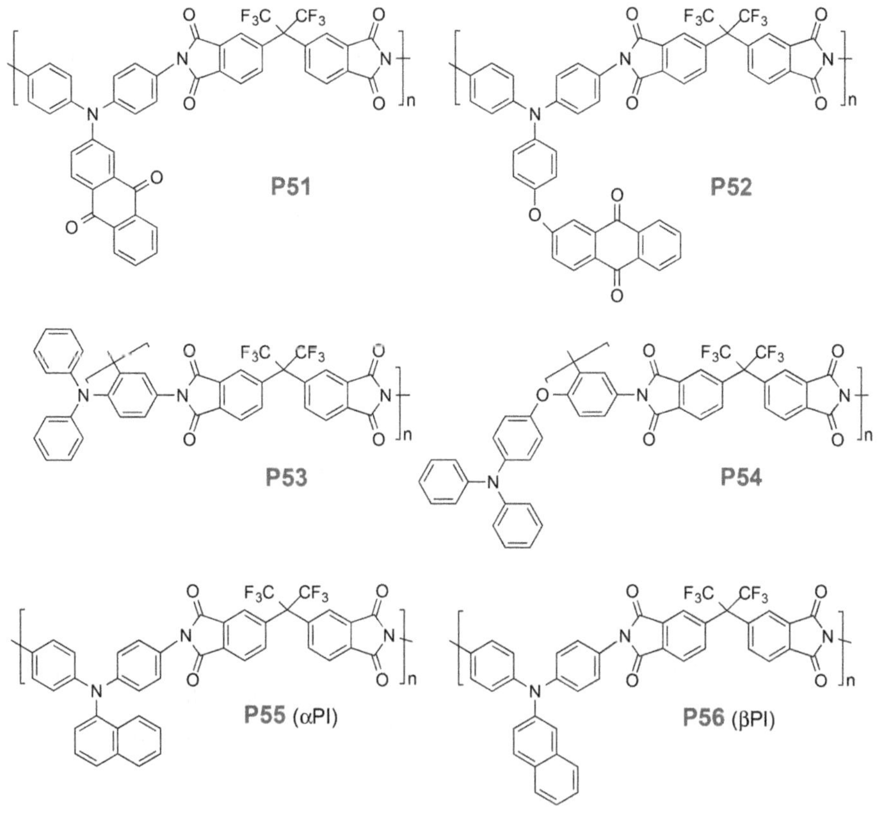

Scheme 4.9 Molecular design of the volatility through the linkage effect.

that the CT mechanism could be used to explain the memory characteristics of these PIs, and the linkage effect between the donor and acceptor was also demonstrated.

Two polyimides, **P53** and **P54**,[66] from N-(2,4-diaminophenyl)-N,N-diphenylamine (DAT) or N-(4-(2′,4′-diaminophenoxy)phenyl)-N,N-diphenylamine (DAPT) and 6FDA were prepared to clarify the structural effect on the resulting memory properties. The memory device based on **P53** showed unstable volatile behavior, while the device based on **P54** with a more bulky donor unit exhibited stable non-volatile flash type memory characteristics with a long retention time of over 10^4 s. Theoretical simulations based on DFT suggested that greater distinct charge separation between the ground and CT states led to highly stable memory behavior. Also, it was clarified that **P54** had a highly twisted conformation compared to **P53** in the ground state, and a more twisted dihedral angle between the donor and acceptor units was induced in the CT state, which led to non-volatile memory characteristics.

Qi et al. synthesized two new diphenylnaphthylamine-based functional aromatic PIs, **P55** (αPI) and **P56** (βPI), for memory device applications.[67] By altering the naphthyl group in diphenylnaphthylamine from α-substitution to β-substitution, the memory behavior of the synthesized PIs can be tuned from irreversible WORM type (**P55**) to programmable flash type (**P56**). The **P55**-based memory device (ITO/PI/Au) exhibits an ON/OFF current ratio as high as about 10^6 and can be switched bi-directionally with comparable positive and negative switch-on threshold voltages, while the **P56**-based device achieves an ON/OFF ratio of up to 10^4 and displays excellent reprogrammable stability during the write–read–erase–reread (WRER) cycles. Theoretical analysis suggests that the electric field-induced donor–acceptor electronic transition and the subsequent formation of CT complexes could be used to explain the electrical switching effects observed in the synthesized PIs. Furthermore, the more non-coplanar conformation of the diphenylnaphthylamine species in **P55**, as compared to **P56**, is suggested to bring about a higher energy barrier that prevents back CT processes under the applied electric field, leading to the irreversible WORM vs. programmable flash memory behaviors. The above results indicate two new functional aromatic PIs with advanced memory device application potential, based on their fast switching and high ON/OFF characteristics.

4.5.4 Thickness Effect

Interestingly, for polymer **P57** reported by Ree and Liou et al. (Scheme 4.10), the film thickness can also influence the memory behavior.[68] It exhibits excellent WORM memory characteristics for films with thicknesses in the range of 34–74 nm, which may be attributed to the formation of stable local filaments. However, it reveals DRAM memory characteristics under a film thickness of 100 nm and the ON/OFF current ratio of the DRAM devices is as high as 10^{11}. This is because the films are too thick to exhibit stable local filament formation, and only local filaments form in the films, which are responsible for the DRAM memory performance.

In addition to TPA, electron-rich sulfur-containing groups can also act as charge donors.[6] Ueda and Chen et al.[56] reported the synthesis and characterization of two polyimides (PIs), **P58** and **P59**, consisting of alternating electron-donating 2,8- or 3,7-phenylenesulfanyl-substituted dibenzothiophene and electron-accepting phthalimide moieties for high-performance memory device applications. The device with the ITO/PI/Al configuration

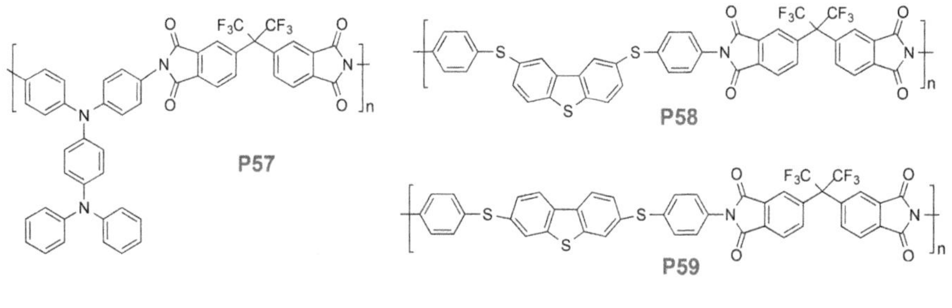

Scheme 4.10 Molecular design of the volatility through the thickness effect.

showed multi-memory characteristics changing from high-conductance ohmic current flow to non-volatile negative differential resistance (NDR), dynamic random access memory, and insulator, with a corresponding film thickness of 12, 20, 25, and 45 nm, respectively. The 25 nm PI device showed reproducible DRAM characteristics with a high ON/OFF current ratio of more than 10^8.

4.6 Polyimide–Inorganic Hybrid Materials

Although several types of single polymer devices have been demonstrated, hybrid composites have also been extensively prepared for memory device applications. The supplementary compounds (NPs or fullerene derivatives) can be viewed as data storage media or can have physical electronic transitions with the organic polymer matrix.

Tian and Qi *et al.* reported the synthesis of a silver-nanoparticle-embedded PI thin film and its electrical bistability.[69] A soluble PI, **P60**, where the 6FDA part serves as an effective electron-accepting moiety, was synthesized as the polymer matrix (Scheme 4.11). Silver nanoparticles (Ag NPs) with diameters less than 7 nm were subsequently generated *in situ* in the parent PI film *via* ultraviolet (UV) reduction of the (1,1,1-trifluoro-2,4-pentadionato) silver(I) complex (AgTFA) previously incorporated in the matrix. Electrical characterization results for the sandwich device (ITO/PI silver nanohybrid film/Au) indicate that the nano-hybrid material possesses electrical bistability and the device exhibits two accessible conductivity states, which can be reversibly switched from the low-conductivity state to the high-conductivity state with an ON/OFF current ratio of around 10^2. The device with the PI silver nano-hybrid film as the active layer shows non-volatile memory behavior. The high- and low-conductivity states of the device can be sustained after removal of the applied voltage.

In addition to Ag NPs, some other kinds of inorganic nanomaterials have also been introduced into polymers to realize rewritable memory behavior. According to our report, novel solution-processable sulfur- or fluorine-containing poly(*o*-hydroxy-imide), **P61**[70] and **P62**,[71] with pendant hydroxyl groups and the corresponding PI–TiO$_2$ hybrids were synthesized and used for memory applications. To enhance the memory behavior, different amounts of TiO$_2$ were introduced into the PIs and the corresponding tunable memory properties were investigated. The hydroxyl groups on the backbone of the PIs can provide reaction sites for organic–inorganic bonding and homogeneous hybrid thin films can therefore be obtained by controlling the molar ratio of titanium butoxide/hydroxyl groups *via* sol–gel reaction. The resulting hybrid films with different TiO$_2$ concentrations from 0 wt% to 50 wt% exhibited electrically programmable digital memory properties from DRAM to SRAM to WORM with a high ON/OFF current ratio (10^8). Furthermore, from the results of the current–voltage *I–V* characteristics, the crystalline phase of titania reveals a higher trapping ability for increasing the retention time in the ON state.

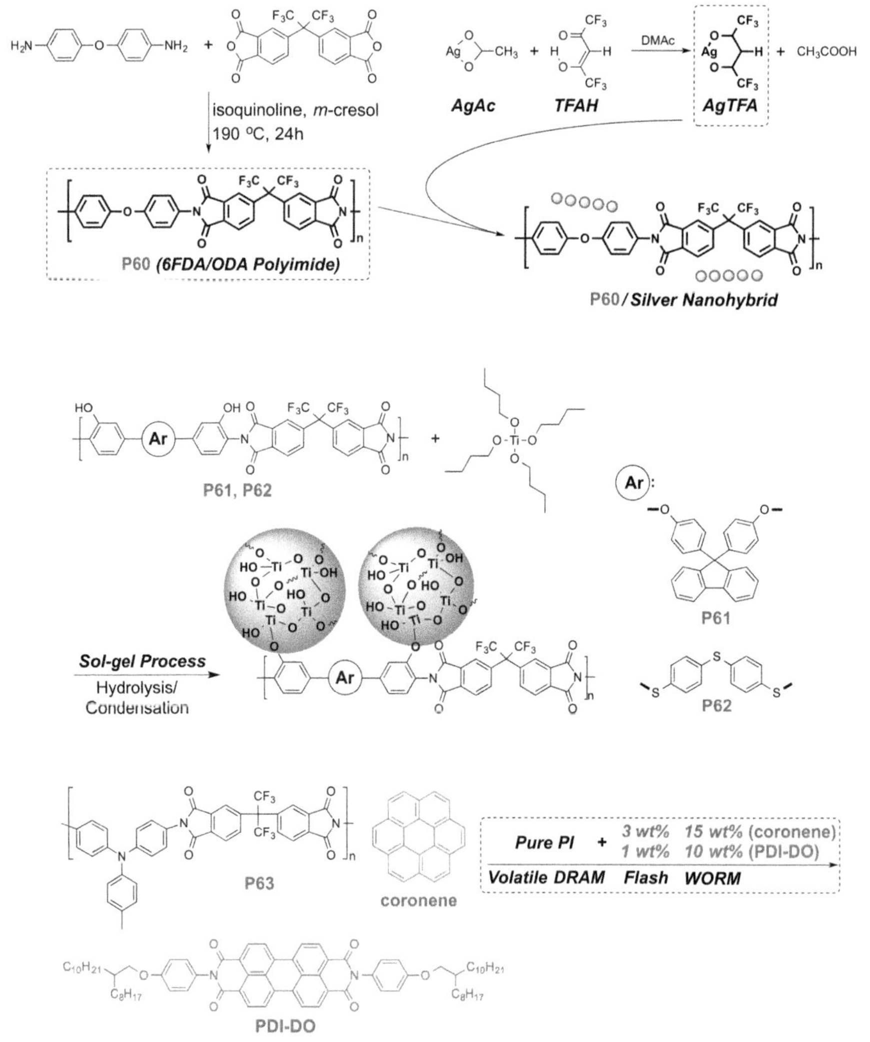

Scheme 4.11 Chemical structures of polyimide–inorganic hybrid materials for memory device applications.

Other resistance switching memory devices based on PI blends were reported by Ueda, Liu, and Chen.[72] The active layers of the PI blend films were prepared from different compositions of **P63** and polycyclic aromatic compounds (coronene or N,N-bis[4-(2-octyldodecyloxy)-phenyl]-3,4,9,10-perylenetetracarboxylic diimide (**PDI-DO**)). The additives of large π-conjugated polycyclic compounds can stabilize the CT complex induced by the applied electric field. Thus, the memory device characteristics change from volatile to non-volatile behavior of flash and WORM as the additive content increases in both blend systems. The main differences between these two blend systems are the threshold voltage values and the additive content required to change the memory behavior. Due to the stronger accepting ability and higher electron affinity of PDI-DO compared to those of coronene, the **P63**:PDI-DO blend-based memory devices show a lower threshold voltage and can change their memory behavior with a smaller additive content.

4.7 Flexible Polyimide Electrical Memory Devices

Similar to other organic electronics, flexible polymer memory devices have also been demonstrated using PI as an active layer.[25,40,73] In 2012, Ueda, Liu, and Chen fabricated a memory device similar to the previous ones on a flexible poly(ethyleneterephthalate) (PET) substrate.[40] The device showed very stable non-volatile behavior even after bending at a radius of curvature of 5 mm and for up to 1000 bending cycles.

In addition, the flexible memory device reported by our group was tested under severe bending at various curvature radii of 11, 9, 7, and 5 mm, respectively, and the flexible device did not crack or deform upon bending. Reliable and reproducible switching memory behavior of the PI film in the device could also be achieved under mechanical bending stress. The development of device fabrication for practical use has been carried out to a large extent. Like other organic electronics, such as organic transistors, organic photovoltaics, *etc.*, the performance of a polymer memory significantly relies on the fabrication process of the device. Therefore, similar to the development of the memory material itself, there is still significant potential in this area.

References

1. M. Ding, *Prog. Polym. Sci.*, 2007, **32**, 623.
2. A. Mochizuki and M. Ueda, *J. Photopolym. Sci. Technol.*, 2001, **14**, 677.
3. H. Kricheldorf, in *Progress in Polyimide Chemistry II*, ed. H. R. Kricheldorf, Springer, Berlin, Heidelberg, 1999, vol. 141, p. 83.
4. P. Lesieur, A. Barraud and M. Vandevyver, *Thin Solid Films*, 1987, **152**, 155.
5. R. Souzy and B. Ameduri, *Prog. Polym. Sci.*, 2005, **30**, 644.
6. Q.-D. Ling, D.-J. Liaw, C. Zhu, D. S.-H. Chan, E.-T. Kang and K.-G. Neoh, *Prog. Polym. Sci.*, 2008, **33**, 917.
7. N.-H. You, C.-C. Chueh, C.-L. Liu, M. Ueda and W.-C. Chen, *Macromolecules*, 2009, **42**, 4456.
8. J. H. A. Smits, S. C. J. Meskers, R. A. J. Janssen, A. W. Marsman and D. M. de Leeuw, *Adv. Mater.*, 2005, **17**, 1169.
9. W. Tang, H. Z. Shi, G. Xu, B. S. Ong, Z. D. Popovic, J. C. Deng, J. Zhao and G. H. Rao, *Adv. Mater.*, 2005, **17**, 2307.
10. F. Verbakel, S. C. J. Meskers and R. A. J. Janssen, *Chem. Mater.*, 2006, **18**, 2707.
11. H.-T. Lin, Z. Pei, J.-R. Chen, G.-W. Hwang, J.-F. Fan and Y.-J. Chan, *IEEE Electron Device Lett.*, 2007, **28**, 951.

12. D. Attianese, M. Petrosino, P. Vacca, S. Concilio, P. Iannelli, A. Rubino and S. Bellone, *IEEE Electron Device Lett.*, 2008, **29**, 44.
13. B. Cho, T.-W. Kim, M. Choe, G. Wang, S. Song and T. Lee, *Org. Electron.*, 2009, **10**, 473.
14. X.-D. Zhuang, Y. Chen, G. Liu, B. Zhang, K.-G. Neoh, E.-T. Kang, C.-X. Zhu, Y.-X. Li and L.-J. Niu, *Adv. Funct. Mater.*, 2010, **20**, 2916.
15. C. W. Chu, J. Ouyang, H. H. Tseng and Y. Yang, *Adv. Mater.*, 2005, **17**, 1440.
16. Q.-D. Ling, F.-C. Chang, Y. Song, C.-X. Zhu, D.-J. Liaw, D. S.-H. Chan, E.-T. Kang and K.-G. Neoh, *J. Am. Chem. Soc.*, 2006, **128**, 8732.
17. Y.-C. Lai, K. Ohshimizu, W.-Y. Lee, J.-C. Hsu, T. Higashihara, M. Ueda and W.-C. Chen, *J. Mater. Chem.*, 2011, **21**, 14502.
18. Y.-Q. Li, R.-C. Fang, A.-M. Zheng, Y.-Y. Chu, X. Tao, H.-H. Xu, S.-J. Ding and Y.-Z. Shen, *J. Mater. Chem.*, 2011, **21**, 15643.
19. B. Zhang, Y.-L. Liu, Y. Chen, K.-G. Neoh, Y.-X. Li, C.-X. Zhu, E.-S. Tok and E.-T. Kang, *Chem.–Eur. J.*, 2011, **17**, 10304.
20. S.-J. Liu, Z.-H. Lin, Q. Zhao, Y. Ma, H.-F. Shi, M.-D. Yi, Q.-D. Ling, Q.-L. Fan, C.-X. Zhu, E.-T. Kang and W. Huang, *Adv. Funct. Mater.*, 2011, **21**, 979.
21. C.-J. Chen, H.-J. Yen, W.-C. Chen and G.-S. Liou, *J. Mater. Chem.*, 2012, **22**, 14085.
22. B. Zhang, G. Liu, Y. Chen, C. Wang, K.-G. Neoh, T. Bai and E.-T. Kang, *ChemPlusChem*, 2012, **77**, 74.
23. C.-L. Liu and W.-C. Chen, *Polym. Chem.*, 2011, **2**, 2169.
24. T. Kuorosawa, C.-C. Chueh, C.-L. Liu, T. Higashihara, M. Ueda and W.-C. Chen, *Macromolecules*, 2010, **43**, 1236.
25. C.-J. Chen, H.-J. Yen, Y.-C. Hu and G.-S. Liou, *J. Mater. Chem. C*, 2013, **1**, 7623.
26. T. Kurosawa, Y.-C. Lai, T. Higashihara, M. Ueda, C.-L. Liu and W.-C. Chen, *Macromolecules*, 2012, **45**, 4556.
27. H. Carchano, R. Lacoste and Y. Segui, *Appl. Phys. Lett.*, 1971, **19**, 414.
28. L. F. Pender and R. J. Fleming, *J. Appl. Phys.*, 1975, **46**, 3426.
29. Y. Segui, B. Ai and H. Carchano, *J. Appl. Phys.*, 1976, **47**, 140.
30. W. Hwang and K. C. Kao, *J. Chem. Phys.*, 1974, **60**, 3845.
31. A. Wierschem, F. J. Niedernostheide, A. Gorbatyuk and H. G. Purwins, *Scanning*, 1995, **17**, 106.
32. W.-J. Joo, T.-L. Choi, J. Lee, S. K. Lee, M.-S. Jung, N. Kim and J. M. Kim, *J. Phys. Chem. B*, 2006, **110**, 23812.
33. Y.-L. Liu, Q.-D. Ling, E.-T. Kang, K.-G. Neoh, D.-J. Liaw, K.-L. Wang, W.-T. Liou, C.-X. Zhu and D. S.-H. Chan, *J. Appl. Phys.*, 2009, **105**, 044501.
34. G. Tian, D. Wu, S. Qi, Z. Wu and X. Wang, *Macromol. Rapid Commun.*, 2011, **32**, 384.
35. Y.-L. Liu, K.-L. Wang, G.-S. Huang, C.-X. Zhu, E.-S. Tok, K.-G. Neoh and E.-T. Kang, *Chem. Mater.*, 2009, **21**, 3391.
36. Y. Liu, Y. Zhang, Q. Lan, S. Liu, Z. Qin, L. Chen, C. Zhao, Z. Chi, J. Xu and J. Economy, *Chem. Mater.*, 2012, **24**, 1212.
37. Y. Liu, Y. Zhang, Q. Lan, Z. Qin, S. Liu, C. Zhao, Z. Chi and J. Xu, *J. Polym. Sci., Part A: Polym. Chem.*, 2013, **51**, 1302.

38. G.-S. Liou, S.-H. Hsiao and H.-W. Chen, *J. Mater. Chem.*, 2006, **16**, 1831.
39. L. Shi, G. Tian, H. Ye, S. Qi and D. Wu, *Polymer*, 2014, **55**, 1150.
40. A.-D. Yu, T. Kurosawa, Y.-C. Lai, T. Higashihara, M. Ueda, C.-L. Liu and W.-C. Chen, *J. Mater. Chem.*, 2012, **22**, 20754.
41. D. M. Kim, S. Park, T. J. Lee, S. G. Hahm, K. Kim, J. C. Kim, W. Kwon and M. Ree, *Langmuir*, 2009, **25**, 11713.
42. Q. Liu, K. Jiang, Y. Wen, J. Wang, J. Luo and Y. Song, *Appl. Phys. Lett.*, 2010, **97**, 253304.
43. K.-L. Wang, Y.-L. Liu, J.-W. Lee, K.-G. Neoh and E.-T. Kang, *Macromolecules*, 2010, **43**, 7159.
44. K. L. Wang, Y. L. Liu, I. H. Shih, K. G. Neoh and E. T. Kang, *J. Polym. Sci., Part A: Polym. Chem.*, 2010, **48**, 5790.
45. S. Park, K. Kim, J. C. Kim, W. Kwon, D. M. Kim and M. Ree, *Polymer*, 2011, **52**, 2170.
46. F. Chen, G. Tian, L. Shi, S. Qi and D. Wu, *Rsc Adv.*, 2012, **2**, 12879.
47. S. G. Hahm, S. Choi, S.-H. Hong, T. J. Lee, S. Park, D. M. Kim, J. C. Kim, W. Kwon, K. Kim, M.-J. Kim, O. Kim and M. Ree, *J. Mater. Chem.*, 2009, **19**, 2207.
48. K. Kim, S. Park, S. G. Hahm, T. J. Lee, D. M. Kim, J. C. Kim, W. Kwon, Y.-G. Ko and M. Ree, *J. Phys. Chem. B*, 2009, **113**, 9143.
49. B. Hu, F. Zhuge, X. Zhu, S. Peng, X. Chen, L. Pan, Q. Yan and R.-W. Li, *J. Mater. Chem.*, 2012, **22**, 520.
50. S. G. Hahm, S. Choi, S.-H. Hong, T. J. Lee, S. Park, D. M. Kim, W.-S. Kwon, K. Kim, O. Kim and M. Ree, *Adv. Funct. Mater.*, 2008, **18**, 3276.
51. S. Park, K. Kim, D. M. Kim, W. Kwon, J. Choi and M. Ree, *ACS Appl. Mater. Interfaces*, 2011, **3**, 765.
52. Y. Li, R. Fang, S. Ding and Y. Shen, *Macromol. Chem. Phys.*, 2011, **212**, 2360.
53. G. Tian, S. Qi, F. Chen, L. Shi, W. Hu and D. Wu, *Appl. Phys. Lett.*, 2011, **98**, 203302.
54. S.-H. Liu, W.-L. Yang, C.-C. Wu, T.-S. Chao, M.-R. Ye, Y.-Y. Su, P.-Y. Wang and M.-J. Tsai, *IEEE Electron Device Lett.*, 2013, **34**, 123.
55. T. J. Lee, S. Park, S. G. Hahm, D. M. Kim, K. Kim, J. Kim, W. Kwon, Y. Kim, T. Chang and M. Ree, *J. Phys. Chem. C*, 2009, **113**, 3855.
56. C.-L. Liu, T. Kurosawa, A.-D. Yu, T. Higashihara, M. Ueda and W.-C. Chen, *J. Phys. Chem. C*, 2011, **115**, 5930.
57. T. J. Lee, Y.-G. Ko, H.-J. Yen, K. Kim, D. M. Kim, W. Kwon, S. G. Hahm, G.-S. Liou and M. Ree, *Polym. Chem.*, 2012, **3**, 1276.
58. Y. Li, H. Xu, X. Tao, K. Qian, S. Fu, Y. Shen and S. Ding, *J. Mater. Chem.*, 2011, **21**, 1810.
59. Y. Li, Y. Chu, R. Fang, S. Ding, Y. Wang, Y. Shen and A. Zheng, *Polymer*, 2012, **53**, 229.
60. D. M. Kim, Y.-G. Ko, J. K. Choi, K. Kim, W. Kwon, J. Jung, T.-H. Yoon and M. Ree, *Polymer*, 2012, **53**, 1703.
61. K. Kim, H.-J. Yen, Y.-G. Ko, C.-W. Chang, W. Kwon, G.-S. Liou and M. Ree, *Polymer*, 2012, **53**, 4135.

62. Y.-G. Ko, W. Kwon, H.-J. Yen, C.-W. Chang, D. M. Kim, K. Kim, S. G. Hahm, T. J. Lee, G.-S. Liou and M. Ree, *Macromolecules*, 2012, **45**, 3749.
63. Y. Li, H. Xu, X. Tao, K. Qian, S. Fu, S. Ding and Y. Shen, *Polym. Int.*, 2011, **60**, 1679.
64. T. Kurosawa, Y.-C. Lai, A.-D. Yu, H.-C. Wu, T. Higashihara, M. Ueda and W.-C. Chen, *J. Polym. Sci., Part A: Polym. Chem.*, 2013, **51**, 1348.
65. Y.-C. Hu, C.-J. Chen, H.-J. Yen, K.-Y. Lin, J.-M. Yeh, W.-C. Chen and G.-S. Liou, *J. Mater. Chem.*, 2012, **22**, 20394.
66. T. Kurosawa, A.-D. Yu, T. Higashihara, W.-C. Chen and M. Ueda, *Eur. Polym. J.*, 2013, **49**, 3377.
67. L. Shi, H. Ye, W. Liu, G. Tian, S. Qi and D. Wu, *J. Mater. Chem. C*, 2013, **1**, 7387.
68. T. J. Lee, C.-W. Chang, S. G. Hahm, K. Kim, S. Park, D. M. Kim, J. Kim, W.-S. Kwon, G.-S. Liou and M. Ree, *Nanotechnology*, 2009, **20**, 135204.
69. G. Tian, D. Wu, L. Shi, S. Qi and Z. Wu, *Rsc Adv.*, 2012, **2**, 9846.
70. C.-L. Tsai, C.-J. Chen, P.-H. Wang, J.-J. Lin and G.-S. Liou, *Polym. Chem.*, 2013, **4**, 4570.
71. C.-J. Chen, C.-L. Tsai and G.-S. Liou, *J. Mater. Chem. C*, 2014, **2**, 2842.
72. A.-D. Yu, T. Kurosawa, Y.-H. Chou, K. Aoyagi, Y. Shoji, T. Higashihara, M. Ueda, C.-L. Liu and W.-C. Chen, *ACS Appl. Mater. Interfaces*, 2013, **5**, 4921.
73. H.-J. Yen, C.-J. Chen and G.-S. Liou, *Adv. Funct. Mater.*, 2013, **23**, 5307.

CHAPTER 5

Nonconjugated Polymers with Electroactive Chromophore Pendants

MOONHOR REE*[a], YONG-GI KO[a], SUNGJIN SONG[a], AND BRIAN J. REE[a]

[a]Department of Chemistry, Division of Advanced Materials Science, Center for Electro-Photo Behaviors in Advanced Molecular Systems, Pohang Accelerator Laboratory, Polymer Research Institute, and BK School of Molecular Science, Pohang University of Science and Technology, Pohang 790-784, Republic of Korea
*E-mail: ree@postech.edu

5.1 Introduction

The properties of organic materials, including small molecules and polymers, can be easily tailored through controlled, well-defined chemical synthesis. They can be processed into miniaturized dimensions and multi-stack layer structures with ease. In general, small organic molecules require more elaborate fabrication processes than inorganic materials, such as vacuum evaporation and deposition. Nevertheless, they often reveal very poor ability to form thin films. Additionally, the relatively low boiling points and low chemical resistances of organic molecules leave the deposited organic layer vulnerable to severe damage resulting from the fabrication of memory devices. In contrast, polymeric materials exhibit great film formation processability, flexibility, high mechanical strength, and good scalability. Also, they can be easily processed at low cost *via* conventional solution processes. Hence, polymers are quite favorable for making the multi-stack layer structures required for high density memory devices.

Due to their advantageous characteristics, great research effort has been devoted to developing electrical memory polymers as alternatives to silicon- and metal-oxide-based memory materials.[1-4] As a result, three major polymer families have been introduced for electrical memory devices.[1-33] The first family consists of polymers based on fully π-conjugated backbones that can trap and transport charge as electroactive chromophores.[5-14] The second family is composed of polymers based on either partially π-conjugated backbones or backbones containing electroactive chromophore units.[15] The last family of electrical memory polymers is based on non-π-conjugated backbones with pendant electroactive chromophores.[16-33] The π-conjugated backbones of the first polymer family limits many options in synthesis. The second polymer family also has a narrow range of options in synthesis because of the partial π-conjugation or electroactive chromophore units in the backbones. The last polymer family, however, has a wide range of synthetic options because it only requires electroactive chromophores to be attached as pendant groups to the backbones of any nonconjugated polymer, and there is a greater variety of nonconjugated polymers than those with π-conjugated backbones. Good candidates for the nonconjugated polymer backbone include polyvinyl, polyimide, polyvinylether, polyester, polyamide, polyurethane, polyurea, and polysulfone. Suitable candidates for the electroactive chromophore are carbazole and its derivatives, triphenylamine and its derivatives, fluorene and its derivatives, anthracene and its derivatives, fullerene and its derivatives, oxadiazole and its derivatives, tetracyanoethylene and its derivatives, and tetracyanoquinodimethane. These electroactive chromophores can be classified into two groups, namely electron-donor and -acceptor groups, depending on the electron (or hole) affinity.

This chapter introduces some representatives of the last family of polymers and discusses their chemical nature, morphological nature, memory modes, device performances, switching mechanisms and related key factors.

5.2 Nonconjugated Polymers with Electron-Donating Chromophore Pendants

5.2.1 Polymers with Carbazole Pendants: Backbone and Molecular Orbital Effects

Aromatic polyimides (PIs) are well known for their high chemical resistance, excellent thermal stability, dimensional stability, and mechanical properties.[34-42] Thus, PIs have been widely used in the microelectronics industry as interdielectric layers, insulators, passivation layers, and flexible carriers.

These properties allow aromatic PIs to be excellent nonconjugated polymers for the development of electrical memory polymers.

Poly(3,3'-bis(N-ethylenyloxycarbazole)-4,4'-biphenylene hexafluoroisopropylidenediphthalimide) (6F-HABCZ) is a good example of a PI with pendant electroactive chromophores (Figure 5.1a).[16,43] The polyimide was synthesized from 2,2'-bis(3,4-dicarboxyphenyl)hexafluoropropane dianhydride (6F) and 3,3'-bis[9-carbazole(ethyloxy)biphenyl]-4,4'-diamine (HABCZ). This polymer contains two electron-donating carbazole moieties per repeat unit as pendant electroactive chromophores. Other HABCZ-based PIs have been synthesized by polycondensation reactions with 3,3',4,4'-biphenyltetracarboxylic dianhydride (BPDA), pyromellitic dianhydride (PMDA), 3,3',4,4'-diphenylethertetracarboxylic dianhydride (ODPA), and 3,3',4,4'-diphenylsulfonyltetracarboxylic dianhydride (DSDA): BPDA-HABCZ, PMDA-HABCZ, ODPA-HABCZ, and DSDA-HABCZ (Figure 5.1a).[17] These PIs are thermally stable up to 325–427 °C,

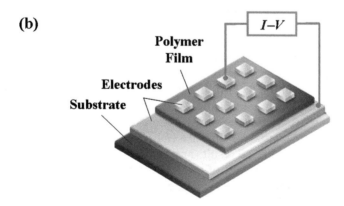

Figure 5.1 (a) Chemical structures of HABCZ-based PIs; (b) a schematic structural diagram of the memory devices fabricated with a PI with pendant electroactive chromophores and metal top and bottom electrodes.

depending on the kind of dianhydride unit. 6F-HABCZ, ODPA-HABCB and DSDA-HABCB PIs show distinctive glass transitions at 225, 219, and 251 °C, respectively. For BPDA-HABCB and PMDA-HABCZ PIs, however, glass transition is not discernible over the temperature range 25–300 °C. These PIs exhibit excellent film formation capability, providing high quality nanoscale thin films with a smooth surface (less than 1.0 nm root-mean-square roughness) via a simple spin-coating process.

Nanoscale thin film layers (20 nm thick) of 6F-HABCZ in devices with aluminum (Al) top and bottom electrodes show write-once-read-many-times (WORM) memory behavior in the positive voltage sweep; the representative current–voltage (I–V) data are presented in Figure 5.2a. The device is switched on at around +3.0 V (= $V_{c,ON}$, the critical voltage). The ON/OFF current ratio is in the range 10^5–10^8, depending on the reading voltage. Nanoscale thin films (30 nm thick) of the other HABCZ-based PIs in devices with Al electrodes also reveal WORM memory behavior with a high ON/OFF current ratio of up to 10^{10} in positive voltage sweeps (Figure 5.2b–e). The $V_{c,ON}$ values range from 2.2 to 3.3 V. These WORM memory behaviors have been observed in negative voltage sweeps. Overall, all HABCZ-based PIs demonstrate excellent unipolar WORM memory behaviors with low electric power consumptions and high ON/OFF current ratios. Therefore, the thermally, dimensionally and chemically stable HABCZ-based PIs are promising materials for the low cost, mass production of high performance, programmable permanent memory devices operating with low power consumption in unipolar switching mode.

The I–V data have been further analyzed.[17,43] The I–V data for the OFF-state of 6F-HABCZ could be fitted well by the trap-limited space-charge limited conduction (SCLC) model while those of the ON-state could be successfully fitted by the ohmic contact model (Figure 5.3a and b). Similar I–V data analysis results are obtained for the other HABCZ-based PIs, as shown in Figure 5.3c and d. These analysis results indicate that the HABCZ-based PI layers in the OFF-state are mainly governed by the trap-limited SCLC mechanism whereas those in the ON-state predominantly operate through an ohmic conduction mechanism.

Poly(3,3′-dihydroxy-4,4′-biphenylene hexafluoroisopropylidene-diphthalimide) (6F-HAB), fabricated into devices using Al electrodes, has also been examined to check for programmable memory behavior.[43] Unlike the HABCZ-based PIs with carbazole pendants, 6F-HAB in nanoscale thin films always exhibits dielectric behavior, even with 5 and 10 nm thickness (Figure 5.2f). In fact, it is well known that conventional PIs are extremely low conductivity materials[44] and that the carbazole unit is a hole-transporting (i.e., electron-donating) moiety.[45] Experimental results and facts lead to the notion that the permanent memory characteristics of the HABCZ-based PIs originate from the carbazole pendants incorporated into the 6F-HAB dielectric, rather than the 6F-HAB polymer itself.

In general, the molecular orbital characteristics (i.e., the highest occupied molecular orbital (E_{HOMO}) and the lowest unoccupied molecular orbital (E_{LUMO})) of electroactive polymers are considered as a critical factor in their

Nonconjugated Polymers with Electroactive Chromophore Pendants 171

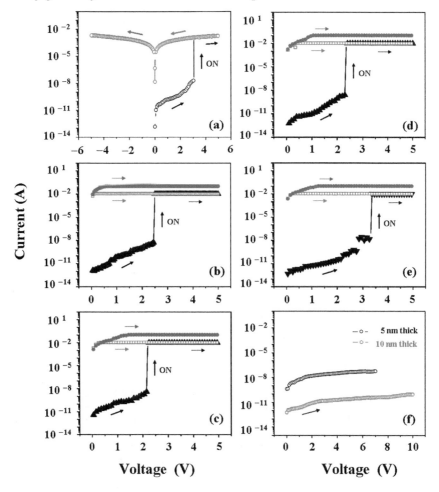

Figure 5.2 *I–V* curves of the Al/HABCZ-based PI/Al devices: (a) 6F-HABCZ (20 nm thick); (b) BPDA-HABCZ (30 nm thick); (c) PMDA-HABCZ (30 nm thick); (d) ODPA-HABCZ (30 nm thick); (e) DSDA-HABCZ (30 nm thick); (f) 6F-HAB (5 and 10 nm thick). The first and second sweeps were conducted with a 0.01 A compliance current, while the third sweep was performed with a 0.1 A compliance current.[16,17,43] The electrode contact area was 0.5 × 0.5 or 0.6 × 0.6 mm². Reproduced with permission. Copyright 2008, Wiley-VCH. Copyright 2011, Elsevier. Copyright 2014, American Chemical Society.

electrical memory behaviors.[2-4] Thus, it is vital to find out how E_{HOMO} and E_{LUMO} contribute to the electrical memory behavior. The E_{HOMO} and E_{LUMO} of the PIs, including their band gaps, have been measured by means of ultraviolet-visible (UV-vis) spectroscopy and cyclic voltammetry.[17,43] The determined E_{HOMO} and E_{LUMO} values are compared with those of the carbazole molecule itself and 6F-HAB in Table 5.1. The E_{HOMO} values are in the range of −4.90 to −5.71 eV while the E_{LUMO} values are in the range of −1.90 to −3.00 eV;

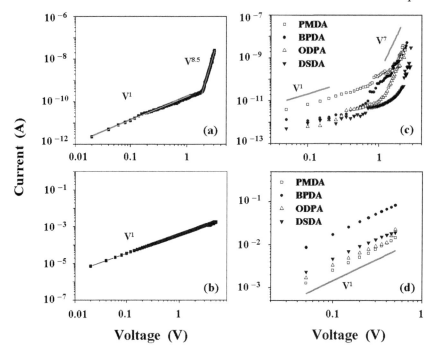

Figure 5.3 Voltage–current plots for the Al/HABCZ-based PI/Al devices in the OFF-state and in the ON-state, which were obtained from the data in Figure 5.2:[16,17,43] (a) 6F-HABCZ (20 nm thick) in the OFF-state; (b) 6F-HABCZ (20 nm thick) in the ON-state; (c) other HABCZ-based PIs (30 nm thick) in the OFF-state; (d) other HABCZ-based PIs (30 nm thick) in the ON-state. The data in the OFF-state: the symbols are the measured data and the solid lines are the data fitted by the trap-limited SCLC model. The data in the ON-state: the symbols are the measured data and the solid line is the data fitted by the ohmic conduction model. Reproduced with permission. Copyright 2008, Wiley-VCH. Copyright 2011, Elsevier. Copyright 2014, American Chemical Society.

the band gap ΔE_{bg} varies in the range of 2.16 to 3.48 eV. The E_{HOMO} value of 6F-HABCZ is lower than those of 6F-HAB and carbazole, whereas the E_{HOMO} values of BPDA-HABCZ and PMDA-HABCZ are higher than those of 6F-HAB and carbazole. The E_{HOMO} values of ODPA-HABCZ and DSDA-HABCZ are comparable to that of 6F-HAB but higher than that of carbazole. In contrast, the E_{LUMO} values of all HABCZ-based PIs are lower than those of 6F-HAB and carbazole. The band gap of 6F-HABCZ is slightly smaller than that of the 6F-HAB-based polymer but much smaller than that of carbazole. The band gaps of the other HABCZ-based PIs are much smaller than those of 6F-HAB and carbazole.

The molecular orbital results collectively provide several key information points. First, the E_{HOMO} and E_{LUMO} values of 6F-HAB are influenced by the incorporation of carbazole as pendants. Second, the E_{HOMO} and E_{LUMO} values of the carbazole moiety are also influenced by its incorporation into 6F-HAB as pendants. Finally, the molecular orbital characteristics of the

Table 5.1 Molecular orbitals, band gaps and energy barriers to Al electrode of HABCZ-based PIs.

Material	E_{HOMO}[a] (eV)	$|E_{HOMO} - \Phi_{Al}|$[b] (eV)	E_{LUMO}[c] (eV)	$|E_{LUMO} - \Phi_{Al}|$[d] (eV)	ΔE_{bg}[e] (eV)	$V_{c,ON}$ (V)	Memory mode
6F-HABCZ	−5.71	1.43	−2.23	2.05	3.48	3.0	WORM
BPDA-HABCZ	−5.06	0.78	−2.90	1.38	2.16	2.5	WORM
PMDA-HABCZ	−4.90	0.62	−1.90	2.38	2.90	2.2	WORM
ODPA-HABCZ	−5.15	0.87	−2.41	1.87	2.74	2.4	WORM
DSDA-HABCZ	−5.16	0.88	−3.00	1.28	2.16	3.4	WORM
6F-HAB	−5.15	0.87	−1.41	2.87	3.74		Dielectric
CZ	−5.35[f]	1.07	−0.55[f]	3.73	4.80		

[a] The highest occupied molecular orbital.
[b] The energy barrier for hole injection into the HOMO from the Al electrode having a work function Φ_{Al} of −4.28 eV.
[c] The lowest unoccupied molecular orbital.
[d] The energy barrier for electron injection into the LUMO from the Al electrode ($\Phi_{Al} = -4.28$ eV).
[e] The band gap: $\Delta E_{bg} = E_{HOMO} - E_{LUMO}$.
[f] Ref. 46 and 47.

HABCZ-based PI are further influenced by changing the dianhydride unit in the backbone. Consequently, the overall molecular orbital characteristics of an electroactive polymer are dependent on all of its chemical components.

Just based on the E_{HOMO} and E_{LUMO} data in Table 5.1, one may easily expect that the 6F-HAB itself can demonstrate electrical memory behavior in devices. This polymer, however, has been confirmed to reveal no electrical memory characteristics at all. This result is good evidence that no one can make any direct prediction about electrical memory characteristics only from the E_{HOMO} and E_{LUMO} levels of polymers. In other words, the information about a polymer's molecular orbitals alone cannot explain its electrical memory behavior. Other critical factors need to be identified and then included to comprehend such phenomena.

However, the E_{HOMO} and E_{LUMO} data may give clues for understanding the efficiency of hole or electron injection into the polymer layer of a device from electrodes. As summarized in Table 5.1, the energy barriers for hole injection into the HABCZ-based PI layers (i.e., the E_{HOMO} levels) from an Al electrode ($\Phi_{Al} = -4.28$ eV, work function) are always lower than those for electron injection into the PI layers from an Al electrode. Similar energy barrier situations are found for carbazole. These facts collectively suggest that the charge traps and conduction processes in the individual HABCZ-based PI layers are predominated by hole injection rather than electron injection. Therefore, these PIs are considered to be p-type memory polymers.

Furthermore, regarding only the energy barriers for hole injection, one may expect that the $V_{c,ON}$ value for WORM memory switching is highest for the 6F-HABCZ layer, intermediate for BPDA-HABCZ, ODPA-HABCZ and DSDA-HABCZ, and lowest for PMDA-HABCZ. Such a correlation could not be found, however. This confirms that the energy barrier of a HABCZ-based polymer to an Al electrode is not the determining factor for the electrical switching behavior. Hence, while the energy barriers and the molecular orbital characteristics do take part in the electrical switching phenomena, the main driving factor(s) behind such phenomena still remains elsewhere.

6F-HAB also shows similar energy barriers to the Al electrode to the HAB-CZ-based PIs. Namely, hole injection from the electrode is also favorable for 6F-HAB. This polymer, however, did not demonstrate any electrical switching behavior. This suggests that 6F-HAB does not have the capability to trap the charges injected from the electrode even with the aid of an applied electric field and, furthermore, could not transport the injected charges efficiently from one electrode to the opposite electrode. In this regard, electroactive chromophore groups such as carbazole are essential in PI polymers for trapping and transporting charges.

As shown in Figure 5.2 and listed in Table 5.1, all HABCZ-based PIs exhibit a single memory mode, namely WORM memory behavior, regardless of the dianhydride units in the polymer backbone and the varying electron-donating or -accepting powers of the different dianhydride units. This result might be attributed to the unique physicochemical environment of the carbazole pendant in the HABCZ-based PIs. As shown in Figure 5.1a, two carbazole pendants are connected to each HAB diamine unit in the PI backbone using nonconjugated ethylenyloxy linkers, placing the individual carbazole units somewhat far from the HAB diamine unit and much further from the dianhydride unit. As a result, any inductive and resonance effects due to variations of the dianhydride unit in the PI backbone cannot easily reach the carbazole pendant. Consequently, the WORM memory mode originating from the carbazole pendants may be retained even with variations in the dianhydride unit in the PI backbone.

Considering the role and the positional distribution of carbazole pendants in the HABCZ-based PIs, the observed electrical switching might take place *via* charge traps on the carbazole pendants and a hopping process of charges using the carbazole pendants as stepping stones.

5.2.2 Polymers with Triphenylamine-Based Pendants: Substituent and Molecular Orbital Effects

Triphenylamine (TPA) and its derivatives are good candidates for incorporation into polymers as electron-donating pendant groups (Figure 5.4). Poly(4,4'-aminotriphenylene hexafluoroisopropylidenediphthalimide) (6F-TPA) thin films (*ca.* 50 nm thick) were reported to exhibit bipolar static random access memory (SRAM) behavior in devices with Al top and indium-tin-oxide (ITO) bottom electrodes.[20]

Poly(3,3'-di(4-(diphenylamino)benzylideniminoethoxy)-4,4'-biphenylene hexafluoroisopropylidenediphthalimide) (6F-HAB-TPAIE) in thin films *ca.* 35 nm thick (which is a 6F-HAB PI with a 4-(diphenylamino)benzylidenimino pendant as a TPA derivative) has been reported to reveal not only unipolar WORM memory behavior but also unipolar and bipolar flash memory

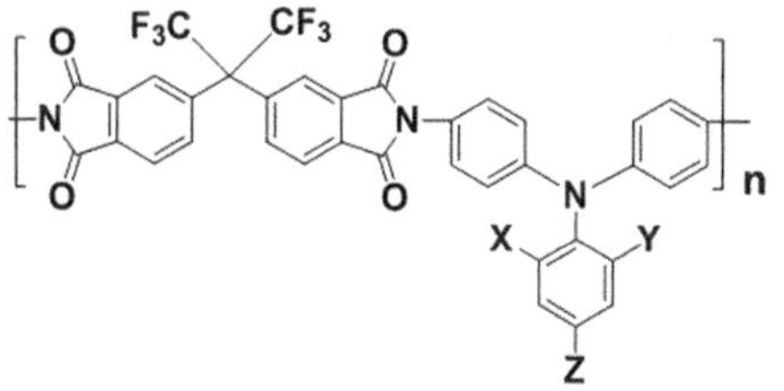

PI	X	Y	Z
6F-TPA	H	H	H
6F-TPA-OH	H	H	OH
6F-TPA-Me₃	Me	Me	Me
6F-TPA-T₂	H	H	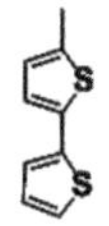 (bithiophene)
6F-TPA-NMe₂	H	H	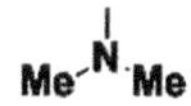 NMe₂
6F-TPA-NPh₂	H	H	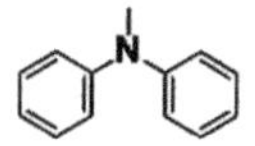 NPh₂
6F-TPA-Ph(CF₃)₂	H	H	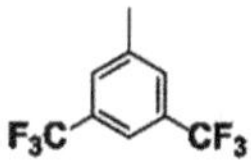 3,5-bis(trifluoromethyl)phenyl

Figure 5.4 Chemical structures of 6F-TPA and its analogues with various substituents.

behaviors in devices with Al electrodes (Figure 5.5).[19] Poly(3,3′-bis(diphenylcarbamyloxy)-4,4′-biphenylene hexafluoroisopropylidenediphthalimide) (6F-HAB-DPC) in thin films (ca. 70 nm thick), which is another 6F-HAB PI with a diphenylcarbamyl pendant (Figure 5.5), also exhibits unipolar WORM memory behavior as well as unipolar and bipolar flash memory behaviors in devices with Al electrodes.[18] Considering these results with the dielectric characteristics of 6F-HAB and poly(p-phenylene hexafluoroisopropylidenediphthalimide) (6F-PDA),[48] the SRAM behavior of 6F-TPA and the various memory modes of 6F-HAB-TPAIE originate from the electron-donating TPA units.

The observation of SRAM behavior suggests low electrical memory stability of the 6F-TPA devices. Such memory instability, however, can be either enhanced or fully resolved by the incorporation of substituents into the aromatic rings of the TPA chromophore.[22–25] Poly(4,4′-amino(4-hydroxyphenyl)diphenylene hexafluoroisopropylidenediphthalimide) (6F-TPA-OH) is a 6F-TPA analogue which has only one additional hydroxyl substituent per TPA unit (Figure 5.4). 6F-TPA-OH in thin films (30 nm thick) shows unipolar WORM memory behavior.[22] The WORM memory mode was retained for devices fabricated with an Al top electrode and Al, ITO, and gold (Au) bottom electrodes (Figure 5.6); the $V_{c,ON}$ is around ±2.5 V and the ON/OFF current ratio is up to 10^6.

Similar WORM memory modes have been demonstrated for 6F-TPA analogues (30 nm thick films) whose TPA unit has a trimethyl or dithiophenyl or dimethylamino or diphenylamino or di(trifluoromethyl)phenyl substituent: 6F-TPA-Me$_3$, 6F-TPA-T$_2$, 6F-TPA-NMe$_2$, 6F-TPA-NPh$_2$, and 6F-TPA-Ph(CF$_3$)$_2$

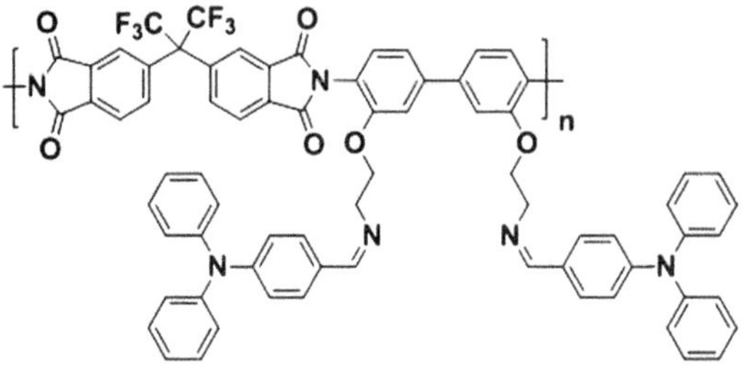

6F-HAB-TPAIE PI

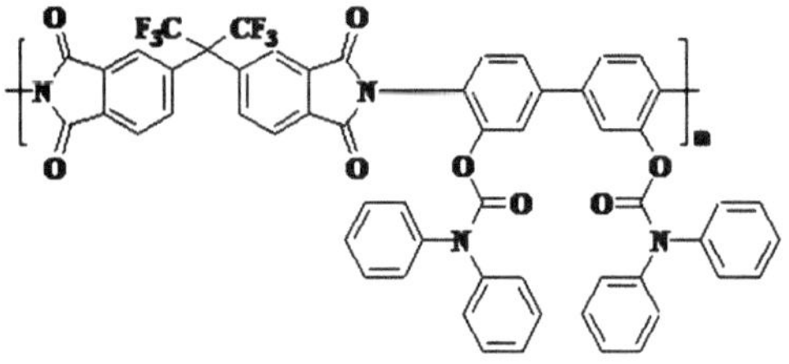

6F-HAB-DPC PI

Figure 5.5 Chemical structures of 6F-HAB-TPAIE and 6F-HAB-DPC.

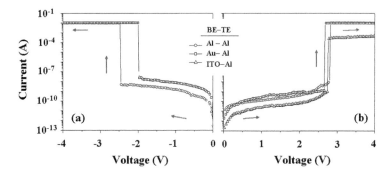

Figure 5.6 *I–V* curves of 54 nm thick 6F-TPA-OH devices fabricated with various bottom electrodes: (a) negative voltage sweeps; (b) positive voltage sweeps.[22] The electrode contact area was 0.5 × 0.5 mm². Reproduced with permission. Copyright 2009, American Chemical Society.

(Figure 5.4).[23–25] As shown in Figure 5.7, 6F-TPA-NMe$_2$ exhibits both unipolar and bipolar switching memory behaviors, which is a unique property amongst organic and polymer memory systems.[25] Overall, the $V_{c,ON}$ ranges from ±2.0 V to ±4.0 V and the ON/OFF current ratio is up to 10^9, depending on the substituents.

For 6F-TPA and its analogues, the E_{HOMO} and E_{LUMO} levels have been measured and summarized in Table 5.2. Both the E_{HOMO} and E_{LUMO} levels of 6F-TPA are influenced by the electron-donating or -withdrawing substituents incorporated into the TPA unit. Regardless of the incorporated substituents, however, the energy barrier for hole injection from the electrode (Al or ITO) is always lower than that for electron injection from the Al electrode. Similar energy barrier results are found for the TPA molecule. These results collectively suggest that the charge traps and conduction processes are predominated by hole injection rather than electron injection for 6F-TPA and its analogues, categorizing them as p-type memory polymers.

Similar trends have been observed for poly(4,4′-aminotriphenylene 3,3′,4,4′-diphenylsulfonyltetracarboximide) (DSDA-TPA) and its derivatives: poly(N-(4-cyanophenyl)-N,N-4,4′-diphenylene 3,3′,4,4′-diphenylsulfonyltetracarboximide) (DSDA-TPA-CN), poly(N-(4-methoxyphenyl)-N,N-4,4′-diphenylene

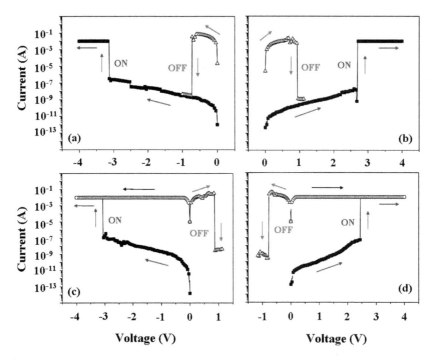

Figure 5.7 I–V curves of the Al/6F-TPA-NMe$_2$(30 nm thick)/Al devices, which were measured with a compliance current of 0.01 A: (a) and (b) unipolar flash memory; (c) and (d) bipolar flash memory.[25] The switching-OFF runs in (a) and (b) were carried out with a compliance current of 0.1 A. The electrode contact area was 0.5 × 0.5 mm^2. Reproduced with permission. Copyright 2012, Royal Society of Chemistry.

Table 5.2 Molecular orbitals, band gaps and energy barriers to the Al electrode of 6F-TPA analogues with various substituents.

Material	$E_{HOMO}{}^a$ (eV)	$\|E_{HOMO} - \Phi_{Al}\|^b$ (eV)	$E_{LUMO}{}^c$ (eV)	$\|E_{LUMO} - \Phi_{Al}\|^d$ (eV)	$\Delta E_{bg}{}^e$ (eV)	$V_{c,ON}$ (V)	Memory mode(s)
6F-TPA	−5.13	0.85	−2.03	2.25	3.10	3.3	SRAM
6F-TPA-OH	−5.26	0.46^f	−2.26	2.02	3.00	1.7^g	WORM
6F-TPA-Me$_3$	−4.75	0.47	−1.60	2.68	3.15	1.8	WORM
6F-TPA-T$_2$	−5.64	1.36	−2.97	1.31	2.67	3.9	WORM
6F-TPA-NMe$_2$	−4.68	0.40	−1.61	2.67	3.07	2.7	WORM Unipolar flash Bipolar flash
6F-TPA-NPh$_2$	−5.10	0.82	−1.88	2.40	3.22	1.2	WORM
6F-TPA-Ph(CF$_3$)$_2$	−5.08	0.80	−1.97	2.31	3.11	3.0	WORM
TPA	−4.85^h	0.57	−0.21^h	4.07	4.64		

^{a} The highest occupied molecular orbital.
^{b} The energy barrier for hole injection into the HOMO from the Al electrode having a work function Φ_{Al} of −4.28 eV.
^{c} The lowest unoccupied molecular orbital.
^{d} The energy barrier for electron injection into the LUMO from the Al electrode (Φ_{Al} = −4.28 eV).
^{e} The band gap: $\Delta E_{bg} = E_{HOMO} - E_{LUMO}$.
^{f} The energy barrier for hole injection into the HOMO from the ITO electrode (Φ_{ITO} = −4.80 eV).
^{g} Measured for ITO/PI/Al devices rather than Al/PI/Al devices.
^{h} Ref. 49.

3,3′,4,4′-diphenylsulfonyltetracarboximide) (DSDA-TPA-OMe), poly(N-(4-dimethylaminophenyl)diphenylene 3,3′,4,4′-diphenylsulfonyltetracarboximide) (DSDA-TPA-NMe$_2$), poly(N,N-bis(4-aminophenyl)-N′,N′-diphenyl-1,4-phenylene 3,3′,4,4′-diphenylsulfonyltetracarboximide) (DSDA-TPA-NPh$_2$), poly(N,N-bis(4-aminophenyl)-N′,N′-di(4-methoxyphenyl)-1,4-phenylene 3,3′,4,4′-diphenylsulfonyltetracarboximide) (DSDA-TPA-N(PhOMe)$_2$), and poly(N,N-bis(4-aminophenyl)-N′,N′-di(4-cyanophenyl)-1,4-phenylene 3,3′,4,4′-diphenylsulfonyltetracarboximide) (DSDA-TPA-N(PhCN)$_2$) (Figure 5.8 and 5.9; Table 5.3).[26,27] DSDA-TPA, DSDA-TPA-OMe, DSDA-TPA-NPh$_2$ and DSDA-N(PhOMe)$_2$ exhibit only unipolar WORM memory characteristics (Figure 5.9a, d and e). For DSDA-TPA-CN, only 80% of device cells show unipolar WORM memory behavior; 20% of device cells reveal unipolar dynamic random access memory (DRAM) behavior (Figure 5.9b). In comparison, DSDA-TPA-N(PhCN)$_2$ exhibits only unipolar DRAM behavior (Figure 5.9f). DSDA-TPA-NMe$_2$ shows unipolar flash memory behavior in addition to unipolar WORM memory characteristics (Figure 5.9c). Considering the dielectric characteristics of poly(m-phenylene 3,3′,4,4′-diphenylsulfonyltetracarboximide) (DSDA-mPDA),[50] the memory behaviors of DSDA-TPA and its analogues are attributed to the TPA unit and its derivatives. In particular, the memory behavior of DSDA-TPA-NMe$_2$ resembles that of 6F-TPA-NMe$_2$. DSDA-TPA-NMe$_2$, however, did not demonstrate bipolar switching memory behavior as 6F-TPA-NMe$_2$ did. This leads to the notion that the DSDA and 6F units play their own roles in the electrical memory behaviors of their PIs. The DSDA unit has a

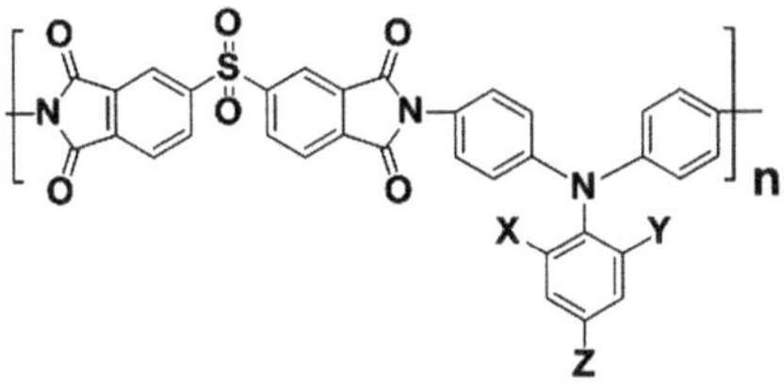

PI	X	Y	Z
DSDA-TPA	H	H	H
DSDA-TPA-OMe	H	H	OMe
DSDA-TPA-NMe_2	H	H	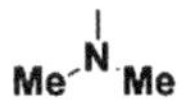
DSDA-TPA-CN	H	H	CN
DSDA-TPA-NPh_2	H	H	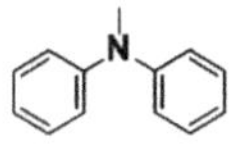
DSDA-TPA-$N(PhOMe)_2$	H	H	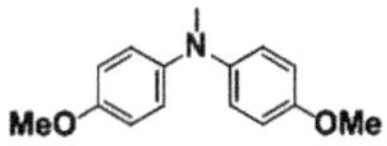
DSDA-TPA-$N(PhCN)_2$	H	H	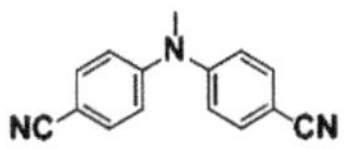

Figure 5.8 Chemical structures of DSDA-TPA and its analogues with various substituents.

relatively high electron-donating power compared to the 6F unit. The difference in the inductive effect may contribute differently in the electrical memory modes.

As summarized in Table 5.3, the E_{HOMO} and E_{LUMO} levels of DSDA-TPA are influenced by the electron-donating or -withdrawing substituents incorporated into the TPA unit. Regardless of the incorporated substituents, however, the energy barrier for hole injection from the electrode is always lower than that for electron injection from the Al electrode. These results suggest that DSDA-TPA and its derivatives are p-type memory polymers.

For 6F-TPA, DSDA-TPA and their analogues, *I–V* data analysis has found that the OFF-states are mainly governed by a trap-limited SCLC mechanism while the ON-states are mainly controlled by an ohmic conduction mechanism.[22–27]

For 6F-TPA, DSDA-TPA and their analogues, it has been found that the $V_{c,ON}$ values show no direct correlations with the molecular orbitals or the energy barriers (Tables 5.2 and 5.3). Moreover, the $V_{c,ON}$ values show no direct correlations with the electron-donating or -accepting powers of the substituents (*i.e.*, pendant groups) in the TPA unit (Tables 5.2 and 5.3). These results

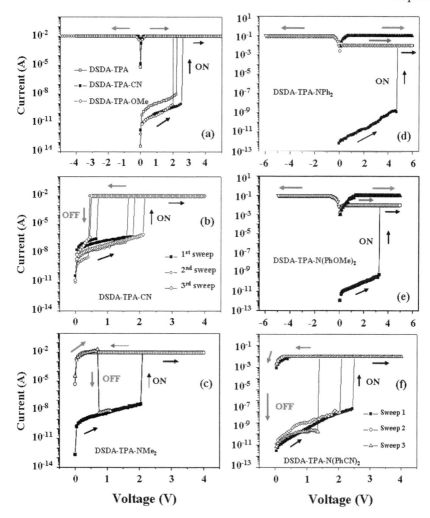

Figure 5.9 I–V curves of the Al/DSDA-TPA derivatives (30 nm thick)/Al devices, which were measured with a compliance current of 0.01 A: (a), (d) and (e) unipolar WORM memory; (b) and (f) unipolar DRAMs; (c) unipolar flash memory.[26,27] The switching-OFF run in (c) was carried out with a compliance current of 0.1 A. The electrode contact area was 0.5 × 0.5 mm². Reproduced with permission. Copyright 2012, American Chemical Society. Copyright 2012, Elsevier.

reiterate that, in addition to the molecular orbitals and energy barriers, some other factors are involved in the electrical switching behaviors of the TPA-based PI analogues.

The memory modes the TPA-based PI analogues have been found to exhibit a strong dependency on both the dianhydride units and the pendant groups of the TPA unit (Tables 5.2 and 5.3). Different from the electron-donating carbazole pendants in the HABCZ-based PIs in Section 5.2.1, the TPA units

Table 5.3 Molecular orbitals, band gaps and energy barriers to Al electrode of DSDA-TPA analogues with various substituents.

Material	E_{HOMO}[a] (eV)	$|E_{HOMO} - \Phi_{Al}|$[b] (eV)	E_{LUMO}[c] (eV)	$|E_{LUMO} - \Phi_{Al}|$[d] (eV)	ΔE_{bg}[e] (eV)	$V_{c,ON}$ (V)	Memory mode(s)
DSDA-TPA	−5.32	1.04	−2.28	2.00	3.04	2.2	WORM
DSDA-TPA-OMe	−5.16	0.88	−2.17	2.11	2.99	2.0	WORM
DSDA-TPA-NMe$_2$	−4.75	0.47	−1.84	2.44	2.91	2.2	WORM Unipolar flash
DSDA-TPA-CN	−5.08	0.80	−1.94	2.34	3.17	2.6	WORM DRAM
DSDA-TPA-NPh$_2$	−4.92	0.64	−1.97	2.31	2.95	4.8	WORM
DSDA-TPA-N(PhOMe)$_2$	−4.83	0.55	−1.78	2.50	3.05	3.2	WORM
DSDA-TPA-N(PhCN)$_2$	−4.98	0.70	−1.91	2.37	3.07	2.4	DRAM
TPA	−4.85[f]	0.57	−0.21[f]	4.07	4.64		

[a] The highest occupied molecular orbital.
[b] The energy barrier for hole injection into the HOMO from the Al electrode having a work function Φ_{Al} of −4.28 eV.
[c] The lowest unoccupied molecular orbital.
[d] The energy barrier for electron injection into the LUMO from the Al electrode ($\Phi_{Al} = -4.28$ eV).
[e] The band gap: $\Delta E_{bg} = E_{HOMO} - E_{LUMO}$.
[f] Ref. 48.

in the TPA-based PI analogues are part of the PI backbone. Namely, the TPA unit is connected directly to the dianhydride units (6F or DSDA units). Furthermore, all considered substituents are directly incorporated into the TPA unit as pendant groups. The direct connection to the dianhydride units and pendant groups can cause a certain degree of inductive effect and resonance effect on the TPA unit's charge trapping power and the trapped charges' stability. Such chemical environmental changes of the TPA unit can affect the electrical memory mode of the TPA-based PI analogues. Some evidence has clearly appeared in devices of TPA-based PIs as discussed in the following paragraphs.

First, 6F-TPA reveals SRAM behavior, and has very poor memory stability, whereas DSDA-TPA exhibits WORM memory behavior, and has the highest memory stability. The difference in the electrical memory stability might arise from the differences in the charge trapping and stabilization powers of 6F-TPA and DSDA-TPA as briefly discussed above. The 6F unit has a relatively higher electron-withdrawing power than the DSDA unit. Thus, the TPA units in 6F-TPA do not have enough electron density to give a suitable level of charge trapping and stabilization power, consequently leading to the SRAM mode. In comparison, the DSDA unit can make a positive contribution or a slightly negative contribution to the charge trapping and stabilization powers of the TPA units, consequently providing a permanent (*i.e.*, WORM) memory mode.

Second, the memory instability (*i.e.*, SRAM) of 6F-TPA has also been improved by incorporating a hydroxyl group (*i.e.*, an electron-donating group) into the TPA as a pendant. The TPA in 6F-TPA is weakened by the nearest neighbor 6F unit (which has an electron-withdrawing power), however, the weakened electron density of the TPA is again made up to a certain level by incorporating an electron-donating hydroxyl group, providing a sufficient level of electron density for a stable WORM memory mode. Considering such positive inductive effects, the WORM memory modes demonstrated in 6F-TPA-Me$_3$, 6F-TPA-T$_2$ and DSDA-TPA-OMe can be understood.

Third, DSDA-TPA-CN, however, again reveals some instability in its electrical memory behavior. Such memory instability is attributed to the electron density of the TPA unit weakened by the incorporation of an electron-withdrawing cyano (CN) group.

Fourth, the memory instability of 6F-TPA has been improved by incorporating a 3,5-di(trifluoromethyl)phenyl (Ph(CF$_3$)$_2$) group even though the pendant has both a positive resonance effect and a negative inductive effect. This indicates that the resonance effect of the extended phenyl linker can enhance the charge trapping and stabilization power of 6F-TPA against the negative inductive effect of two trifluoromethyl groups. Such a low contribution from the two trifluoromethyl groups might be attributed to their position relatively far from the TPA unit.

Fifth, the dimethylamino (NMe$_2$) substituent, as an electron-donating pendant, has been found to have a significant impact on the electrical memory mode and performance of 6F-TPA and DSDA-TPA. Due to the NMe$_2$ pendant, two or three different memory modes (unipolar WORM and flash memory; unipolar and bipolar flash and unipolar WORM memory) with high performance are demonstrated in 6F-TPA-NMe$_2$ and DSDA-TPA-NMe$_2$. Such an interesting memory mode variety might be attributed to a positive synergetic effect of the TPA unit and the NMe$_2$ pendant. Overall, the role of the NMe$_2$ pendant is very unique in controlling the electrical memory mode and performance of TPA-based PIs.

Sixth, diphenylamino (NPh$_2$) and di(4-methoxyphenyl)amino (N(PhOMe)$_2$) pendants, as electron-donating groups having a resonance effect, have also improved the memory instability of 6F-TPA. However, they have led to only a single memory mode, unipolar WORM memory, which is quite different from the modes observed in TPA-based PIs having NMe$_2$ pendants.

Finally, DSDA-TPA-N(PhCN)$_2$ exhibits unipolar DRAM behavior. The observation of such a DRAM mode indicates that the memory instability of 6F-TPA has gotten worse even though the 6F unit has been replaced by a DSDA unit and an N(PhCN)$_2$ group is incorporated as a pendant. Thus, the result suggests that the negative inductive effect of the two CN groups can override any positive inductive and resonance contributions of the DSDA unit and NPh$_2$ groups to improve the memory instability. In comparison, the two CN groups in DSDA-TPA-N(PhCN)$_2$ seem to have a greater negative impact on the memory mode and performance than does the CN group in DSDA-TPA-CN.

5.3 Nonconjugated Polymers with Electron-Accepting Chromophore Pendants

A number of electrical memory polymer systems have been reported so far.[1-33,43] Additional memory polymers have been discussed in other chapters. The majority of the memory polymers are made of π-conjugated and nonconjugated backbones bearing electron-donating chromophores; so, they can be classified as p-type memory systems. Nevertheless, memory polymers bearing only electron-accepting chromophores have rarely been introduced. A few examples are discussed in this section.

Recently, two nonconjugated polymers with pendant fullerene (C_{60}) chromophores have been introduced as electrical memory polymers, in which the C_{60} moiety is heavily loaded: poly(1,2,3-triazol-1,4-diylmethylene-1,3-(5-methoxycarbonyl-2,4,6-tris(dodecyloxy)phenylene)methylene-1,2,3-triazol-1,4-diyl(dipropylene 1,2-methano[60]fullerene-61,61-dicarboxylate)) (P1-C_{60}) and poly(1,2,3-triazol-1,4-diylmethylene-1,3-(5-methoxycarbonyl-2,4,6-tris(dodecyloxy)phenylene)-methylene-1,2,3-triazol-1,4-diyl(diethyldipropylene bismethano[60]fullerene-61,61,62,62-tetra-carboxylate)) (P2-C_{60}) (Figure 5.10).[28] Interestingly, nanoscale thin films (60 nm thick) of these C_{60}-based polymers in devices with Al top and ITO bottom electrodes reveal excellent bipolar flash memory behaviors: $V_{c,ON}$ = −1.10 V to −1.44 V, $V_{c,OFF}$ = +0.90 V to +0.95 V (critical switch-OFF voltage), and the ON/OFF current ratio ranges from 10^4 to 10^7 (Figure 5.11). The C_{60}-based polymers additionally exhibit unipolar WORM and flash memory characteristics in devices with Al electrodes (Figure 5.12). I–V data analysis has found that trap-limited SCLC

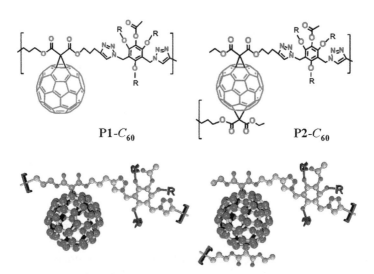

Figure 5.10 Chemical structures of P1-C_{60} and P2-C_{60}.[28] Reproduced with permission. Copyright 2014, American Chemical Society.

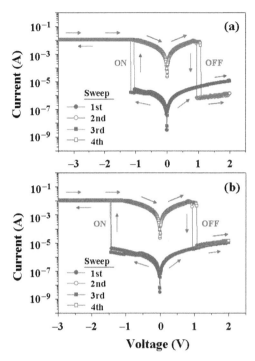

Figure 5.11 Bipolar I–V curves of Al/polymer(60 nm thick)/ITO devices: (a) P1-C_{60}; (b) P2-C_{60}.[28] The electrode contact area was 1.0 × 1.0 mm². Reproduced with permission. Copyright 2014, American Chemical Society.

and hopping processes govern the outstanding switching behaviors of the P1-C_{60} and P2-C_{60} devices.

Considering the molecular orbitals (E_{HOMO} = −7.11 eV for P1-C_{60} and −6.87 eV for P2-C_{60}; E_{LUMO} = −3.91 eV for P1-C_{60} and −3.77 eV for P2-C_{60}) and the electrodes' work function Φ, the energy barriers for electron injection into the polymer layers are always lower than those for hole injection into the polymer layers. These considerations collectively indicate that the electrical switching behaviors of P1-C_{60} and P2-C_{60} in the devices with an Al top electrode and ITO, Al or Au bottom electrodes can be operated favorably by electron injection. Based on the best knowledge of ours so far, P1-C_{60} and P2-C_{60} are the first n-type bipolar and unipolar digital memory polymers operating in flash and WORM memory modes.

Poly(5-phenyl-1,3,4-oxadiazol-2-yl-[1,1′-biphenyl]carboxyloxy-n-nonyl acrylate) (PPOXBPA) is another nonconjugated polymer that contains electron-accepting pendants (Figure 5.13a).[29,30] This polymer has E_{HOMO} = −4.58 eV, E_{LUMO} = −1.18 eV, and ΔE_{bg} = 3.40 eV. As-cast and thermally annealed films of PPOXBPA exhibit crystalline-like grazing incidence X-ray scattering (GIXS) patterns, as shown in Figure 5.13c and e. The as-cast films, however, show ring-like scattering, whereas the thermally annealed films reveal highly anisotropic and spot-like scattering. Quantitative analysis of the GIXS patterns reveals the following: the as-cast films form vertically-oriented

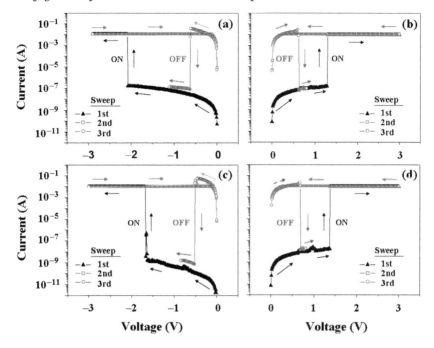

Figure 5.12 Unipolar $I–V$ curves of Al/polymer (60 nm thick)/Al devices: (a) and (b) P1-C_{60}; (c) and (d) P2-C_{60}.[28] The first and second voltage sweeps were performed with a compliance current of 0.01 A, whereas the third voltage sweep was carried out with a compliance current of 0.10 A. The electrode contact area was 1.0 × 1.0 mm². Reproduced with permission. Copyright 2014, American Chemical Society.

multibilayer structures with partial interdigitation between bristles from different layers occurring *via* the π–π stacking of the oxadiazole mesogen units. Its orientation distribution is very broad (Figure 5.13c and g), however, with an overall crystallinity X_c of 55%. On the other hand, thermally annealed films form well ordered, horizontally-oriented multibilayer structures with X_c = 85.9% (Figure 5.13f and g). These multibilayer structures maintain their structural integrity up to 220 °C before completely collapsing.

The morphology differences were found to impact the electrical memory behavior. The as-cast films 30 nm thick reveal unipolar SRAM behavior in their devices with Al top and ITO bottom electrodes (Figure 5.14a), whereas the thermally annealed films (30 nm thick) exhibit unipolar WORM memory characteristics (Figure 5.14b). These results collectively indicate that the formation of a well ordered, horizontally-oriented multibilayer structure is necessary for PPOXBPA films to demonstrate highly stable memory behavior like WORM memory. Considering the dielectric nature of the polymer backbone, the excellent SRAM and WORM memory characteristics of PPOXBPA are thought to originate from the electron-accepting oxadiazole pendants.

$I–V$ data analysis confirms that the electrical switching behaviors in the as-cast films and the thermally annealed films are governed by trap-limited

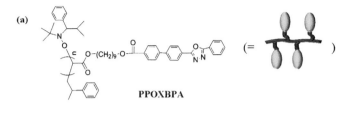

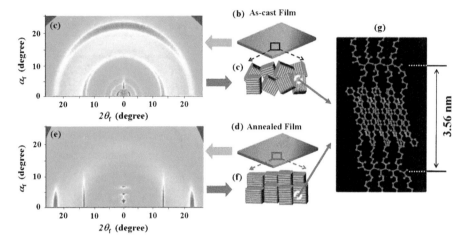

Figure 5.13 (a) Chemical structure of PPOXBPA; (b) as-cast film (65 nm thick); GIXS image of the as-cast film; (c) vertically-oriented multilayer structure with a broad orientation distribution, which was obtained from analysis of the GIXS pattern in (c); (d) thermally annealed film (65 nm thick; annealed at 115 °C for 1 day *in vacuo*); (e) GIXS image of the thermally annealed film; (f) horizontally-oriented multilayer structure with a very narrow orientation distribution, which was obtained from analysis of the GIXS pattern in (e); (g) schematic representation of the multibilayer structure of PPOXBPA molecules where the atoms are colored: C, gray; O, red; N, blue; hydrogen atoms are omitted for clarity. In the multibilayer structure the bristles are partially interdigitated *via* π–π stacking of the mesogenic part containing oxadiazole units.[29] Reproduced with permission. Copyright 2011, American Chemical Society.

SCLC and hopping processes, regardless of the difference in memory mode and in the morphology and orientation (Figure 5.14c–f).

Considering the molecular orbitals of PPOXBPA and the work functions of the top and bottom electrodes, the energy barrier for hole injection into the polymer layer from either the Al or ITO electrode is always lower than that for electron injection into the polymer layer from the electrodes. These suggest that the PPOXBPA-based devices undergo conduction processes governed mainly by hole injection. In general, the oxadiazole moiety is well known as an electron-acceptor. Thus, one may easily expect that the electrical switching of PPOXBPA is driven by electron injection. This prediction, however, could not be realized for the PPOXBPA devices showing hole injection driven

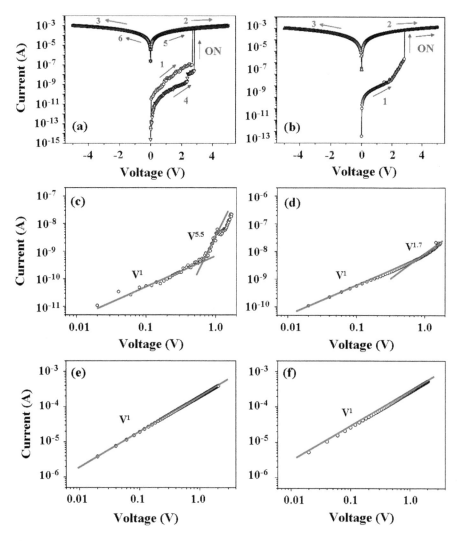

Figure 5.14 I–V curves of ITO/PPOXBPA (30 nm thick)/Al devices, which were measured with a compliance current level of 0.01 A:[29] (a) unipolar DRAMs of the as-cast film; (b) unipolar WORM memory behavior of the thermally annealed film. The size of the top Al electrodes was $100 \times 100\ \mu m^2$. The I–V data for the OFF- and ON-states in (a) and (b) were analyzed: (c) and (d) the OFF-state fitted with a combination of the ohmic model and the trap-limited SCLC model; (e) and (f) the ON-state fitted with the ohmic model. The symbols are the measured data and the solid lines are the fits obtained using the models. Reproduced with permission. Copyright 2011, American Chemical Society.

switching behavior. Hence, the oxadiazole unit in the pendants of PPOXBPA can be classified as an electron-donor rather than an electron-acceptor. Therefore, PPOXBPA is a p-type memory polymer.

Another nonconjugated polymer bearing electron-accepting pendants is poly(3,3′-di(9-anthracenemethoxy)-4,4′-biphenylene hexafluoroisopropylidenediphthalimide) (6F-HAB-AM) which has two anthracene moieties per repeat unit (Figure 5.15a).[31] This polymer is amorphous but thermally stable up to approximately 410 °C. This polymer in 20 nm thick film devices with Al electrodes exhibits unipolar WORM memory behavior (Figure 5.15b and c). In addition, the polymer demonstrates excellent unipolar and bipolar

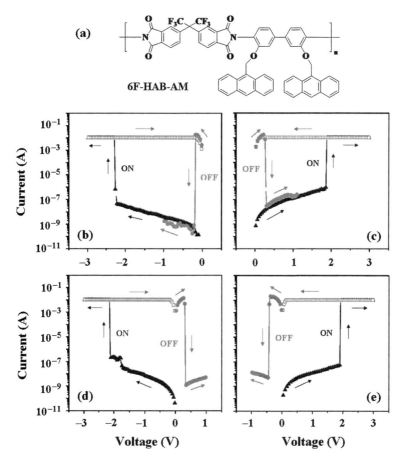

Figure 5.15 (a) Chemical structure of 6F-HAB-AM. I–V curves of Al/6F-HAB-AM (20 nm thick)/Al devices, which were measured with a compliance current level of 0.01 A:[31] (b) and (c) unipolar WORM and flash memory devices; (d) and (e) bipolar flash memory devices. The switching-OFF runs in (b) and (c) were carried out with a 0.1 A compliance current. The electrode contact area was 0.2 × 0.2 mm². Reproduced with permission. Copyright 2011, American Chemical Society.

switching behaviors over a very small voltage range less than ±2 V (Figure 5.15b–e). The 6F-HAB-AM films show repeatable writing, reading, and erasing ability with high reliability and a high ON/OFF current ratio (up to 10^7) under ambient air conditions. The excellent electrical memory behavior has been confirmed to occur through trap-limited SCLC and hopping processes *via* the utilization of anthracene pendants. Moreover, considering its molecular orbitals (E_{HOMO} = −5.09 eV and E_{LUMO} = −1.50 eV) and the electrodes' work function, the energy barrier for hole injection into the polymer layer from the electrodes is lower than that for electron injection into the polymer layer from the electrodes. This indicates that 6F-HAB-AM in devices can undergo electrical switching processes *via* hole injection. Thus, the anthracene pendant in 6F-HAB-AM can be classified as an electron-donor rather than an electron-acceptor. Therefore, 6F-HAB-AM also is also a p-type memory polymer.

Overall, more research effort is still required for developing new n-type memory polymer systems demonstrating a variety of memory modes with high stability and performance.

5.4 Nonconjugated Polymers with Electron-Donor and -Acceptor Chromophore Pendants

In recent years, research effort has been made to develop nonconjugated polymers consisting of electron-donating and -accepting units either in the backbone or in the pendants for electrical memory devices. As a result, some polymer systems have been reported.[2–4] Their electrical memory behaviors have been interpreted mainly by using the charge transfer (CT) mechanism.[2–4] To support the proposed CT mechanism, density functional theory (DFT) calculations, UV-vis spectra, *in-situ* fluorescence (FL) spectra and transmission electron microscope (TEM) images have been provided.[2–4,20,51–56]

The DFT approach can only calculate local electron density on an electron-donating unit, an electron-accepting unit, and their conjugated units, thus providing their HOMO and LUMO levels and further indicating their CT complex formation. UV-vis and FL spectroscopies and TEM images can also give some information on the CT complex formation between electron-donating and -accepting units. Nevertheless, these analyses could not provide any direct evidence for the CT mechanism because such CT complex formation in polymer film layers does not soundly explain the mechanism behind electrical switching phenomena. Instead, it is more reasonable to consider CT complexed donor–acceptor units as another class of electroactive chromophores. A CT complex as a single electroactive chromophore unit can be interpreted to have its own molecular orbitals (E_{HOMO}, E_{LUMO}, and ΔE_{bg}) and a certain level of capability to trap and transport charge with the aid of an applied electric field. Otherwise, the electron-donating and -accepting units may have separate roles as charge trapping sites. The electron-donating units can trap holes favorably; the trapped holes can be interrupted in part or fully to transport between neighboring donors through the electron-accepting

Table 5.4 Molecular orbitals, band gaps and energy barriers to Al electrode of PVPK$_m$BEOXD$_n$ in various compositions.

Material	E_{HOMO}[a] (eV)	$\|E_{HOMO} - \Phi_{Al}\|$[b] (eV)	E_{LUMO}[c] (eV)	$\|E_{LUMO} - \Phi_{Al}\|$[d] (eV)	ΔE_{bg}[e] (eV)	$V_{c,ON}$ (V)	Memory mode(s)
PVPK	−4.88	0.60	−1.42	2.86	3.46		Conductive
PVPK$_8$BEOXD$_2$	−5.05	0.77	−1.63	2.65	3.42	2.3	WORM
PVPK$_5$BEOXD$_5$	−5.19	0.91	−1.79	2.49	3.40	3.0	WORM Unipolar flash
PVPK$_2$BEOXD$_8$	−5.07	0.79	−1.70	2.58	3.37	2.8	DRAM
PBEOXD	−5.23	0.95	−1.90	2.38	3.33		Dielectric

[a] The highest occupied molecular orbital.
[b] The energy barrier for hole injection into the HOMO from the Al electrode having a work function Φ_{Al} of −4.28 eV.
[c] The lowest unoccupied molecular orbital.
[d] The energy barrier for electron injection into the LUMO from the Al electrode ($\Phi_{Al} = -4.28$ eV).
[e] The band gap: $\Delta E_{bg} = E_{HOMO} - E_{LUMO}$.

units. In contrast, the electron-accepting units can trap electrons favorably; the trapped electrons can be interrupted in part or fully to transport between the neighboring acceptors through the electron-donating units.

Overall, the mechanism(s) of electrical switching in polymer systems consisting of both electron-donating and -accepting chromophore units may be much more complex than those of polymers composed of either electron-donating or -accepting chromophores.

The series poly(9-(4-vinylphenyl)carbazole-*ran*-2-(4-vinylbiphenyl)-5-(4-ethoxyphenyl)-1,3,4-oxadiazole) (PVPK$_m$BEOXD$_n$: m and n are in mol%) is a good model memory polymer system consisting of both electron-donating and -accepting chromophore pendants.[32] Synchrotron GIXS analysis has found that all thin films of PVPK$_m$BEOXD$_n$ are found to be fully amorphous, which is attributed to the random copolymer characteristics and the irregularities in electron-donating and -accepting chromophore bristle size and shape. The molecular orbitals have been calculated and summarized in Table 5.4. The results indicate that the molecular orbitals and band gaps can be tuned by changing the composition of VPK and BEOXD pendants in the copolymer.

For the PVPK$_m$BEOXD$_n$ copolymers, devices with Al/polymer (40 nm thick)/Al sandwich structures were fabricated and electrically evaluated. The results are summarized in Figure 5.16b–f and Table 5.4. Surprisingly, the PVPK homopolymer was found to immediately reach a low resistance state (*i.e.*, ON-state) in the voltage sweep. The PVPK film layer retains the ON-state in an electric field (Figure 5.16b). In contrast, the PBEOXD film layer initially exhibits high resistance (*i.e.*, OFF-state) and no switching behavior during the voltage sweep, *i.e.* it exhibits dielectric characteristics (Figure 5.16f). The copolymer films also initially enter an OFF-state but exhibit various switching behaviors depending on their chemical composition. The PVPK$_8$BEOXD$_2$ film layer in the device exhibits permanent nonvolatile memory behavior,

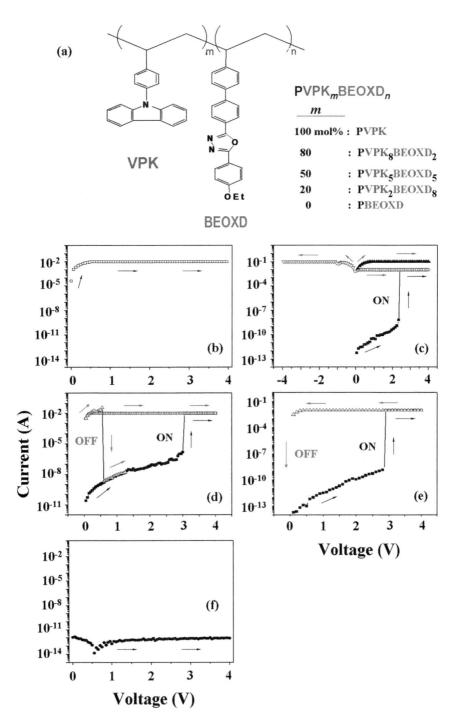

Figure 5.16 (a) Chemical structures of PVPK$_m$BEOXD$_n$ random copolymers. I–V curves of the Al/PVPK$_m$BEOXD$_n$ (40 nm thick)/Al devices, which were measured using voltage sweeps with a compliance current of 0.01 A:[32] (b) PVPK; (c) PVPK$_8$BEOXD$_2$; (d) PVPK$_5$BEOXD$_5$; (e) PVPK$_2$BEOXD$_8$; (f) PBEOXD. The switching-OFF run in (d) was carried out with a compliance current of 0.1 A. The electrode contact area was 0.5 × 0.5 mm^2. Reproduced with permission. Copyright 2013, Royal Society of Chemistry.

namely unipolar WORM memory characteristics (Figure 5.16c). Interestingly, the PVPK$_5$BEOXD$_5$ layer reveals unipolar flash memory behavior in addition to unipolar WORM memory characteristics (Figure 5.16d). In contrast, the PVPK$_2$BEOXD$_8$ film layer exhibits unipolar DRAM behavior (Figure 5.16e). Hence, the BEOXD-rich copolymers show relatively poor memory stability compared to the VPK-rich copolymers.

Overall, the PVPK$_m$BEOXD$_n$ copolymers exhibited excellent and versatile unipolar electrical memory tunable *via* chemical composition. The memory versatility might be attributed to the CT-complexed VPK–BEOXD pairs and the isolated VPK and BEOXD chromophores. Considering the molecular orbitals of the copolymers and the electrodes' work functions, all memory behaviors are thought to occur *via* hole injection from the electrode, and are governed by trap-limited SCLC and hopping processes, regardless of the compositions of the VPK and BEOXD pendants. Collectively, PVPK$_m$BEOXD$_n$ copolymers of any composition behave like electron-donating polymers and are p-type memory systems.

Poly(2-(*N*-carbazolyl)ethyl methacrylate) end-capped with fullerene (PCzMA-C$_{60}$) is another good model memory polymer system consisting of both electron-donating and -accepting chromophore pendants (Figure 5.17a).[33] Poly(2-(*N*-carbazolyl)ethyl methacrylate) (PCzMA) in 60 nm thick films, which contains one electron-donating carbazole moiety per repeat unit but no C$_{60}$ end group, exhibits unipolar flash and WORM memory behaviors, as shown in Figure 5.17b and c. Such WORM memory behavior has also been demonstrated in HABCZ-based PIs containing two carbazole pendants per repeat unit,[16,43] as discussed in Section 5.2.1. The flash memory mode could not be demonstrated in devices fabricated using HABCZ-based PIs. The C$_{60}$-ended PCzMA (*i.e.*, PCzMA-C$_{60}$) in 60 nm thick films nicely demonstrates both unipolar flash and WORM memory behaviors as PCzMA exhibits (Figure 5.17d and e). Interestingly, the $V_{c,ON}$ value is increased to ±4.0 V from ±1.9 V through the incorporation of only one electron-accepting C$_{60}$ at the polymer chain end. Furthermore, PCzMA-C$_{60}$ additionally shows bipolar flash memory characteristics (Figure 5.17f and g), which could not be realized in PCzMA. *I–V* data analysis indicates that all memory characteristics in PCzMA-C$_{60}$, and in PCzMA are governed by trap-limited SCLC conduction and hopping processes. Moreover, considering the molecular orbitals (E_{HOMO} = −5.36 eV and E_{LUMO} = −3.56 eV) and the electrodes' work function, the versatile memory behavior takes place by hole injection, rather than electron injection.

Thus, the incorporation of only one C$_{60}$ at the PCzMA backbone end impacts the electrical memory mode and characteristics. For PCzMA-C$_{60}$, the C$_{60}$-end group may possibly play two different roles. A fraction of the C$_{60}$-end groups may make complexes with electron-donating carbazole pendants and then participate in electrical memory phenomena as charge (in particular, hole) trappers and transporters. Another fraction of the C$_{60}$-end groups may participate solely as electron-acceptors and interrupt the transportation of holes between neighboring C$_{60}$–carbazole complexes, between C$_{60}$–carbazole

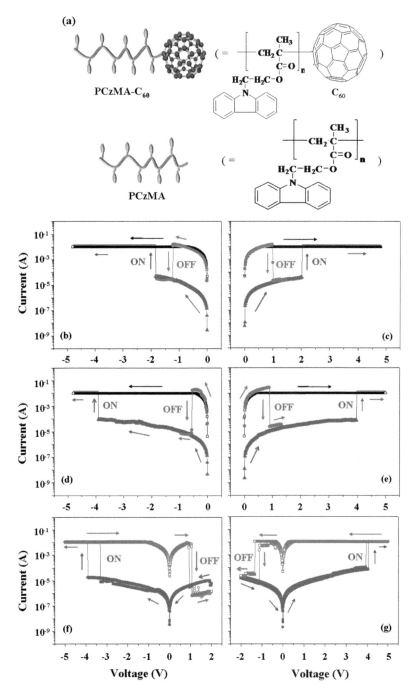

Figure 5.17 (a) Chemical structures of PCzMA-C_{60} and PCzMA. I–V curves of the Al/polymer (60 nm thick)/ITO devices, which were measured using voltage sweeps with a compliance current of 0.01 A:[33] (b) and (c) unipolar WORM and flash memory behavior of PCzMA; (d) and (e) unipolar WORM and flash memory behavior of PCzMA-C_{60}; (f) and (g) bipolar flash memory behavior of PCzMA-C_{60}. The switching-OFF runs in (b–e) were carried out with a compliance current of 0.1 A. The electrode contact area was 0.6 × 0.6 mm². Reproduced with permission. Copyright 2012, Wiley-VCH.

complexes neighboring carbazole pendants, and between neighboring carbazole pendants.

As discussed above, the incorporation of electron-donating and -accepting chromophores into polymers as pendants and as backbone components is just one of the synthetic ways to develop electrical memory polymer systems, regardless of CT complex formation. Again, CT complex formation is not the electrical memory mechanism; instead, CT complexes can be comprehended as a kind of electroactive unit giving rise to electrical memory behaviors by functioning as charge-trapping and hopping sites. In principle, their role resembles that of electron-donors or electron-acceptors. Therefore, the electrical memory mode and performance of a polymer based on electron-donating and -accepting chromophores are dependent on the types of electron-donating and -accepting chromophores and their combinations.

5.5 Stability and Reliability of Electrical Memory in Polymer-Based Devices

Some electrical memory polymers have been discussed in the previous sections, which include polymers containing electron-donating or/and -accepting chromophore pendants. They have demonstrated versatile electrical memory behavior with high ON/OFF current ratios and low power consumption.

For the polymers described in the sections above, their devices have been confirmed to reveal a long retention time up to around 10 hours in the ON-state as well as in the OFF-state even under ambient air conditions.[16–19,22–33,43,48] The retention times of 6F-HABCZ and 6F-HAB-TPAIE devices in the ON- and OFF-states, as two representative examples, are shown in Figure 5.18.[16,19] Moreover, memory polymer devices have been confirmed to be operable for several years. For example, P1-C_{60} and P2-C_{60} memory devices in both the bipolar and unipolar modes have functioned without error after being kept in ambient air for 1.5 years (Figure 5.19).[28] Another example is the PCzMA-C_{60} memory device in unipolar flash and WORM memory modes and in bipolar flash mode that functions properly after being kept in ambient air conditions for 3 years (Figure 5.20).[33] Overall, these devices exhibit excellent reliability.

In addition, memory polymer devices have been examined in the aspect of thermal stability. Figure 5.21 shows the thermal stability of 6F-TPA-OH devices in a nitrogen atmosphere.[22] The devices nicely show WORM memory characteristics at various temperatures up to 150 °C, which is the middle point of the glass transition of the polymer. The devices exhibit lower $V_{c,ON}$ at higher temperature and furthermore their ON-state current level is slightly decreased with increasing temperature, whereas the OFF-state current increases with increasing temperature. The I–V data for the OFF-state is still governed by trap-limited SCLC. Thus, the increases in the OFF-state current might result from the increase in hopping rate due to high thermal excitation. In contrast, above 160 °C (which is near to the end point of the

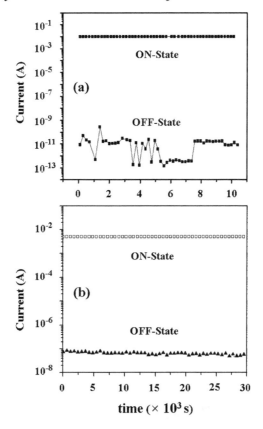

Figure 5.18 Long-term responses (*i.e.*, retention times) of the ON- and OFF-states of Al/polymer/Al devices: (a) 6F-HABCZ (70 nm thick), which was probed with a reading voltage of +0.7 V;[16] (b) 6F-HAB-TPAIE (35 nm thick), which was probed with a reading voltage of +0.5 V.[19] Reproduced with permission. Copyright 2008, Wiley-VCH. Copyright 2009, American Chemical Society.

glass transition of the polymer), the devices failed to show nice WORM memory behavior. This failure of WORM memory behavior might be attributed to several different reasons. The PI molecules become mobile above the glass transition, decreasing the stability to the dimensions of the polymer layer of the device. The mobile polymer chains then may further mobilize the charge trapping sites in the polymer layer, destabilizing the hopping paths formed in the device under the applied electric field and ultimately rupturing the hopping paths.

Figure 5.22 presents the thermal stability of 6F-HAB-TPAIE devices in a nitrogen atmosphere.[19] The ON-state current level of the devices does not change when the device temperature is raised up to 130 °C. Furthermore, the switching characteristics (*i.e.*, nonvolatile memory behavior) of the device are retained up to 130 °C. The turn-ON voltage of the device varies little with

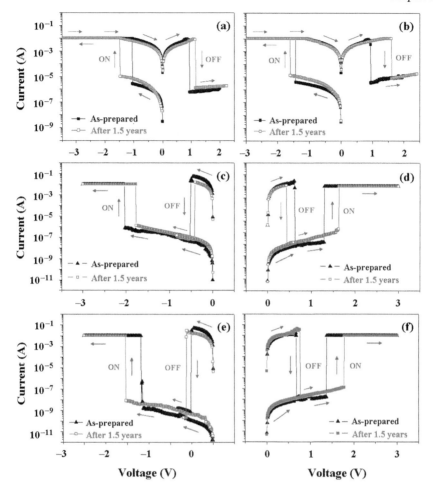

Figure 5.19 Reliability of polymer devices:[28] (a) bipolar Al/P1-C_{60} (60 nm thick)/ITO and (b) Al/P2-C_{60} (60 nm thick)/ITO devices; (c) and (d) unipolar Al/P1-C_{60} (60 nm thick)/Al and (e) and (f) Al/P2-C_{60} (60 nm thick)/Al devices. The devices used in the reliability test were stored in ambient air for 1.5 years. The electrode contact area was 1.0 × 1.0 mm². Reproduced with permission. Copyright 2014, American Chemical Society.

temperature, with no apparent correlation with the temperature variation. The turn-OFF voltage of the device increases slightly with increasing temperature, however, resulting in a slightly increased turn-OFF current level. These results indicate that the ON-state resistance increases with increasing temperature. When the temperature is increased to 140 °C and above, the ON-state of the device, however, is no longer retained, *i.e.* DRAM characteristics are found. On the other hand, the OFF-state current level of the device was found to decrease with increasing temperature. The OFF-state current decreases slowly with increasing temperature until around 70 °C and then

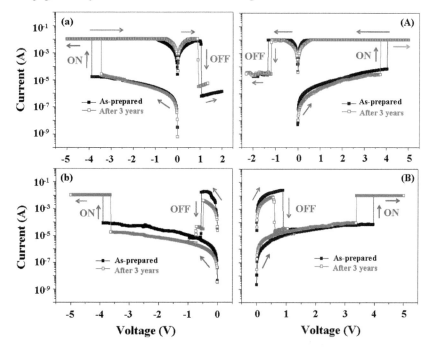

Figure 5.20 Reliability of Al/PCzMA-C$_{60}$ (60 nm thick)/ITO devices:[33] (a) and (A) bipolar flash memory behavior before and after 3 years; (b) and (B) unipolar flash memory behavior before and after 3 years. The devices used in the reliability test have been stored in ambient air for 3 years. The electrode contact area was 0.6 × 0.6 mm^2. Reproduced with permission. Copyright 2012, Wiley-VCH.

drops rapidly with further increases in temperature. Overall, the 6F-HAB-TPAIE PI devices have high temperature stability and retain excellent nonvolatile memory performance up to 130 °C.

As discussed above, the memory devices of a polymer can be rationalized to function properly up to the polymer's glass transition temperature, T_g. In general, all consumer electronic products, including logic and memory devices, are accepted to function with good stability and reliability up to 80 °C or higher under ambient air conditions. Therefore, thermal and dimensional stability up to ≥80 °C should be considered as a mandatory requirement for the development of high performance memory polymer materials and their devices.

5.6 Memory Mechanisms in Polymer-Based Devices

The electrical memory behaviors of polymers need to be thoroughly understood to develop polymers with high performance and utilize them in the production of advanced high density memory devices. In particular, an exact comprehension of the various switching mechanisms is essential. The switching mechanisms of polymer devices, however, yield complexity

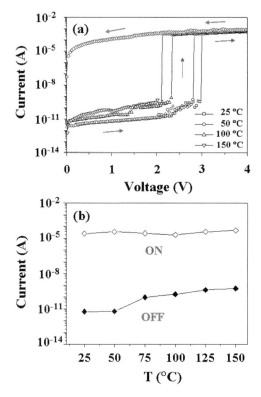

Figure 5.21 (a) *I–V* curves measured for ITO/6F-TPA-OH (54 nm thick)/Al devices at various temperatures under a nitrogen atmosphere; (b) variations of the ON- and OFF-state currents with temperature for ITO/6F-TPA-OH (54 nm thick)/Al devices, as probed under a constant bias of +0.8 V.[22] The electrode contact area was 0.5 × 0.5 mm². Reproduced with permission. Copyright 2009, American Chemical Society.

because multiple properties of the polymer and the architecture of the device combine to produce the memory behavior. Significant factors include the molecular orbital levels, band gap, the electron-donor group and its population and donor ability, the electron-accepting group and its population and accepting ability, the morphological structure and orientation and their changes in an applied electric field, the charge-trapping ability and capacity, the stabilization of trapped charges, the trapped-charge-induced space charge field and its effects, mobile charges and their population, charge transport, polymer layer thickness, the electrode's work function, interfacial contact, electric field strength (voltage and current), and voltage sweep direction. Thus, despite numerous studies on the switching mechanisms of polymer memory devices, more quantitative research is absolutely necessary to reach a definitive conclusion on the switching mechanisms.

Several switching mechanisms have been reported for polymer memory devices: ohmic conduction, Schottky conduction, thermionic conduction, SCLC, trap-limited SCLC, hopping conduction, tunneling conduction,

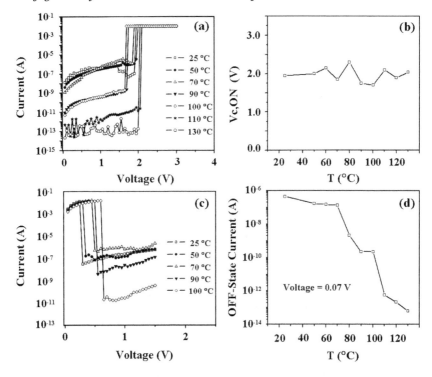

Figure 5.22 I–V curves for Al/6F-HAB-TPAIE (35 nm thick)/Al devices at various temperatures up to 130 °C:[19] (a) the turn-ON process was performed with a compliance current of 0.01 A from 0 to +3.0 V; (b) turn-ON voltage levels, which are plotted as a function of temperature; (c) the turn-OFF process was carried out with a compliance current of 0.1 A from 0 to +1.5 V; (d) OFF-state current levels, which are plotted as a function of temperature. The OFF-state level was determined to be 0.7 V. The electrode contact area was 0.2 × 0.2 mm². Reproduced with permission. Copyright 2009, American Chemical Society.

local filamentary conduction (which is similar to hopping conduction in its physical meaning), CT conduction, conformation change conduction, and polymer fuse conduction.[1–33,43,51–73] The majority of the mechanisms (ohmic conduction, Schottky conduction, thermionic conduction, SCLC, trap-limited SCLC, and tunneling conduction) were proposed to explain the correlations between the current levels and applied voltages.[1,5–19,21–33,43,72,73] In general, the current levels of polymer memory devices in ON-states are relatively high, and the voltage variation is simple: current flow (the flow of mobile charges) mainly occurs in the ON-state. The ohmic conduction mechanism appears appropriate for the current–voltage behaviors of ON-states. In contrast, other conduction mechanisms have been proposed to explain the current–voltage relationship for OFF-states. Current levels in OFF-states are much lower than those of ON-states and exhibit no simple relationship with the voltage levels. In fact, for polymers in OFF-states under an electric field, various electrical phenomena can arise, including charge (hole or electron or both) injection from electrodes and charge traps, and charge flow relatively lower than that of trapped charges. Overall, these mechanisms need to be quantized into the individual factors associated with the switching phenomena for better comprehension.

Other switching mechanisms (*i.e.*, local filamentary (*i.e.*, hopping) conduction, tunneling conduction, CT conduction, and conformation change) have been proposed as explanations for electrical switching phenomena at the molecular level.[1–33,43,51–73]

The local filamentary conduction (*i.e.*, hopping conduction) mechanism arises from charge-trapped sites in the polymer layer of memory devices.[1,5–19,21–33,43,72,73] Experimental evidence for local filament formation has been found for several polymer memory systems.[1,5–8,58] Local filaments in polymer memory devices are commonly investigated using current-sensing atomic force microscopy (CS-AFM); representative CS-AFM images are shown in Figure 5.23.[1,6,8] In general, any electroactive chromophore components in the polymer, such as electron-donor groups, electron-accepting groups and their combined units, can act as charge-trapping sites.[1,5–19,21–33,43,72,73] When an electric field is applied to the polymer layer *via* electrodes, charges are trapped by the polar chemical moieties in the polymer. Such charge-trapping continues until the sites are filled completely (that is, until the applied electric field reaches the $V_{c,ON}$); the majority of these trapped charges are immobile, although some can be mobile, resulting in the low levels of current flow that are commonly observed in the OFF-state under an electric field. Above the threshold voltage, charges are excessively injected into the polymer layer and then the excessively trapped charges are able to move through the charge-trapped sites, leading to high current flow in the ON-state. Here, the charge-trapped sites in the polymer layer can function as stepping stones, providing conduction paths, namely conductive filaments, for current flow. However, the charge-trapping sites in the polymer medium cannot be in physical contact.[1,5–19,21–33,43,72,73] Charges, however, are known to be capable of hopping a

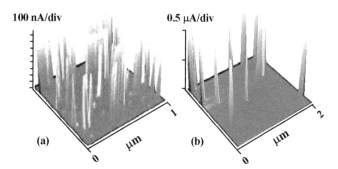

Figure 5.23 CS-AFM images showing local filaments formed in memory polymer layers: (a) poly(*o*-anthranilic acid);[6] (b) hyperbranched copper phthalocyanine polymer.[8] Reproduced with permission. Copyright 2007, American Chemical Society. Copyright 2008, Wiley-VCH.

certain distance, several nanometers or shorter.[43,72,73] Therefore, when the charge-trapped sites are positioned with interdistances of several nanometers or shorter, they can act as conductive filaments, *i.e.*, mobile charges can move between the charge-trapped sites by hopping.[43,72,73] Overall, the memory type and performance of a polymer is primarily dependent upon its trapping sites' characteristics, including the type and number of electron-donor groups, the type and number of electron-accepting groups, the conjugated or nonconjugated links between electron-donor and electron-accepting groups, charge-trapping power and capacity, the trapped charge-induced electric field and its power and volume, and the stabilization of trapped charges. In addition, polymer layer morphology, conformation, orientation, thickness, and the molecular orbitals of the polymer, the applied electric field strength (voltage and current, often compliance current controlled) and the electrode's work function are factors that need to be considered to understand the memory type and performance of systems with the formation of local filaments (*i.e.*, hopping paths) based on charge-trap sites.[16,22,23,26,43,72,73]

The CT conduction mechanism has also been proposed for some polymers that exhibit resistive memory behaviors.[2–4,20,51–56,60] CT takes place in electron-donor and -acceptor complexes and is defined as a partial electronic charge transfer from the donor moiety to the acceptor moiety. The CT mechanism basically relies on the molecular orbitals of the electron-donor and -acceptor moieties and electric field induced charge transportation from the HOMO to the LUMO state. Information about the HOMO and LUMO levels of polymers has generally been obtained experimentally. Some calculations of the HOMO and LUMO levels of electron-donor and -acceptor units have been performed by using DFT. DFT simulations, however, provide no direct evidence of the CT switching mechanism. They only support the experimentally measured overall HOMO and LUMO levels of the polymer. Such charge transportation may be favorable in single π-conjugated polymers containing electron-donor and -acceptor moieties or in polymers consisting of alternating electron-donor and -acceptor moieties, but not in nonconjugated polymers containing only electron-donor moieties, only electron-acceptor moieties or both moieties with large interdistances. Also, even when charge transportation is feasible in single π-conjugated polymers, it may be limited to very short lengths or distances. Moreover, charges can undergo recombination that degrades memory performance. However, there is no clear experimental evidence for the CT mechanism in polymer memory devices.

The conformation change mechanism has also been proposed to explain the behavior of polymer memory devices.[60,69,71] As discussed above, memory polymers are composed of electron-donating or/and -accepting electroactive chromophores as backbones or/and pendants. When an electric field is applied to the polymer layer of a memory device, such electroactive chromophore units may respond to the electric field and induce conformation changes inside the polymer chain in part or completely, which contribute positively or negatively to the memory behavior. Thus, such electric field induced conformation changes in the polymer layer are another aspect to be considered in understanding polymer memory and mechanism, rather than providing a complete switching mechanism. Additionally, there is no experimental evidence proving that conformation change alone is the electrical switching mechanism. Such induced conformation changes might essentially be needed for single molecule-based memory devices, however.[74,75]

It is conclusive that charge-trapping and hopping conduction of electroactive chromophore units (appropriate electron-donating and -accepting units and their combinations, and π-conjugated backbones) is the most prominent switching mechanism for the resistive type of polymer memory. Charge-trapping and hopping conduction might be associated with many other factors, however. Hence, more effort needs to be put into comprehending the associated factors and quantizing their individual roles in charge-trapping and hopping conduction, and ultimately the electrical switching mechanism.

5.7 Concluding Remarks

There has been substantial research into polymer memory technology over the last two decades because of its potential as an alternative to conventional silicon- and metal-oxide-based memory technologies, which do not appear to be adequate for the required reduction in device size from the micro- to the nano-scale. As a result, a number of memory polymers have been reported. In particular, some nonconjugated polymers with electroactive chromophore pendants have been reported for high performance memory devices. In addition, there are extensive investigations being carried out on these polymers in an attempt to provide insight into the switching mechanisms. p-Type polymers outnumber n-type polymers in the literature, however. Thus, more research effort is necessary to develop high performance n-type memory polymers. Furthermore, more studies on the switching mechanisms are needed supported by quantitative experimental evidence. Thorough comprehension of the relationships between chemical and morphological structures and electrical memory performance remains an urgent challenge that must be solved in order to continue the development of high performance memory polymers.

This chapter has been prepared with financial support from the National Research Foundation (NRF) of Korea (Doyak Program 2011-0028678 and Center for Electro-Photo Behaviors in Advanced Molecular Systems (2010-0001784)), the Ministry of Science, ICT & Future Planning (MSIP) and the Ministry of Education (BK21 Plus Program and Global Excel Program).

References

1. S. G. Hahm, Y.-G. Ko, W. Kwon and M. Ree, *Curr. Opin. Chem. Eng.*, 2013, **2**, 79.
2. T. Kurosawa, T. Higashihara and M. Ueda, *Polym. Chem.*, 2013, **4**, 16.
3. Q. D. Ling, D. J. Liaw, C. Zhu, D. S. H. Chan, E. T. Kang and K. G. Neoh, *Prog. Polym. Sci.*, 2008, **33**, 917.

4. W. P. Lin, S. J. Liu, T. Gong, Q. Zhao and W. Huang, *Adv. Mater.*, 2014, **26**, 570.
5. S. Baek, D. Lee, J. Kim, S. Hong, O. Kim and M. Ree, *Adv. Funct. Mater.*, 2007, **17**, 2637.
6. J. Kim, S. Cho, S. Choi, S. Baek, D. Lee, O. Kim, S.-M. Park and M. Ree, *Langmuir*, 2007, **23**, 9024.
7. S.-H. Hong, O. Kim, S. Choi and M. Ree, *Appl. Phys. Lett.*, 2007, **91**, 093517.
8. S. Choi, S.-H. Hong, S. H. Cho, S. Park, S.-M. Park, O. Kim and M. Ree, *Adv. Mater.*, 2008, **20**, 1766.
9. D. Lee, S. Baek, M. Ree and O. Kim, *IEEE Electron Device Lett.*, 2008, **29**, 694.
10. D. Lee, S. Baek, M. Ree and O. Kim, *Jpn. J. Appl. Phys.*, 2008, **47**, 5665.
11. S. Park, T. J. Lee, D. M. Kim, J. C. Kim, K. Kim, W. Kwon, Y.-G. Ko, H. Choi, T. Chang and M. Ree, *J. Phys. Chem. B*, 2010, **114**, 10294.
12. Y.-G. Ko, W. Kwon, D. M. Kim, Y.-S. Gal and M. Ree, *Polym. Chem.*, 2012, **3**, 2028.
13. S. Nam, Y.-G. Ko, S. G. Hahm, S. Park, J. Seo, H. Lee, H. Kim, M. Ree and Y. Kim, *NPG Asia Mater.*, 2013, **5**, e33.
14. Y.-G. Ko, D. M. Kim, K. Kim, S. Jung, D. Wi, T. Michinobu and M. Ree, *ACS Appl. Mater. Interfaces*, 2014, **6**, 8415.
15. S. G. Hahm, T. J. Lee, D. M. Kim, W. Kwon, Y.-G. Ko, T. Michinobu and M. Ree, *J. Phys. Chem. C*, 2011, **115**, 21954.
16. S. G. Hahm, S. Choi, S.-H. Hong, T. J. Lee, S. Park, D. M. Kim, W.-S. Kwon, K. Kim, O. Kim and M. Ree, *Adv. Funct. Mater.*, 2008, **18**, 3276.
17. S. Park, K. Kim, J. C. Kim, W. Kwon, D. M. Kim and M. Ree, *Polymer*, 2011, **52**, 2170.
18. S. G. Hahm, S. Choi, S.-H. Hong, T. J. Lee, S. Park, D. M. Kim, J. C. Kim, W.-S. Kwon, K. Kim, M.-J. Kim, O. Kim and M. Ree, *J. Mater. Chem.*, 2009, **19**, 2207.
19. K. Kim, S. Park, S. G. Hahm, T. J. Lee, D. M. Kim, J. C. Kim, W. Kwon, Y. G. Ko and M. Ree, *J. Phys. Chem. B*, 2009, **113**, 9143.
20. Q.-D. Ling, F.-C. Chang, Y. Song, C.-X. Zhu, D.-J. Liaw, D. S.-H. Chan, E.-T. Kang and K.-G. Neoh, *J. Am. Chem. Soc.*, 2006, **128**, 8732.
21. K.-J. Lee, D. M. Kim, K. Ihm, M. Ree, T.-H. Kang and S. Chung, *Appl. Phys. Lett.*, 2012, **100**, 053306.
22. D. M. Kim, S. Park, T. J. Lee, S. G. Hahm, K. Kim, J. C. Kim, W. Kwon and M. Ree, *Langmuir*, 2009, **25**, 11713.
23. T. J. Lee, C.-W. Chang, S. G. Hahm, K. Kim, S. Park, D. M. Kim, J. Kim, W.-S. Kwon, G.-S. Liou and M. Ree, *Nanotechnology*, 2009, **20**, 135204.
24. D. M. Kim, Y.-G. Ko, J. K. Choi, K. Kim, W. Kwon, J. Jung, T.-H. Yoon and M. Ree, *Polymer*, 2012, **53**, 1703.
25. T. J. Lee, Y.-G. Ko, H.-J. Yen, K. Kim, D. M. Kim, W. Kwon, S. G. Hahm, G.-S. Liou and M. Ree, *Polym. Chem.*, 2012, **3**, 1276.
26. Y.-G. Ko, W. Kwon, H.-J. Yen, C.-W. Chang, D. M. Kim, K. Kim, S. G. Hahm, T. J. Lee, G.-S. Liou and M. Ree, *Macromolecules*, 2012, **45**, 3749.

27. K. Kim, H.-J. Yen, Y.-G. Ko, C.-W. Chang, D. M. Kim, W. Kwon, S. G. Hahm, G.-S. Liou and M. Ree, *Polymer*, 2012, **53**, 4135.
28. Y.-G. Ko, S. G. Hahm, K. Murata, Y. Y. Kim, B. J. Ree, S. Song, T. Michinobu and M. Ree, *Macromolecules*, 2014, **47**, 8154.
29. W. Kwon, B. Ahn, D. M. Kim, Y.-G. Ko, S. G. Hahm, Y. Kim, H. Kim and M. Ree, *J. Phys. Chem. C*, 2011, **115**, 19355.
30. W. Kwon, B. Ahn, Y. Kim, Y. Y. Kim, B. J. Ree, Y.-G. Ko, J. Lee and M. Ree, *Sci. Adv. Mater.*, 2014, **6**, 2289.
31. S. Park, K. Kim, D. M. Kim, W. Kwon, J. Choi and M. Ree, *ACS Appl. Mater. Interfaces*, 2011, **3**, 765.
32. K. Kim, Y.-K. Fang, W. Kwon, S. Pyo, W.-C. Chen and M. Ree, *J. Mater. Chem. C*, 2013, **1**, 4858.
33. S. G. Hahm, N.-G. Kang, W. Kwon, K. Kim, Y.-G. Ko, S. Ahn, B.-G. Kang, T. Chang, J.-S. Lee and M. Ree, *Adv. Mater.*, 2012, **24**, 1062.
34. T. J. Shin and M. Ree, *J. Phys. Chem. B*, 2007, **111**, 13894.
35. S. G. Hahm, S. W. Lee, J. Suh, B. Chae, S. B. Kim, S. J. Lee, K. H. Lee, J. C. Jung and M. Ree, *High Perform. Polym.*, 2006, **18**, 549.
36. M. Ree, *Macromol. Res.*, 2006, **14**, 1.
37. J. Wakita, S. Jin, T. J. Shin, M. Ree and S. Ando, *Macromolecules*, 2010, **43**, 1930.
38. S. G. Hahm, T. J. Lee and M. Ree, *Adv. Funct. Mater.*, 2007, **17**, 1359.
39. S. G. Hahm, T. J. Lee, T. Chang, J. C. Jung, W. C. Zin and M. Ree, *Macromolecules*, 2006, **39**, 5385.
40. S. W. Lee, S. J. Lee, S. G. Hahm, T. J. Lee, B. Lee, B. Chae, S. B. Kim, J. C. Jung, W. C. Zin, B. H. Sohn and M. Ree, *Macromolecules*, 2005, **38**, 4331.
41. T. J. Shin, B. Lee, H. Y. Youn, K.-B. Lee and M. Ree, *Langmuir*, 2001, **17**, 7842.
42. A. Viehbeck, M. J. Goldberg and C. A. Kovac, *J. Electrochem. Soc.*, 1990, **137**, 1460.
43. B. J. Ree, W. Kwon, K. Kim, Y.-G. Ko, Y. Y. Kim, H. Lee and M. Ree, *ACS Appl. Mater. Interfaces*, 2014, **6**, 21692.
44. A. Viehbeck, M. J. Goldberg and C. A. Kovac, *J. Electrochem. Soc.*, 1990, **137**, 1460.
45. H. Miyasaka, S. R. Kahn and A. Itaya, *J. Phys. Chem. A*, 2002, **106**, 2192.
46. D. R. Lide, *CRC Handbook of Chemistry and Physics*, CRC, New York, 8th edn, 2007, pp. 3–86.
47. http://www.nanoniele.jp.
48. M. Ree, K. Kim, S. H. Woo and H. Chang, *J. Appl. Phys.*, 1997, **81**, 698.
49. W. Shi, S. Fan, F. Huang, W. Yang, R. Liu and Y. Cao, *J. Mater. Chem.*, 2006, **16**, 2387.
50. S. I. Kim, M. Ree, T. J. Shin, C. Lee, T.-H. Woo and S. B. Rhee, *Polymer*, 2000, **41**, 5173.
51. Y. Q. Li, R. C. Fang, A. M. Zheng, Y. Y. Chu, X. Tao, H. H. Xu, S. J. Ding and Y. Z. Shen, *J. Mater. Chem.*, 2011, **21**, 15643.
52. B. Zhang, Y. L. Liu, Y. Chen, K. G. Neoh, Y. X. Li, C. X. Zhu, E. S. Tok and E. T. Kang, *Chem.–Eur. J.*, 2011, **17**, 10304.

53. Y. C. Lai, K. Ohshimizu, W. Y. Lee, J. C. Hsu, T. Higashihara, M. Ueda and W. C. Chen, *J. Mater. Chem.*, 2011, **21**, 14502.
54. A. D. Yu, W. Y. Tung, Y. C. Chiu, C. C. Chueh, G. S. Liou and W. C. Chen, *Macromol. Rapid Commun.*, 2014, **35**, 1039.
55. L. C. Lin, H. J. Yen, C. J. Chen, C. L. Tsai and G. S. Liou, *Chem. Commun.*, 2014, **50**, 13917.
56. C. J. Chen, H. J. Yen, Y. C. Hu and G. S. Liou, *J. Mater. Chem. C*, 2013, **1**, 7623.
57. Y. Q. Li, R. C. Fang, A. M. Zheng, Y. Y. Chu, X. Tao, H. H. Xu, S. J. Ding and Y. Z. Shen, *J. Mater. Chem.*, 2011, **21**, 15643.
58. B. Hu, F. Zhuge, X. Zhu, S. Peng, X. Chen, L. Pan, Q. Yan and R. W. Li, *J. Mater. Chem.*, 2011, **22**, 520.
59. Y. S. Lai, C. H. Tu, D. L. Kwong and J. S. Chen, *Appl. Phys. Lett.*, 2005, **87**, 122101.
60. E. Teo, Q. Ling, Y. Song, Y. Tan, W. Wang, E. Kang, D. Chan and C. Zhu, *Org. Electron.*, 2006, **7**, 173.
61. Y. S. Lai, C. H. Tu, D. L. Kwong and J. S. Chen, *IEEE Electron Device Lett.*, 2006, **27**, 451.
62. M. Kim, S. Choi, M. Ree and O. Kim, *IEEE Electron Device Lett.*, 2007, **28**, 967.
63. Q. Liu, K. Jiang, L. Wang, Y. Wen, J. Wang, Y. Ma and Y. Song, *Appl. Phys. Lett.*, 2010, **96**, 213305.
64. D. Lee, S. Baek, M. Ree and O. Kim, *Electron. Lett.*, 2008, **44**, 596.
65. S. Moller, C. Perlov, W. Jackson, C. Taussig and S. R. Forrest, *Nature*, 2003, **426**, 166.
66. Y. Li, H. Xu, X. Tao, K. Qian, S. Fu, Y. Shen and S. Ding, *J. Mater. Chem.*, 2011, **21**, 1810.
67. S. Yoshida, T. Ono and M. Esashi, *Nanotechnology*, 2011, **22**, 335302.
68. M. Lauters, B. McCarthy, D. Sarid and G. Jabbour, *Appl. Phys. Lett.*, 2006, **89**, 013507.
69. B. Zhang, Y. Chen, Y. Zhang, X. Chen, Z. Chi, J. Yang, J. Ou, M. D. Zhang, D. Li, D. Wang, M. Liu and J. X. Zhou, *Phys. Chem. Chem. Phys.*, 2012, **14**, 4640.
70. D. Braun, *J. Polym. Sci., Part B: Polym. Phys.*, 2003, **41**, 2622.
71. S. L. Lim, Q. Ling, E. Y. H. Teo, C. X. Zhu, D. S. H. Chan, E.-T. Kang and K. G. Neoh, *Chem. Mater.*, 2007, **19**, 5148.
72. B. Ahn, D. M. Kim, J.-C. Hsu, Y.-G. Ko, T. J. Shin, J. Kim, W.-C. Chen and M. Ree, *ACS Macro Lett.*, 2013, **2**, 555.
73. K. Kim, Y. Y. Kim, S. Park, Y.-G. Ko, Y. Rho, W. Kwon, T. J. Shin, J. Kim and M. Ree, *Macromolecules*, 2014, **47**, 4397.
74. Z. Donhauser, B. Mantooth, K. Kelly, L. Bumm, J. Monnell and J. Stapleton, *Science*, 2001, **292**, 2303.
75. J. Vandendriessche, P. Palmans, S. Toppet, N. Boens, S. F. De and H. Masuhara, *J. Am. Chem. Soc.*, 1984, **106**, 8057.

CHAPTER 6

Polymer Composites for Electrical Memory Device Applications

CHENG-LIANG LIU*[a] AND WEN-CHANG CHEN*[b]

[a]Department of Chemical and Materials Engineering, National Central University, Taoyuan 32001, Taiwan; [b]Department of Chemical Engineering, National Taiwan University, Taipei 10617, Taiwan
*E-mail: clliu@ncu.edu.tw, chenwc@ntu.edu.tw

6.1 Introduction

Computers and many electronic gadgets usually rely on stored information memory devices. Devices with electrical bistability have been widely reported because they can be used as switches and memory elements which are important for digital electric technology. Resistive switching memory devices have been widely investigated using correlated inorganic materials. However, electric field controlled switching devices based on organic polymer materials have undergone great growth recently. Organic/polymeric materials have advantages over their inorganic counterparts because of the inherent flexibility of the compounds, their versatility in chemical structure, low cost, and simplified processing. Several representative reviews have examined the use of organic or polymeric materials in memory devices.[1–8] Although many types of single organic/polymeric molecule-based devices have been demonstrated, organic–inorganic hybrid blends and multilayer composite systems have also been extensively prepared for resistor memory devices because of their easy and diverse modulation of the host polymers and guest additives to control the resulting memory characteristics. The supplementary compounds can be viewed as data storage media or can provide physical electronic transitions within the organic polymer matrix. The general fabrication processes for polymer composites include solution-processable techniques (such as spin-coating or dip-coating) for organic–inorganic blends and layer-by-layer deposition to form multilayer devices. Functional organic compounds, inorganic nanomaterials, allotropes of carbon, and ionic liquids embedded into the organic polymer matrix or fabricated into multilayer architectures as sandwiched resistance switching materials between two metal electrodes will be discussed in this chapter.

6.2 Polymer–Organic Molecule Composites

Lai et al. developed soft electronics information modules, which were based on an elastic graphene bottom electrode grown by chemical vapor deposition and a poly(methyl methacrylate) (PMMA) and poly(3-butylthiophene) (P3BT) composite chosen as the active layer in the sandwiched device.[9–11] The current–voltage curve of the PMMA:P3BT memory device demonstrates electrical bistability with typical write-once-read-many times (WORM) features and a high ON/OFF ratio of >10^5. Even under repetitive bending/stretching, the polymer composite memory still exhibits excellent electrical functions and memory effects. Similar results in a WORM memory device fabricated using a phase-separated poly(3-hexylthiophene) (P3HT) and PMMA hybrid layer were also reported by Kim et al.[12,13] Organic composite bistable memory devices were made by employing size-controlled conducting polypyrrole (PPy) nanoparticle-embedded poly(vinyl alcohol) (PVA) as active layer.[14] Monodisperse PPy nanoparticles with different diameters (20, 60, and 100 nm) were fabricated via dispersion polymerization in order to evaluate their size-dependent bistable memory characteristics. As-prepared devices with 20 nm PPy nanoparticles demonstrated a high ON/OFF ratio over 100, and stable multilevel switching behavior over more than 100 cycles was observed.

Ferroelectric polarization is an attractive physical property as the mechanism for nonvolatile switching, because the two polarizations can be used as two binary levels. Blom et al. presented a bistable rectifying diode by blending semiconducting polymers (such as poly(3-hexylthiophene) (P3HT) or poly(9,9-dioctylfluorene) (PFO)) and ferroelectric polymers (such as poly(vinylidene fluoride-trifluoroethylene) (PVDF-TrFE)) into phase-separated networks based on a simple cross-bar array.[15–19] During the ON state the ferroelectric polarization points towards the anode, and the injection barriers for holes and electrons are simultaneously lowered by the ferroelectric stray fields such that they can effectively be disregarded. On the other hand, in the OFF state the ferroelectric polarization points towards the cathode. Effectively, both injection barriers then increase and no current flows, irrespective of the bias. The resistance of the diode can be read out nondestructively at any bias lower than the coercive field.

Solution-processable donor–acceptor blends between a polymer and small molecule[20–23] or two small molecules[24] have also been reported for electrical memory devices with fine-tunable storage performance. The electrical conductivity modulation exhibited by this class of material system is understood to be due to electric field-induced charge transfer from the HOMO of the electron-donating material (donor) to the LUMO of the electron-accepting material (acceptor), which is related to charge separation, charge transfer, and charge storage occurring within the active components.[1] Salaoru et al. demonstrated bistable nonvolatile behavior from a blend of two small organic molecules where one was an electronic donor tetrathiafulvalene (TTF) and the second one was an electron acceptor tetracyanoethylene (TCNE) in poly(vinyl acetate) (PVAc).[20] Huang and Pyo et al. used an active layer made by blending a polystyrene (PS) insulating matrix, poly[10-(2′-ethylhexyl)phenothiazine-3,7-diyl] (PPTZ) acting as donor and tetracyanoquinodimethane (TCNQ) acting as acceptor.[21] The PS:PPTZ:TCNQ device exhibits good switching and memory effects such as a large ON/OFF ratio and a long retention time. Lee et al. discussed the donor–acceptor charge transfer characteristics between electron-donating P3HT and molecular hexaazatriphenylene derivative acceptors where the molecular acceptor was selected because of its heterocyclic structure which would likely promote close interaction with the conjugated chains of P3HT.[22] Through slight modification of the acceptors, variation in the LUMO can be achieved, which alters the threshold voltage of the resistive switching device. Memory devices fabricated using these spin-coated blends demonstrated electrical bistability with a conductivity modulation of ~2 orders of magnitude and with negligible data loss after a stress test of 500 consecutive read-outs conducted at 0.2 V. Our group reported tunable electrical memory characteristics using polyimide:polycyclic aromatic compound blends on flexible substrates.[23] The active layers of the polyimide blend films were prepared from different compositions of poly[4,4′-diamino-4″-methyltriphenylamine-hexafluoroisopropylidene diphthalimide] (PI(AMTPA)) and polycyclic aromatic compounds (coronene or N,N-bis[4-(2-octyldodecyloxy)-phenyl]-3,4,9,10-perylenetetracarboxylic diimide (PDI-DO)), as shown in Figure 6.1(a). Thus, the memory device characteristics changed from volatile to nonvolatile behavior of flash and WORM as the additive contents increased in both blend systems. The main differences between these two blend systems were the threshold voltage values and the additive content needed to change the memory behavior. Besides, the memory devices fabricated on a flexible substrate exhibited excellent durability under bending conditions. The above results clearly showed that the introduction of coronene or PDI-DO additives dramatically affected the resistive memory performance (Figure 6.1(b)).

Small molecule donor–acceptor blends, using the 9,9″-dibutyl-9′-(2-ethylhexyl)-9H,9′H,9″H-3,3′:6′,3″-tercarbazole (TCz) donor and the 2,9-di(tridecan-7-yl)anthra[2,1,9-def:6,5,10-$d'e'f'$]-diisoquinoline-1,3,8,10-(2H,9H)-tetraone (PI)

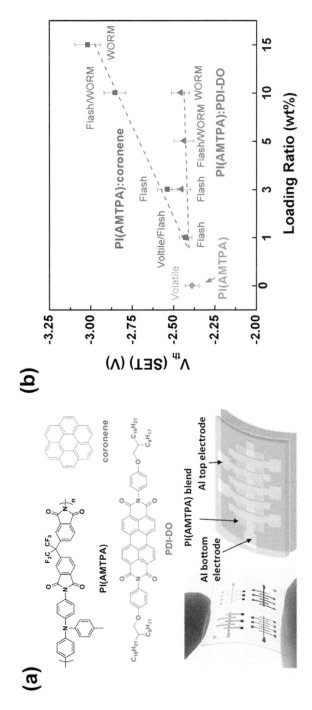

Figure 6.1 (a) Schematic diagram of polymer composite memory device configuration and the chemical structures of the polyimide matrix and polycyclic aromatic compound additives. (b) Variation of the threshold voltage (V_{th}(SET)) and memory performance of memory devices with the loading ratio of coronene or PDI-DO. Reproduced with permission from ref. 23. Copyright 2013 American Chemical Society.

acceptor, were designed for electrical memory devices. As the blend ratios were adjusted, dually tunable electrical data storage performance with volatility and reversibility was achieved. When the film had a donor:acceptor ratio of 5:1, the acceptor could not hold the transferred charges well, leading to an unstable ON state. Therefore, a corresponding DRAM behavior was observed. As for the films with donor:acceptor ratios of 2:1 and 1:1, the transferred charges could be stably held and even a reverse voltage could not dissociate the formed CT complex. Correspondingly, films of both ratios exhibited WORM behavior. When the donor:acceptor ratio is 1:2, the formed CT complex might have intermediate stability and could be dissociated with a reverse bias, exhibiting flash behavior.

6.3 Organic Polymer–Inorganic Nanomaterial Composites

Nanometer-sized materials exhibit particularly interesting electrical properties due to the interesting quantum effects of the electron. Organic polymer–inorganic composite materials have attracted much interest in attempts to use nanomaterials for applications such as light-emitting diodes, photovoltaics, transistors, bio-detection, data storage, *etc.* In this section, organic polymer–inorganic nanohybrid memory systems are demonstrated using organic polymer–discrete nanoparticle blends in single layer or multilayer structures where the inorganic nanomaterial is sandwiched between two organic layers.

6.3.1 Multistacked Composite with Metallic Nanoparticles Embedded

Yang and coworkers first proposed a new multilayer memory structure consisting of a layered organic/metal/organic deposited between Al electrodes.[25-31] Semiconducting small molecule, 2-amino-4,5-imidazoledicarbonitrile (AIDCN), was selected for the organic layer whereas the thermal evaporation of a thin Al middle layer forms the metal nanoclusters (see Figure 6.2(a)). It was suggested that the nanocluster layer was polarized after the applied bias was above a threshold, consequently making the organic layers undergo a conductance change, switching the whole device to an ON state that was confirmed by the theoretical single band Hubbard model.[29] The device remained in the ON state even when the power was turned off, making it a candidate for nonvolatile memory cells. The OFF state could be recovered by a reverse bias to neutralize the stored charge and to restore the device to its original state (Figure 6.2(a)). One million writing/erasing cycles for the electrically bistable device were tested in ambient atmosphere without significant device degradation (Figure 6.2(b)).[27] Charge storage occurred through the redistribution of charges within the Al nanocluster layer and hence the switching process was quite fast with nanoseconds of write-in

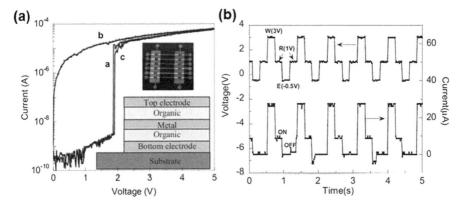

Figure 6.2 (a) Current–voltage curves of the memory device with the nominal structure Al/AIDCN (50 nm)/Al nanoclusters (20 nm)/AIDCN (50 nm)/Al. The bottom inset shows the device structure, the top inset shows an image of a typical memory device. (b) Current responses of the device during the write–read–erase voltage cycle. Reproduced with permission from ref. 27. Copyright 2002, AIP Publishing LLC.

speed for the device.[27] It was also found that the high yield memory device must be fabricated with a slow deposition rate for the middle metal layer to allow the formation of small-sized nanoclusters.[28] The thickness-related electrical characteristics in organic/metal/organic systems were also studied.[31] The ON state current showed an exponential decrease with increasing thickness of the organic layer, suggesting that the electron transmission probability of device decreased with increasing thickness of the organic layer as the thickness of the metal nanocluster layer was fixed. However, the current step shifted toward low voltage with increasing metal nanocluster layer thickness. A further study used the third contacts to the middle cluster layer to provide a three-terminal organic memory device (two two-terminal devices).[30] Through appropriate interface engineering, both top and bottom devices showed electrical bistability and memory effects and this could double the data storage density of the memory device.[30]

Bozano et al. fabricated similar multilayer structures[32,33] and proposed the activation mechanism in these organic devices to be essentially the same as the concept of electroformed metal–insulator–metal (MIM) diodes interpreted by the Simmons and Verderber (SV) model.[34] The devices were fabricated with Alq_3 as the semiconducting medium, granular Al as charge trapping sites and Al as bottom and top electrodes, exhibiting the bistability and multistability inherent to the SV mechanism of metastable charge storage.[33] The results showed that the current switched between the OFF and ON states at the threshold, reached a maximum, then went through the negative differential resistance (NDR) region to a minimum and finally increased again exponentially.[32,33] Metals with different work functions such as Cr, Cu, ITO, Au and Ni can be used as the bottom electrodes. Again, Mg, Ag, Al, Cr, Au, and CuPc were evaporated as the middle materials. Other polymeric semiconductors or crosslinking polymers have been selected as semiconducting hosts.[33] Suh and *et al.* fabricated a nonvolatile memory device with an embedded Al layer between poly(*N*-vinylcarbazole) (PVK) to observe its charge behavior.[35] The conductance of the device was controlled by the trapping and detrapping of charges at the effective trap sites in the embedded Al layer, and PVK as an organic active material merely carried out the charge function. The effective trap site also preferred to be charged at a high conductivity state and took relatively low energy states, which were affected by thermal stimulation.

Other chemically self-assembled or thermally-deposited nanocrystals/nanoparticles (such as Cu_2O,[36] MnO_3,[37,38] Ni–NiO,[39-41] CdSe/ZnS,[42] and $Ni_{1-x}Fe_x$[43]) were surrounded by an organic polymer matrix in multilayer structures. The fabricated devices utilizing embedded nanocrystals/nanoparticles exhibited significantly enhanced information storage capability and nonvolatile behavior on the basis of current–voltage results. The free charge injected from the electrode into the organic polymer layer could be captured by the ground electronic sub-band of the nanocrystals/nanoparticles which resulted in the formation of a trapped charge space field. Charge transport could be facilitated since charge traps were filled with holes. The surface morphology of the nanostructure trapping layer also plays an important role.[31] Polarization resulting from charge trapping of an MoO_3 nanocluster-like layer was more random than that of an MoO_3 nanoparticle layer. This means that the effective polarization effect on the carrier transportation of the MoO_3 nanocluster-like layer was weaker than that of the MoO_3 nanoparticle layer.[37,38] Therefore, after switching from the low conductance state to the high conductance state, the device remained in the high conductance state without NDR.[37,38]

The layer-by-layer (LBL) assembly method offers diverse opportunities to prepare organic/metallic multilayer composites with tailored electrical properties and has proved useful for fabricating composite based memory devices.[44-51] Cho and coworkers reported electrostatic LBL assembled multilayer films containing inorganic nanoparticles,[44-47] metallic oxides,[46-48] and enzymes[50,51] that were employed as active layers for charge-trap flash and resistance switching nonvolatile memory devices requiring high memory performance with a good memory stability and a high ON/OFF ratio.

6.3.2 Organic Polymer–Metal Nanoparticle Hybrids

In this approach, the memory material consists of an organic polymer matrix with a sufficient amount of well-dispersed capped-metal nanoparticles to provide the memory environment. The nanoparticles are blended into polymers that determine the electrical bistability of the device. Based on this approach, Yang's group extended the organic phase from pure PS[52-55] and PMMA[55] to PS doped with organic species[55-57] and conjugated polymers[58-60] as host with a capping agent on Au nanoparticles as guest.

PS films were preferred to be selected as the matrix with 2-naphthalenethiol (2NT)[52-55] or 2-benzeneethanethiol (BET)-capped Au nanoparticles sandwiched between two Al electrodes.[55] The lower current difference and higher threshold voltage for Al/Au-BET+PS/Al than that of Al/Au-2NT+PS/Al might be due to the less conjugated π-electrons of BET than those of 2NT, which were the capping molecules on the Au nanoparticles. The current increased with the increase of the Au-2NT nanoparticle concentration, and the threshold voltage increased linearly with the increase of the film thickness. These results suggested that the Au-2NT nanoparticles were the media for charge transport through the film due to the conjugated naphthalene structures, and electric field-induced charge transfer between the Au nanoparticles and the capping 2NT induces a polarization along the applied electric field that results in the formation of the film in two stable conductivity states. Besides, the resistive switching voltage was related to the work function of the electrode.[54,61,62] The electrode sensitivity toward resistive switching was attributed to the contact potential between the nanoparticles and metal electrode. These observations and Au nanoparticles' concentration-dependent electrical transition also evidenced that the resistive switching of the polymer:metal nanoparticle films was not due to filament formation in the polymer composite films. Another group synthesized 1-dodecanethiol (DT) protected Au and proposed that Au nanoparticles could be treated as the nano-trap in the PS[63] or 4-cyano-2,4,4-trimethyl-2-methylsulfonylthiocarbonyl sulfonyl-poly(butyric acid 1-adamantan-1-yl-1-methyl-ethyl ester) (PCm) through a carrier hopping process.[64–67] A trap-filled space-charge-limited-current (SCLC) model was supported by the experimental data and theoretical calculations[68] to explain the charge trapping/detrapping mechanism in this memory system. A similar explanation that treats Au nanoparticles as traps in PVK was investigated by estimating the density of state calculations for pristine and Au-interacting PVK.[69,70]

A device based on a mixture of capped-Au nanoparticles and an aromatic 8-hydroxyquinoline (8-HQ) compound in a PS matrix was also proposed by Yang's group.[55–57] Before the electrical transition, the device exhibited a low conductivity controlled by the charge injection. After the voltage bias was greater than the threshold voltage, the current increased by more than four orders of magnitude. PS acted as an inert matrix for the Au nanoparticles and did not play a part in the electrical transition.[55,56] Charge transfer between the 8HQ and Au nanoparticles as donor and acceptor, respectively, under a high electric field for the electronic transition was observed. The tunnelling process between the neighboring 8HQ molecules was analyzed well by a combination of direct tunnelling and Fowler–Nordheim (F–N) tunnelling.[55,56] Direct integration of the Au nanoparticles + 8-HQ + PS bistable layer with a light-emitting polymer layer showed unique light-emitting properties modulated by electrical bistability.

Based on similar strategies, Au nanoparticles could form stable charge transfer complexes with thiol end-capped PS,[71] conjugated polymers (polyaniline[72] or P3HT[73] or crosslinked poly(4-n-hexylphenyldiphenylamine)[74]), and other pendent polymers (such as linear homopolymers[75–79] and ordered

PS-b-PMMA block copolymers[80]) where the polymers served both as the supporting matrix and active component in the memory device. The conjugated polymer:Au nanoparticle device exhibited a higher current in the OFF state, a more gradual transition between the two conductivity states and better stability in the ON state than the PS + 8HQ + Au nanoparticle device since the conjugated polymer had a higher HOMO and lower LUMO than 8HQ and higher charge mobility than the PS film dispersed with 8HQ.[53] Moreover, ordered PS-b-PMMA block copolymer:Au nanoparticle nanocomposites were employed to prevent particle aggregation and produce uniform sized nanoparticles (Figure 6.3(a)).[80] A relatively reduced threshold voltage was obtained (Figure 6.3(b)) and only a small amount of expensive Au nanoparticle material was needed. This system may be used to scale down the active size of the defined device for real applications.

In addition, different dispersed metal nanoparticles such as Pt, Ag and Al nanoparticles were incorporated into organic polymeric matrices.[81–90] A bio-based information processing device derived from tobacco mosaic virus (TMV) nanowires coated with Pt nanoparticles showed an ON/OFF ratio of 10^3 as a result of charge trapping of the nanoparticles for data storage and tunnelling processes in the high conductivity state.[81] However, there were two proposed switching mechanisms discussed in Ag nanoparticle–polymer composites. If the memory device was fabricated using nanocomposite films

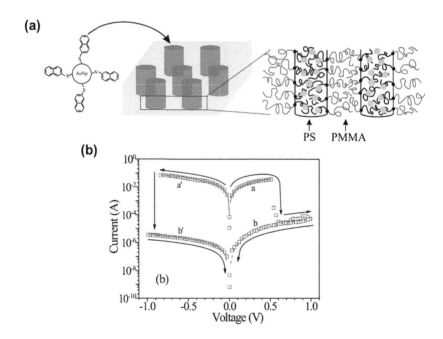

Figure 6.3 (a) A cartoon showing ordered block copolymer:Au nanoparticle nanocomposites. (b) Current–voltage curve of the Au/block copolymer:Au nanoparticles/Au device. Reproduced with permission from ref. 80. Copyright 2010 WILEY-VCH Verlag GmbH & Co. KGaA, Weinheim.

consisting of Ag nanoparticles and a semiconducting PVK, electrical field-induced charge transfer occurred with PVK positively charged and the Ag nanoparticles negatively charged when the external field was high enough.[82] When a reverse voltage was applied to the device, the charge transfer complex could return back to its original state. However, the current bistability and the carrier transport mechanism of the device using Ag nanoparticles with insulating polymer composites were probably due to charge trapping during the OFF state (thermionic emission) and trap-filling during the ON state (SCLC model).[83–89]

Besides metal nanoparticles, Winey and coworkers reported resistive switching in bulk and glassy polymer nanocomposites, specifically a series of Ag nanowire–PS composites.[91,92] At compositions near the electrical percolation threshold, resistive switching in the Ag nanowire–PS nanocomposites occurred upon increasing the voltage and subsequently recovered as the voltage decreased. The resistive switching in these Ag nanowire–PS devices was likely caused by field-induced formation of Ag filaments that bridge between adjacent nanowire clusters, extending the percolation network and decreasing bulk resistivity.[91–93]

6.3.3 Organic Polymer–Other Inorganic Nanomaterial Composites

To date, a single layer polymer embedded with various kinds of inorganic nanomaterials fabricated utilizing a solution method has been investigated in device structures based on polymer composites sandwiched between two metal electrodes. Inorganic nanomaterials randomly dispersed inside the polymer layer without aggregation may have a number of advantages such as higher density and strong charge confinement which allow the memory devices to be operated under better conditions. Various research groups introduced semiconductor inorganic nano-species such as Cu_2S,[94,95] CdS,[96,97] $CdTe$,[98] ZnO,[99–114] ZrO_2,[114] Al_2O_3,[114] CeO_2,[114] Y_2O_3,[114] TiO_2,[114–117] SnO_2,[118] MoS_2,[119] FeS_2,[120] $BaTiO_3$,[121] and $CdSe-ZnS$,[122–126] $InP-ZnS$,[127] $CuInS_2-ZnS$,[128] and $CdSe-InP$[129] core–shell materials, embedded in an organic polymer matrix to fabricate electrically bistable devices.

It is generally believed that the resistance switching in polymer–inorganic hybrid nanocomposite devices is related to the charge storage of the inorganic nanomaterials. Electric field-induced charge transfer and charge trapping in the inorganic nanomaterials have been proposed although the switching mechanism in hybrid composite devices is under debate. Sulfur-containing[116] or fluorene-based[117] aromatic polyimides with different TiO_2 (domain size: 3–5 nm) concentrations have been prepared, resulting in tunable memory properties from volatile dynamic random access memory (DRAM) to static random access memory (SRAM) to WORM with a high ON/OFF current ratio of 10^8. Introduction of TiO_2 as electron acceptor into polyimides can stabilize the charge transfer complex and thus functional polyimide–TiO_2 composite materials with higher TiO_2 content reveal longer retention ability.[116] Furthermore, the effect of the crystalline phase of TiO_2 in the PI composite films on the memory behavior was also investigated, and anatase TiO_2 exhibited higher trapping ability to increase the retention time in the ON state compared to amorphous TiO_2.[116] In addition, Zhang et al. proposed a facile method for exfoliation and dispersion of MoS_2 in ethanol with the aid of polyvinylpyrrolidone (PVP).[119] The resultant PVP-coated MoS_2 was used for flexible nonvolatile rewritable memory devices with reduced graphene oxide (rGO) electrodes, as shown in Figure 6.4. Initially, when the device was swept positively (with rGO as the anode) from 0 to 5 V, the current increased progressively with the voltage increase and an abrupt current increase occurred at a switching threshold voltage of 3.5 V. This indicated that the device transition switched from a high resistance state to a low resistance state, equivalent to a write process in digital memory. The high resistance state could be recovered by applying a reverse voltage at −4.5 V (as the erasing process of a digital memory device). The electrical transition of the device might be attributed to the charge trapping and detrapping behavior of MoS_2 in the PVP matrix under an electrical field. The memory effect was strongly correlated to the presence of inorganic nanomaterials in the hybrid composite memory device. Murata et al. reported on the fabrication of nonvolatile WORM memory devices with the ITO/PVP:ZnO/Al structure with different concentrations of ZnO nanoparticles.[108] The electrical characteristics of the device should be dominated by the size and number density of ZnO. Indeed, the ON and OFF currents and ON/OFF ratio strongly depend on the weight ratio of ZnO to PMMA (x). The currents tend to increase with increasing concentration of ZnO nanoparticles. When $x > 2 \times 10^{-1}$, the devices were immediately short-circuited after fabrication owing to the fact that the active layer was mainly nanoparticles, resulting in very large currents. However, when $x < 10^{-4}$, a conducting path could not be generated due to the negligible appearance of ZnO nanoparticles in PMMA, causing the currents to be very low. Kim et al. fabricated a series of

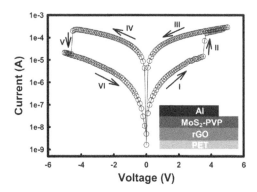

Figure 6.4 Current–voltage curve of the rGO/MoS_2-PVP/Al flexible memory device. Reproduced with permission from ref. 119. Copyright 2012 WILEY-VCH Verlag GmbH & Co. KGaA, Weinheim.

core–shell type quantum dots blended with a polymer layer using a spin-coating method.[122–129] Fittings to the current–voltage curves and their temperature dependences are often used to study charge transport during device operation. Ohmic conduction, thermionic emission, SCLC and Fowler–Nordheim tunneling are common mechanisms used to explain charge transport behavior.[7] Electrical bistability in such systems has been explained in terms of charge trapping or carrier confinement in the inorganic nanomaterials. These classes of composite memory device could be switched between two conductivity states to form inorganic nanomaterials acting as charging and discharging media.

6.4 Polymer–Carbon Allotrope Composites

Allotropes of carbon such as buckyballs, carbon nanotubes (CNTs) or graphene derivatives in the polymer matrix have been proposed for possible applications in resistive memory materials.

6.4.1 Polymer–Buckyball Cluster Composites

Fullerene or its derivatives blended into polymers as composite thin films have been prepared for nonvolatile memory applications. The polymers used include tetracyanoquinodimethane (TCNQ),[130] PS,[131–135] PMMA,[136–140] PI,[141–149] P3HT,[150–153] poly(triphenylamine),[154] poly(4-vinylphenol),[155–157] natural rubber,[158] poly[4-(diphenylamino)benzyl methacrylate] (PTPMA),[159] PVDF-TrFE ferroelectric,[160] and other block copolymers.[161–165] The switching mechanisms based on such blended composites and configurations include electric field-induced charge transfer, the tunneling effect, charge trapping/detrapping, and field-induced polarization.[3] Generally, the electrical switching characteristics of memory devices fabricated using polymer–fullerene nanocomposites can be modulated by controlling the fullerene concentration. Ikkala and coworkers showed that the nanostructure of [6,6]-phenyl-C_{61}-butyric acid methyl ester (PCBM) and PS allowed facile tuning of the switching behavior at low PCBM concentration upon annealing above the glass transition temperature of PS.[131] By increasing the PCBM concentration from 2 to 6 wt%, the switching voltage from OFF to ON state during the first voltage sweep systematically decreased and a consistent nonlinear NDR was observed, suggesting the polarization of the PCBM clusters separated by the tunneling barrier of PS and the generation of a stronger electric field between adjacent clusters. However, above ca. 7 wt%, chains of PCBM clusters coupled to the electrodes led to ohmic behavior. Yang and coworkers observed that the electrical bistability of PCBM and tetrathiofulvalene (TTF) dispersed in a PS matrix was due to a charge transfer effect between the TTF donor and PCBM acceptor.[133] A strong electric field facilitated electron transfer from the HOMO of TTF to the LUMO of PCBM, producing a positively charged TTF and negatively charged PCBM. A sharp increase in conductivity occurred after the charge transfer.

Lee et al. prepared an organic resistor memory component consisting of polyimide:PCBM composites sandwiched between two electrodes.[141–148] Typical unipolar memory switching behavior was achieved by successive application of voltages with the same polarity. A polyimide:PCBM composite resistor memory device was fabricated based on a vertically stacked three-dimensional (3D) architecture to increase the memory cell density,[144] for patterning through a direct metal transfer (DMT) method to downscale the cell size to 100 μm,[143] for a one diode–one resistor (1D–1R) device to prevent crosstalk reading interference,[141] and for an 8 × 8 cross-bar array-type flexible organic memory with a transparent multilayer graphene electrode.[148]

Recently, Huang et al. designed a unique device structure with a configuration of rGO/P3HT:PCBM/Al for polymer nonvolatile memory devices, as shown in Figure 6.5(a).[150,151] The current–voltage characteristics of the fabricated device showed electrical bistability with a WORM memory effect (Figure 6.5(b)). The memory device exhibited a high ON/OFF ratio (10^4–10^5) and a low switching threshold voltage (0.5–1.2 V), which were dependent on the sheet resistance of the rGO electrode. The experimental results confirmed that the carrier transport mechanisms in the OFF and ON states were dominated by the thermionic emission current (Figure 6.5(c)) and the ohmic current (Figure 6.5(d)), respectively. The polarization of the PCBM domains and the localized internal electrical field formed among the adjacent domains were proposed to explain the electrical transition of the memory device.

The design and fabrication of nanostructured materials are expected to lead to advanced high-density memory based on nanoscale element cells. Block copolymers that consist of two or more covalently bonded blocks could produce various morphologies depending on the nature of the block copolymers (block ratio or chemical structures) which can be used to effectively disperse fillers and prevent their aggregation for electrical memory device applications. Our group reported the fabrication and characterization of polymer switching memory devices prepared from supramolecular composite thin films of block copolymers (such as poly-(fluorenylstyrene)-*block*-poly-(2-vinylpyridine) (P(St-Fl)-*b*-P2VP),[161] poly(3-hexylthiophene)-*block*-poly(3-phenoxymethylthiophene) (P3HT-*b*-P3PT),[162] poly[2,7-(9,9-dihexylfluorene)]-*block*-poly-(2-vinylpyridine) (PF-*b*-P2VP),[163] and poly(styrene-*block*-4-vinylpyridine) (PS-*b*-P4VP)[164]) and PCBM. For example, the electrical characteristics for the PCBM:asymmetric PS-*b*-P4VP composite device showed electrically insulating, bistable switching or conducting behavior which strongly depended on the loading of the PCBM, the block length ratio and the solvent annealing conditions.[164] PS-*b*-P4VP with a longer PS block:PCBM memory device showed nonvolatile WORM-like memory behavior but PS-*b*-P4VP with a longer P4VP block:PCBM device exhibited a volatile SRAM nature. Electric field-induced conductance transitions depended on efficient charge transport ability in the co-existing charge-trapping environment in the composite thin film.

Another possible approach to make polymer–buckyball cluster composites for memory switching materials is to covalently incorporate the fullerene

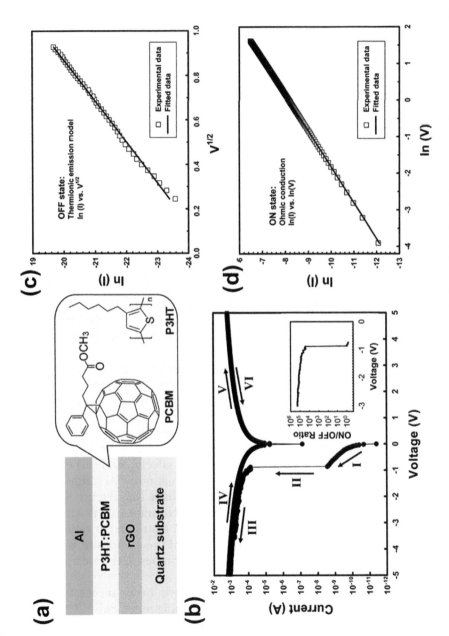

Figure 6.5 (a) Schematic diagram of the configuration of the memory device and the chemical structures of P3HT and PCBM. (b) Current–voltage curve of rGO/P3HT:PCBM/Al device. The inset shows the ON/OFF ratio as a function of applied voltage in the negative sweep. Experimental results and theoretical fitted data of current–voltage curves in the (c) OFF and (d) ON state. Reproduced with permission from ref. 151. Copyright 2010 American Chemical Society.

molecules into the polymers as side chain/backbone/end group compounds to produce fullerene-containing polymers.[166–170] Kang et al. first reported PVK bearing fullerene covalently (PVK-C_{60}) which was found to reveal bipolar flash memory behavior.[166] Later, Ree and coworkers directly attached a C_{60} electron acceptor to the end of poly(2-(N-carbazolyl)ethyl methacrylate) (PCzMA) which contained one electron-donating carbazole moiety per repeat unit, resulting in excellent bipolar and unipolar nonvolatile memory performance with a high ON/OFF current ratio, low power consumption and long-term reliability even under ambient air conditions.[167] A high loading of fullerene molecules in the polymer was achieved without aggregation problems, and the covalent incorporation of fullerene into the polymer resulted in superior nonvolatile memory performance with excellent reliability.[166–170]

6.4.2 Polymer–Carbon Nanotube Composites

Quasi one-dimensional systems such as carbon nanotubes are potential candidates for the inter-connectors of nano-devices and active electrical circuit elements. Pal and coworkers observed electrical bistability and high conductance switching in functionalized carbon nanotube-conjugated P3HT composites at room temperature.[171] The transition to the ON state was explained in terms of charge transfer from the carbon nanotubes to the conjugated polymer chains. The ratio between the conductance of the high and low states increased with an increase in carbon nanotube concentration inside the polymer matrix. The transition between the two states was rewritable in nature and associated with memory phenomena.

Kang et al. reported controllable electrical conductance switching and nonvolatile memory effects in PVK and carbon nanotube composite thin films.[172] The electrical and bistable switching behaviors of the PVK–carbon nanotube composite films could be tuned by varying the carbon nanotube content (doping level) in the composite films. Unique device behaviors, including (i) insulator behavior, (ii) bistable electrical conductance switching effects (WORM memory and rewritable memory effects), and (iii) conductor behavior, were realized in an ITO/PVK–carbon nanotube/Al sandwich structure by increasing the CNT content in the composite film (Figure 6.6). The nonvolatile memory devices were stable under a constant voltage stress or a continuous pulse voltage stress, with an ON/OFF state current ratio in excess of 10^3. These conductance switching effects of the composite films are attributed to electron trapping in the carbon nanotubes of the electron-donating/hole-transporting PVK matrix.

Carbon nanotubes doped with heteroatoms have been successfully employed as charge trap materials for resistive switching memory.[173,174] The controlled work function and high dispersibility of substitutionally doped carbon nanotubes significantly improved the resistive memory performance of carbon nanotube-containing nanocomposite devices.[173] Consequently, doped carbon nanotube devices demonstrated greatly enhanced nonvolatile memory performance (ON/OFF ratio >10^2, endurance cycle >10^2, retention

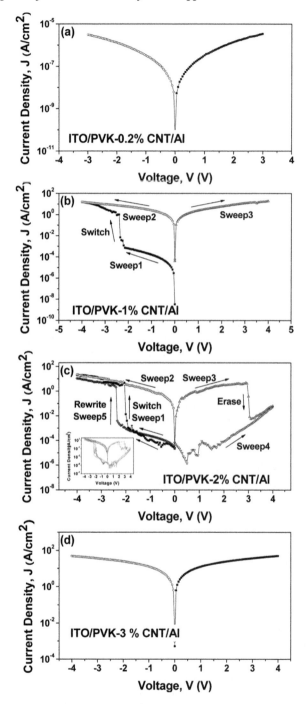

Figure 6.6 Current–voltage curve of the ITO/PVK–carbon nanotubes/Al memory device containing (a) 0.2%, (b) 1%, (c) 2% and (d) 3% carbon nanotubes. Reproduced with permission from ref. 172. Copyright 2009 American Chemical Society.

time >10^5) compared to undoped carbon nanotube devices. More significantly, a device employing both B- and N-doped carbon nanotubes with different charge trap levels exhibited multilevel resistive switching with a discrete and stable intermediate state.[173] In addition, carbon nanotubes embedded into conjugated poly(styrene-*block-para*phenylene) (PS-*b*-PPP) block copolymers,[175] PEDOT:PSS,[176] poly(dimethylsiloxane) (PDMS)[177] and PMMA[178] were reported by different groups.

6.4.3 Polymer–Graphene Sheet Composites

Graphene is a two-dimensional (2D) planar sheet of sp^2-bonded carbon atoms densely packed in a honeycomb crystal lattice. The majority of the graphene literature involves electrically insulating graphene oxide (GO) with chemically reactive oxygen functionalities.[179] Both small molecules and polymers have been covalently attached to the reactive oxygen functionalities or noncovalently attached to the graphitic surfaces. As a result, the transport of charge carriers can be effectively confined, and polymer–graphene nanocomposite have been utilized for electronic memory materials and switching devices.[179]

Recent results have clearly shown the electrical resistance switching behavior of polymer (or functional small molecule)–graphene (or GO) blended nanocomposite thin films.[180–189] Shang *et al.* fabricated nonvolatile memory devices based on electrical conductance tuning in thin films of PVK–graphene nanocomposites.[180] The electrical characteristics of the fabricated device exhibited different conductance behaviors such as insulating, WORM memory, rewritable memory effect and conducting behavior, depending on the doping level of graphene in the nanocomposites. Under ambient conditions, both the OFF and ON states of the bistable memory devices were stable under a constant voltage stress and survived up to 10^6 read cycles, with an ON/OFF current ratio in excess of 10^3. The conductance switching effects of the composites were attributed to electron trapping in the graphene nanosheets of the electron-donating/hole-transporting PVK matrix. Other graphene-containing nanocomposite memory devices based on a film of exfoliated graphene nanoflakes[182] or graphene quantum dots[181] directly blended with an insulating polymer matrix have also been demonstrated for bistable electrical switching behavior and nonvolatile rewritable memory effects.

GO bearing oxygen functionalities on its basal planes and edges can exhibit physical properties for device applications. Therefore, the morphology of the well-dispersed GO hybrid is a critical factor for endowing enhanced properties for electronic memory applications. Both our group[185] and Park's group[186] used approaches with block copolymer nanocomposites to effectively disperse the GO through noncovalent interactions. A bistable memory device based on PS-*b*-P4VP block copolymer micelles and GO sheet composite thin films was obtained using a supramolecular hybrid route.[185] The incorporation of GO with oxygeneous groups facilitated hydrogen bonding with pyridine in the P4VP block that allowed the production of nano-memory objects.

An ITO/7 wt% GO composite/Al device exhibited a WORM memory effect with an ON/OFF ratio of 10^5 at −1.0 V, a retention time of 10^4 s and a 10^8 read pulse at −1.0 V. Highly efficient control of RGO in nonpolar liquid and solid media was achieved by nondestructive modification of RGO with a conjugated block copolymer, poly(styrene-*block-para*phenylene) (PS-*b*-PPP).[186] Good dispersion of RGO was achieved because the PPP blocks strongly adhered to the basal planes of RGO through π–π interactions, while the PS blocks provided good solubility in a variety of nonpolar environments. The nanohybrid film exhibited reliable bistable current switching behavior upon a voltage sweep with excellent data retention and endurance cycling properties which further confirmed such interactions. GO noncovalently functionalized with tetracene[188] or thionine[189] small molecules could also be realized in the form of a homogenous solution and film for making electrical memory devices.

Another approach to fabricate graphene-based polymer memory devices is to modify GO by using a chemical method.[190–198] Kang et al. synthesized a single layer polymer hybrid memory containing GO sheets covalently grafted with triphenylamine-based polyazomethine (TPAPAM),[190] PVK,[191,193] poly(*tert*-butyl acrylate) (PtBA),[192] conjugated polymers containing fluorene and carbazole/triphenylamine units,[194–197] and polyaniline.[198] For example, a TPAPAM conjugate polymer–GO was synthesized and successfully used to fabricate a nonvolatile memory device using a spin-coating technique.[190] At the switching threshold voltage, electrons transitioned from the HOMO of the hole-transporting (electron donating) polymer TPAPAM into the LUMO of the graphene monoatomic layer *via* intramolecular charge transfer interaction. This device exhibited a typical bistable electrical switching and rewritable memory effect, with a turn-on voltage of about −1 V and an ON/OFF current ratio of more than 10^3. Both the ON and OFF states were stable under a constant voltage stress and survived up to 10^8 read cycles at a read voltage of −1.0 V.

Multilayer-structured memory devices based on graphene sandwiched between two insulating polymer layers could also be assembled for graphene-based nanocomposite memory devices with good reproducibility.[199–203] Nanocomposite bistable devices based on PMMA/graphene/PMMA trilayers were fabricated by Kim and coworkers.[199] The maximum ON/OFF ratio of the current bistability was as large as 10^7, and the endurance number of ON/OFF switchings is 1.5×10^5 cycles, and an ON/OFF ratio of 4.4×10^6 was maintained for retention times longer than 10^5 s. The bistable behavior of the devices made from graphene embedded in an insulating PMMA polymer layer was attributed to conducting filament formation in the hybrid layer at the state transition.

6.5 Polymer–Ionic Liquid Composites

Nishide and coworkers designed a radical polymer layer employed as active layer for a nonvolatile memory device.[204,205] To prepare the well-defined nanostructured matrices, they selected a poly(styrene-*block*-ethylene oxide) (PS-*b*-PEO) block copolymer in different segment ratios doped with a redox-active stable 2,2,6,6-tetramethylpiperidinyl-*N*-oxy (TEMPO) radical-substituted ionic liquid (TEMPO-IL).[205] The devices comprising PS-*b*-PEO with 50 and 20 wt% TEMPO-IL, corresponding to cylindrical and spherical morphologies, respectively, exhibited resistive switching behavior under the application of increasing voltage (0 to −5 V). The current density drastically increased at *ca.* −3 V with an ON/OFF ratio of >10^3. Upon application of a positive bias (0 to 5 V), the device switched back to the high resistance state at −3.5 V. Among various morphologies, the spherical morphology was considered as the most simple and ideal configuration for achieving high bit crosspoint memory, because spheres were easily close-packed to form uniform stacks for memory cells.

6.6 Conclusion

This chapter highlights various types of electrical memory devices based on polymer composite materials where the role of these materials in the memory functionality provides helpful indications for further establishing the relationship between the polymer composite materials and their performances. However, the consideration of the resulting memory effects and their mechanisms includes a lot of speculation. All of the issues including device reproducibility, cycling endurance and retention ability must be optimized for further improvement. It is hoped that polymer composite memory devices will be used for real applications in the future after resolving all of the scientific and technical problems discussed above.

References

1. Y. Yang, J. Ouyang, L. P. Ma, R. J. H. Tseng and C. W. Chu, *Adv. Funct. Mater.*, 2006, **16**, 1001.
2. J. C. Scott and L. D. Bozano, *Adv. Mater.*, 2007, **19**, 1452.
3. Q.-D. Ling, D.-J. Liaw, C. Zhu, D. S.-H. Chan, E.-T. Kang and K.-G. Neoh, *Prog. Polym. Sci.*, 2008, **33**, 917.
4. P. Heremans, G. H. Gelinck, R. Muller, K.-J. Baeg, D.-Y. Kim and Y.-Y. Noh, *Chem. Mater.*, 2011, **23**, 341.
5. B. Cho, S. Song, Y. Ji, T.-W. Kim and T. Lee, *Adv. Funct. Mater.*, 2011, **21**, 2806.
6. C.-L. Liu and W.-C. Chen, *Polym. Chem.*, 2011, **2**, 2169.
7. T. W. Kim, Y. Yang, F. Li and W. L. Kwan, *NPG Asia Mater.*, 2012, **4**, e18.
8. W.-P. Lin, S.-J. Liu, T. Gong, Q. Zhao and W. Huang, *Adv. Mater.*, 2014, **26**, 570.
9. Y.-C. Lai, F.-C. Hsu, J.-Y. Chen, J.-H. He, T.-C. Chang, Y.-P. Hsieh, T.-Y. Lin, Y.-J. Yang and Y.-F. Chen, *Adv. Mater.*, 2013, **25**, 2733.
10. Y.-C. Lai, Y.-C. Huang, T.-Y. Lin, Y.-X. Wang, C.-Y. Chang, Y. Li, T.-Y. Lin, B.-W. Ye, Y.-P. Hsieh, W.-F. Su, Y.-J. Yang and Y.-F. Chen, *NPG Asia Mater.*, 2014, **6**, e87.

11. Y.-C. Lai, Y.-X. Wang, Y.-C. Huang, T.-Y. Lin, Y.-P. Hsieh, Y.-J. Yang and Y.-F. Chen, *Adv. Funct. Mater.*, 2014, **24**, 1430.
12. K. K. Park, J. H. Jung and T. W. Kim, *Appl. Phys. Lett.*, 2011, **98**, 193301.
13. W. S. Song, H. Y. Yang, C. H. Yoo, D. Y. Yun and T. W. Kim, *Org. Electron.*, 2012, **13**, 2485.
14. J.-Y. Hong, S. O. Jeon, J. Jang, K. Song and S. H. Kim, *Org. Electron.*, 2013, **14**, 979.
15. K. Asadi, D. M. De Leeuw, B. De Boer and P. W. M. Blom, *Nat. Mater.*, 2008, **7**, 547.
16. K. Asadi, M. Li, N. Stingelin, P. W. M. Blom and D. M. de Leeuw, *Appl. Phys. Lett.*, 2010, **97**, 193308.
17. C. R. McNeill, K. Asadi, B. Watts, P. W. M. Blom and D. M. de Leeuw, *Small*, 2010, **6**, 508.
18. K. Asadi, M. Li, P. W. M. Blom, M. Kemerink and D. M. de Leeuw, *Mater. Today*, 2011, **14**, 592.
19. M. Li, N. Stingelin, J. J. Michels, M.-J. Spijkman, K. Asadi, R. Beerends, F. Biscarini, P. W. M. Blom and D. M. de Leeuw, *Adv. Funct. Mater.*, 2012, **22**, 2750.
20. I. Salaoru and S. Paul, *Thin Solid Films*, 2010, **519**, 559.
21. J.-E. Park, J.-H. Eom, T. Lim, D.-H. Hwang and S. Pyo, *J. Polym. Sci., Part A: Polym. Chem.*, 2012, **50**, 2188.
22. R. Sim, W. Ming, Y. Setiawan and P. S. Lee, *J. Phys. Chem. C*, 2013, **117**, 677.
23. A.-D. Yu, T. Kurosawa, Y.-H. Chou, K. Aoyagi, Y. Shoji, T. Higashihara, M. Ueda, C.-L. Liu and W.-C. Chen, *ACS Appl. Mater. Interfaces*, 2013, **5**, 4921.
24. S. Miao, Y. Zhu, Q. Bao, H. Li, N. Li, S. Ji, Q. Xu, J. Lu and L. Wang, *J. Phys. Chem. C*, 2014, **118**, 2154.
25. L. P. Ma, J. Liu, S. Pyo, Q. F. Xu and Y. Yang, *Mol. Cryst. Liq. Cryst.*, 2002, **378**, 185.
26. L. P. Ma, J. Liu, S. M. Pyo and Y. Yang, *Appl. Phys. Lett.*, 2002, **80**, 362.
27. L. P. Ma, J. Liu and Y. Yang, *Appl. Phys. Lett.*, 2002, **80**, 2997.
28. L. P. Ma, S. Pyo, J. Ouyang, Q. F. Xu and Y. Yang, *Appl. Phys. Lett.*, 2003, **82**, 1419.
29. J. H. Wu, L. P. Ma and Y. Yang, *Phys. Rev. B*, 2004, **69**, 115321.
30. J. He, L. P. Ma, J. H. Wu and Y. Yang, *J. Appl. Phys.*, 2005, **97**, 064507.
31. S. Pyo, L. P. Ma, J. He, Q. F. Xu, Y. Yang and Y. L. Gao, *J. Appl. Phys.*, 2005, **98**, 054303.
32. L. D. Bozano, B. W. Kean, V. R. Deline, J. R. Salem and J. C. Scott, *Appl. Phys. Lett.*, 2004, **84**, 607.
33. L. D. Bozano, B. W. Kean, M. Beinhoff, K. R. Carter, P. M. Rice and J. C. Scott, *Adv. Funct. Mater.*, 2005, **15**, 1933.
34. J. G. Simmons and R. R. Verderber, *Proc. R. Soc. London, Ser. A*, 1967, **301**, 77.
35. J.-S. Choi, Y. S. Cho, J. Y. Yook and D. H. Suh, *Polym. Adv. Technol.*, 2010, **21**, 780.

36. J. H. Jung, J.-H. Kim, T. W. Kim, M. S. Song, Y.-H. Kim and S. Jin, *Appl. Phys. Lett.*, 2006, **89**, 122110.
37. K. S. Yook, S. O. Jeon, C. W. Joo, J. Y. Lee, S. H. Kim and J. Jang, *Org. Electron.*, 2009, **10**, 48.
38. T.-Y. Chang, Y.-W. Cheng and P.-T. Lee, *Appl. Phys. Lett.*, 2010, **96**, 094330.
39. J.-G. Park, W.-S. Nam, S.-H. Seo, Y.-G. Kim, Y.-H. Oh, G.-S. Lee and U.-G. Paik, *Nano Lett.*, 2009, **9**, 1713.
40. S.-H. Seo, W.-S. Nam, J.-S. Kim, S.-Y. Lee, T.-H. Shim and J.-G. Park, *Curr. Appl. Phys.*, 2010, **10**, E32.
41. W.-S. Nam, S.-H. Seo, K.-H. Park, S.-H. Hong, G.-S. Lee and J.-G. Park, *Curr. Appl. Phys.*, 2010, **10**, E37.
42. F. Li, D.-I. Son, J.-H. Ham, B.-J. Kim, J. H. Jung and T. W. Kim, *Appl. Phys. Lett.*, 2007, **91**, 162109.
43. S. H. Cho, W. T. Kim, J. H. Jung, T. W. Kim, C. S. Yoon and Y.-H. Kim, *Jpn. J. Appl. Phys.*, 2010, **49**, 01AD03.
44. J.-S. Lee, J. Cho, C. Lee, I. Kim, J. Park, Y.-M. Kim, H. Shin, J. Lee and F. Caruso, *Nat. Nanotechnol.*, 2007, **2**, 790.
45. Y. Ko, H. Baek, Y. Kim, M. Yoon and J. Cho, *ACS Nano*, 2013, **7**, 143.
46. C. Lee, I. Kim, H. Shin, S. Kim and J. Cho, *Nanotechnology*, 2010, **21**, 185704.
47. S. Cheong, Y. Kim, T. Kwon, B. J. Kim and J. Cho, *Nanoscale*, 2013, **5**, 12356.
48. Y. Kim, C. Lee, I. Shim, D. Wang and J. Cho, *Adv. Mater.*, 2010, **22**, 5140.
49. B. Koo, H. Baek and J. Cho, *Chem. Mater.*, 2012, **24**, 1091.
50. H. Baek, C. Lee, K.-i. Lim and J. Cho, *Nanotechnology*, 2012, **23**, 155604.
51. H. Baek, C. Lee, J. Park, Y. Kim, B. Koo, H. Shin, D. Wang and J. Cho, *J. Mater. Chem.*, 2012, **22**, 4645.
52. J. Y. Ouyang, C. W. Chu, D. Sieves and Y. Yang, *Appl. Phys. Lett.*, 2005, **86**, 123507.
53. A. Prakash, J. Ouyang, J.-L. Lin and Y. Yang, *J. Appl. Phys.*, 2006, **100**, 054309.
54. J. Ouyang and Y. Yang, *Appl. Phys. Lett.*, 2010, **96**, 063506.
55. J. Y. Ouyang, C. W. Chu, R. J. H. Tseng, A. Prakash and Y. Yang, *Proc. IEEE*, 2005, **93**, 1287.
56. J. Y. Ouyang, C. W. Chu, C. R. Szmanda, L. P. Ma and Y. Yang, *Nat. Mater.*, 2004, **3**, 918.
57. R. J. Tseng, J. Ouyang, C. W. Chu, J. S. Huang and Y. Yang, *Appl. Phys. Lett.*, 2006, **88**, 123506.
58. R. J. Tseng, J. X. Huang, J. Ouyang, R. B. Kaner and Y. Yang, *Nano Lett.*, 2005, **5**, 1077.
59. R. J. Tseng, C. O. Baker, B. Shedd, J. Huang, R. B. Kaner, J. Ouyang and Y. Yang, *Appl. Phys. Lett.*, 2007, **90**, 053101.
60. C. O. Baker, B. Shedd, R. J. Tseng, A. A. Martinez-Morales, C. S. Ozkan, M. Ozkan, Y. Yang and R. B. Kanert, *ACS Nano*, 2011, **5**, 3469.
61. J. Ouyang, *Org. Electron.*, 2013, **14**, 665.
62. J. Ouyang, *Org. Electron.*, 2013, **14**, 1458.

63. H.-T. Lin, Z. Pei and Y.-J. Chan, *IEEE Electron Device Lett.*, 2007, **28**, 569.
64. H.-T. Lin, Z. Pei, J.-R. Chen, G.-W. Hwang, J.-F. Fan and Y.-J. Chan, *IEEE Electron Device Lett.*, 2007, **28**, 951.
65. J.-R. Chen, H.-T. Lin, G.-W. Hwang, Y.-J. Chan and P.-W. Li, *Nanotechnology*, 2009, **20**, 255706.
66. H.-T. Lin, Z. Pei, J.-R. Chen and Y.-J. Chan, *IEEE Electron Device Lett.*, 2009, **30**, 18.
67. H.-T. Lin, C.-Y. Lin, Z. Pei, J.-R. Chen, Y.-J. Chan, Y.-H. Yeh and C.-C. Wu, *Org. Electron.*, 2011, **12**, 1632.
68. F. Santoni, A. Gagliardi, M. A. D. Maur and A. Di Carlo, *Org. Electron.*, 2014, **15**, 2792.
69. D. I. Son, D. H. Park, J. Bin Kim, J.-W. Choi, T. W. Kim, B. Angadi, Y. Yi and W. K. Choi, *J. Phys. Chem. C*, 2011, **115**, 2341.
70. J. Lin, D. Li, J.-S. Chen, J.-H. Li and D.-G. Ma, *Chin. Phys. Lett.*, 2007, **24**, 3280.
71. S.-k. Bae, S.-Y. Lee and S. C. Hong, *React. Funct. Polym.*, 2011, **71**, 187.
72. D. Wei, J. K. Baral, R. Osterbacka and A. Ivaska, *J. Mater. Chem.*, 2008, **18**, 1853.
73. Q. Chen, L. Zhao, C. Li and G. Shi, *J. Phys. Chem. C*, 2007, **111**, 18392.
74. D. T. Simon, M. S. Griffo, R. A. DiPietro, S. A. Swanson and S. A. Carter, *Appl. Phys. Lett.*, 2006, **89**, 133510.
75. Y. Song, Q. D. Ling, S. L. Lim, E. Y. H. Teo, Y. P. Tan, L. Li, E. T. Kang, D. S. H. Chan and C. Zhu, *IEEE Electron Device Lett.*, 2007, **28**, 107.
76. P. Y. Lai and J. S. Chen, *Appl. Phys. Lett.*, 2008, **93**, 153305.
77. P. Y. Lai and J. S. Chen, *Org. Electron.*, 2009, **10**, 1590.
78. P. Y. Lai and J.-S. Chen, *IEEE Electron Device Lett.*, 2011, **32**, 387.
79. S. Patil, S. Datar, N. Rekha, S. K. Asha and C. V. Dharmadhikari, *Nanoscale*, 2013, **5**, 4404.
80. C. De Rosa, F. Auriemma, R. Di Girolamo, G. P. Pepe, T. Napolitano and R. Scaldaferri, *Adv. Mater.*, 2010, **22**, 5414.
81. R. J. Tseng, C. Tsai, L. Ma and J. Ouyang, *Nat. Nanotechnol.*, 2006, **1**, 72.
82. A. Tang, S. Qu, Y. Hou, F. Teng, H. Tan, J. Liu, X. Zhang, Y. Wang and Z. Wang, *J. Appl. Phys.*, 2010, **108**, 094320.
83. A. Kiesow, J. E. Morris, C. Radehaus and A. Heilmann, *J. Appl. Phys.*, 2003, **94**, 6988.
84. B. Mukherjee and M. Mukherjee, *Appl. Phys. Lett.*, 2009, **94**, 173510.
85. W. T. Kim, J. H. Jung, T. W. Kim and D. I. Son, *Appl. Phys. Lett.*, 2010, **96**, 253301.
86. J. C. Ribierre, T. Aoyama, T. Muto and P. Andre, *Org. Electron.*, 2011, **12**, 1800.
87. Y.-C. Hung, W.-T. Hsu, T.-Y. Lin and L. Fruk, *Appl. Phys. Lett.*, 2011, **99**, 253301.
88. G. Tian, D. Wu, L. Shi, S. Qi and Z. Wu, *RSC Adv.*, 2012, **2**, 9846.
89. Y. Wang, X. Yan and R. Dong, *Org. Electron.*, 2014, **15**, 3476.
90. O. V. Molodtsova, I. M. Aristova, S. V. Babenkov, O. V. Vilkov and V. Y. Aristov, *J. Appl. Phys.*, 2014, **115**, 164310.

91. S. I. White, P. M. Vora, J. M. Kikkawa, J. E. Fischer and K. I. Winey, *J. Phys. Chem. C*, 2010, **114**, 22106.
92. S. I. White, P. M. Vora, J. M. Kikkawa and K. I. Winey, *Adv. Funct. Mater.*, 2011, **21**, 233.
93. R. M. Mutiso, J. M. Kikkawa and K. I. Winey, *Appl. Phys. Lett.*, 2013, **103**, 223302.
94. A. Tang, F. Teng, L. Qian, Y. Hou and Y. Wang, *Appl. Phys. Lett.*, 2009, **95**, 143115.
95. A. Tang, F. Teng, J. Liu, Y. Wang, H. Peng, Y. Hou and Y. Wang, *J. Nanopart. Res.*, 2011, **13**, 7263.
96. A. Tang, F. Teng, Y. Hou, Y. Wang, F. Tan, S. Qu and Z. Wang, *Appl. Phys. Lett.*, 2010, **96**, 163112.
97. G. Nenna, S. Masala, V. Bizzarro, M. Re, E. Pesce, C. Minarini and T. Di Luccio, *J. Appl. Phys.*, 2012, **112**, 044508.
98. J. Liu, X. Qi, T. Jiang, Z. Lin, S. Chen, L. Xie, Q. Fan, Q. Ling, H. Zhang and W. Huang, *Sci. China: Chem.*, 2010, **53**, 2324.
99. F. Verbakel, S. C. J. Meskers and R. A. J. Janssen, *Appl. Phys. Lett.*, 2006, **89**, 102103.
100. F. Verbakel, S. C. J. Meskers and R. A. J. Janssen, *J. Appl. Phys.*, 2007, **102**, 083701.
101. B. Pradhan, S. K. Majee, S. K. Batabyal and A. J. Pal, *J. Nanosci. Nanotechnol.*, 2007, **7**, 4534.
102. F. Li, T. W. Kim, W. Dong and Y.-H. Kim, *Appl. Phys. Lett.*, 2008, **92**, 011906.
103. K. H. Park, J. H. Jung, F. Li, D. I. Son and T. W. Kim, *Appl. Phys. Lett.*, 2008, **93**, 132104.
104. J. H. Jung, H. J. Kim, B. J. Kim, T. W. Kim and Y. H. Kim, *Appl. Phys. Lett.*, 2007, **91**, 182107.
105. D. I. Son, C. H. You, W. T. Kim, J. H. Jung and T. W. Kim, *Appl. Phys. Lett.*, 2009, **94**, 132103.
106. D. I. Son, C. H. You, J. H. Jung and T. W. Kim, *Appl. Phys. Lett.*, 2010, **97**, 013304.
107. K. Onlaor, T. Thiwawong and B. Tunhoo, *Org. Electron.*, 2014, **15**, 1254.
108. T. T. Dao, T. V. Tran, K. Higashimine, H. Okada, D. Mott, S. Maenosono and H. Murata, *Appl. Phys. Lett.*, 2011, **99**, 233303.
109. J. Jang, W. Park, K. Cho, H. Song and T. Lee, *Curr. Appl. Phys.*, 2013, **13**, 1237.
110. A. Kathalingam and J.-K. Rhee, *J. Electron. Mater.*, 2012, **41**, 2162.
111. C. V. V. Ramana, M. K. Moodley, V. Kannan and A. Maity, *Solid-State Electron.*, 2013, **81**, 45.
112. D. Y. Yun, J. K. Kwak, J. H. Jung, T. W. Kim and D. I. Son, *Appl. Phys. Lett.*, 2009, **95**, 143301.
113. F. Verbakel, S. C. J. Meskers and R. A. J. Janssen, *J. Phys. Chem. C*, 2007, **111**, 10150.
114. F. Verbakel, S. C. J. Meskers, D. M. de Leeuw and R. A. J. Janssen, *J. Phys. Chem. C*, 2008, **112**, 5254.

115. B. Cho, T.-W. Kim, M. Choe, G. Wang, S. Song and T. Lee, *Org. Electron.*, 2009, **10**, 473.
116. C.-J. Chen, C.-L. Tsai and G.-S. Liou, *J. Mater. Chem. C*, 2014, **2**, 2842.
117. C.-L. Tsai, C.-J. Chen, P.-H. Wang, J.-J. Lin and G.-S. Liou, *Polym. Chem.*, 2013, **4**, 4570.
118. J. K. Kwak, D. Y. Yun, D. I. Son, J. H. Jung, D. U. Lee and T. W. Kim, *J. Nanosci. Nanotechnol.*, 2010, **10**, 7735.
119. J. Liu, Z. Zeng, X. Cao, G. Lu, L.-H. Wang, Q.-L. Fan, W. Huang and H. Zhang, *Small*, 2012, **8**, 3517.
120. C. W. Lin, D. Y. Wang, Y. Tai, Y. T. Jiang, M. C. Chen, C. C. Chen, Y. J. Yang and Y. F. Chen, *J. Phys. D: Appl. Phys.*, 2011, **44**, 292002.
121. I. Salaoru and S. Paul, *Philos. Trans. R. Soc., A*, 2009, **367**, 4227.
122. F. Li, D.-I. Son, S.-M. Seo, H.-M. Cha, H.-J. Kim, B.-J. Kim, J. H. Jung and T. W. Kim, *Appl. Phys. Lett.*, 2007, **91**, 122111.
123. D.-I. Son, J.-H. Kim, D.-H. Park, W. K. Choi, F. Li, J. H. Ham and T. W. Kim, *Nanotechnology*, 2008, **19**, 055204.
124. D. I. Son, C. H. Yoo, J. H. Jung and T. W. Kim, *Jpn. J. Appl. Phys.*, 2010, **49**, 01AD01.
125. F. Li, D. I. Son, B. J. Kim and T. W. Kim, *Appl. Phys. Lett.*, 2008, **93**, 021913.
126. D.-I. Son, D.-H. Park, S.-Y. Ie, W.-K. Choi, J.-W. Choi, F. Li and T.-W. Kim, *Nanotechnology*, 2008, **19**, 233303.
127. J. H. Ham, D. H. Oh, S. H. Cho, J. H. Jung, T. W. Kim, E. D. Ryu and S. W. Kim, *Appl. Phys. Lett.*, 2009, **94**, 112101.
128. J. H. Shim, J. H. Jung, M. H. Lee, T. W. Kim, D. I. Son, A. N. Han and S. W. Kim, *Org. Electron.*, 2011, **12**, 1566.
129. D. Y. Yun, W. S. Song, T. W. Kim, S. W. Kim and S. W. Kim, *Appl. Phys. Lett.*, 2012, **101**, 103305.
130. Z. C. Liu, F. L. Xue, Y. Su and K. Varahramyan, *IEEE Electron Device Lett.*, 2006, **27**, 151.
131. A. Laiho, H. S. Majumdar, J. K. Baral, F. Jansson, R. Osterbacka and O. Ikkala, *Appl. Phys. Lett.*, 2008, **93**, 203309.
132. J. K. Baral, H. S. Majumdar, A. Laiho, H. Jiang, E. I. Kauppinen, R. H. A. Ras, J. Ruokolainen, O. Ikkala and R. Osterbacka, *Nanotechnology*, 2008, **19**, 035203.
133. C. W. Chu, J. Ouyang, H. H. Tseng and Y. Yang, *Adv. Mater.*, 2005, **17**, 1440.
134. H. S. Majumdar, J. K. Baral, R. Osterbacka, O. Ikkala and H. Stubb, *Org. Electron.*, 2005, **6**, 188.
135. S. Paul, *IEEE Trans. Nanotechnol.*, 2007, **6**, 191.
136. M. H. Lee, J. H. Jung, J. H. Shim and T. W. Kim, *Org. Electron.*, 2011, **12**, 1341.
137. M. H. Lee, D. Y. Yun, H. M. Park and T. W. Kim, *Appl. Phys. Lett.*, 2011, **99**, 183301.
138. S. Qi, H. Iida, L. Liu, S. Irle, W. Hu and E. Yashima, *Angew. Chem., Int. Ed.*, 2013, **52**, 1049.
139. C. H. Yoo, S. H. Ko and T. W. Kim, *Jpn. J. Appl. Phys.*, 2012, **51**, 06FG12.

140. S. H. Ko, C. H. Yoo and T. W. Kim, *J. Electrochem. Soc.*, 2012, **159**, G93.
141. B. Cho, T.-W. Kim, S. Song, Y. Ji, M. Jo, H. Hwang, G.-Y. Jung and T. Lee, *Adv. Mater.*, 2010, **22**, 1228.
142. Y. Ji, B. Cho, S. Song, T.-W. Kim, M. Choe, Y. H. Kahng and T. Lee, *Adv. Mater.*, 2010, **22**, 3071.
143. J. J. Kim, B. Cho, K. S. Kim, T. Lee and G. Y. Jung, *Adv. Mater.*, 2011, **23**, 2104.
144. S. Song, B. Cho, T.-W. Kim, Y. Ji, M. Jo, G. Wang, M. Choe, Y. H. Kahng, H. Hwang and T. Lee, *Adv. Mater.*, 2010, **22**, 5048.
145. B. Cho, S. Song, Y. Ji and T. Lee, *Appl. Phys. Lett.*, 2010, **97**, 063305.
146. B. Cho, K. H. Nam, S. Song, Y. Ji, G.-Y. Jung and T. Lee, *Curr. Appl. Phys.*, 2012, **12**, 940.
147. S. Song, J. Jang, Y. Ji, S. Park, T.-W. Kim, Y. Song, M.-H. Yoon, H. C. Ko, G.-Y. Jung and T. Lee, *Org. Electron.*, 2013, **14**, 2087.
148. Y. Ji, S. Lee, B. Cho, S. Song and T. Lee, *ACS Nano*, 2011, **5**, 5995.
149. T.-W. Kim, D. F. Zeigler, O. Acton, H.-L. Yip, H. Ma and A. K. Y. Jen, *Adv. Mater.*, 2012, **24**, 828.
150. J. Liu, Z. Lin, T. Liu, Z. Yin, X. Zhou, S. Chen, L. Xie, F. Boey, H. Zhang and W. Huang, *Small*, 2010, **6**, 1536.
151. J. Liu, Z. Yin, X. Cao, F. Zhao, A. Lin, L. Xie, Q. Fan, F. Boey, H. Zhang and W. Huang, *ACS Nano*, 2010, **4**, 3987.
152. S. Gao, C. Song, C. Chen, F. Zeng and F. Pan, *J. Phys. Chem. C*, 2012, **116**, 17955.
153. S. Gao, F. Zeng, C. Chen, G. Tang, Y. Lin, Z. Zheng, C. Song and F. Pan, *Nanotechnology*, 2013, **24**, 335201.
154. C.-J. Chen, Y.-C. Hu and G.-S. Liou, *Chem. Commun.*, 2013, **49**, 2804.
155. W. S. Machado, M. A. Mamo, N. J. Coville and I. A. Huemmelgen, *Thin Solid Films*, 2012, **520**, 4427.
156. S. Paul, A. Kanwal and M. Chhowalla, *Nanotechnology*, 2006, **17**, 145.
157. A. Kanwal and M. Chhowalla, *Appl. Phys. Lett.*, 2006, **89**, 203103.
158. P. Predeep, D. Devasia, J. Aneesh and N. M. Faseena, *Microelectron. Eng.*, 2013, **107**, 54.
159. J.-C. Hsu, Y. Chen, T. Kakuchi and W.-C. Chen, *Macromolecules*, 2011, **44**, 5168.
160. M. A. Khan, U. S. Bhansali, D. Cha and H. N. Alshareef, *Adv. Funct. Mater.*, 2013, **23**, 2145.
161. J.-C. Hsu, C.-L. Liu, W.-C. Chen, K. Sugiyama and A. Hirao, *Macromol. Rapid Commun.*, 2011, **32**, 528.
162. Y.-C. Lai, K. Ohshimizu, W.-Y. Lee, J.-C. Hsu, T. Higashihara, M. Ueda and W.-C. Chen, *J. Mater. Chem.*, 2011, **21**, 14502.
163. S.-L. Lian, C.-L. Liu and W.-C. Chen, *ACS Appl. Mater. Interfaces*, 2011, **3**, 4504.
164. J.-C. Chen, C.-L. Liu, Y.-S. Sun, S.-H. Tung and W.-C. Chen, *Soft Matter*, 2012, **8**, 526.
165. H. Jo, J. Ko, J. A. Lim, H. J. Chang and Y. S. Kim, *Macromol. Rapid Commun.*, 2013, **34**, 355.

166. Q.-D. Ling, S.-L. Lim, Y. Song, C.-X. Zhu, D. S.-H. Chan, E.-T. Kang and K.-G. Neoh, *Langmuir*, 2007, **23**, 312.
167. S. G. Hahm, N.-G. Kang, W. Kwon, K. Kim, Y.-G. Ko, S. Ahn, B.-G. Kang, T. Chang, J.-S. Lee and M. Ree, *Adv. Mater.*, 2012, **24**, 1062.
168. D. Yue, R. Cui, X. Ruan, H. Huang, X. Guo, Z. Wang, X. Gao, S. Yang, J. Dong, F. Yi and B. Sun, *Org. Electron.*, 2014, **15**, 3482.
169. C.-J. Chen, J.-H. Wu and G.-S. Liou, *Chem. Commun.*, 2014, **50**, 4335.
170. Y.-G. Ko, S. G. Hahm, K. Murata, Y. Y. Kim, B. J. Ree, S. Song, T. Michinobu and M. Ree, *Macromolecules*, 2014, **47**, 8154.
171. B. Pradhan, S. K. Batabyal and A. J. Pal, *J. Phys. Chem. B*, 2006, **110**, 8274.
172. G. Liu, Q.-D. Ling, E. Y. H. Teo, C.-X. Zhu, D. S.-H. Chan, K. G. Neoh and E.-T. Kang, *ACS Nano*, 2009, **3**, 1929.
173. S. K. Hwang, J. M. Lee, S. Kim, J. S. Park, H. I. Park, C. W. Ahn, K. J. Lee, T. Lee and S. O. Kim, *Nano Lett.*, 2012, **12**, 2217.
174. M. A. Mamo, A. O. Sustaita, Z. N. Tetana, N. J. Coville and I. A. Huemmelgen, *Nanotechnology*, 2013, **24**, 125203.
175. S. K. Hwang, J. R. Choi, I. Bae, I. Hwang, S. M. Cho, J. Huh and C. Park, *Small*, 2013, **9**, 831.
176. J. A. Avila-Nino, W. S. Machado, A. O. Sustaita, E. Segura-Cardenas, M. Reyes-Reyes, R. Lopez-Sandoval and I. A. Huemmelgen, *Org. Electron.*, 2012, **13**, 2582.
177. C. Hu, C. Liu, Y. Zhang, L. Chen and S. Fan, *ACS Appl. Mater. Interfaces*, 2010, **2**, 2719.
178. S. ChandraKishore and A. Pandurangan, *RSC Adv.*, 2014, **4**, 9905.
179. Y. Chen, B. Zhang, G. Liu, X. Zhuang and E.-T. Kang, *Chem. Soc. Rev.*, 2012, **41**, 4688.
180. Q. Zhang, J. Pan, X. Yi, L. Li and S. Shang, *Org. Electron.*, 2012, **13**, 1289.
181. L. Kou, F. Li, W. Chen and T. Guo, *Org. Electron.*, 2013, **14**, 1447.
182. Y.-C. Lai, D.-Y. Wang, I. S. Huang, Y.-T. Chen, Y.-H. Hsu, T.-Y. Lin, H.-F. Meng, T.-C. Chang, Y.-J. Yang, C.-C. Chen, F.-C. Hsu and Y.-F. Chen, *J. Mater. Chem. C*, 2013, **1**, 552.
183. L. Zhang, Y. Li, J. Shi, G. Shi and S. Cao, *Mater. Chem. Phys.*, 2013, **142**, 626.
184. C. Wu, F. Li and T. Guo, *Vacuum*, 2014, **101**, 246.
185. A.-D. Yu, C.-L. Liu and W.-C. Chen, *Chem. Commun.*, 2012, **48**, 383.
186. D. B. Velusamy, S. K. Hwang, R. H. Kim, G. Song, S. H. Cho, I. Bae and C. Park, *J. Mater. Chem.*, 2012, **22**, 25183.
187. M. A. Mamo, A. O. Sustaita, N. J. Coville and I. A. Huemmelgen, *Org. Electron.*, 2013, **14**, 175.
188. S. Wang, K. K. Manga, M. Zhao, Q. Bao and K. P. Loh, *Small*, 2011, **7**, 2372.
189. B. Hu, R. Quhe, C. Chen, F. Zhuge, X. Zhu, S. Peng, X. Chen, L. Pan, Y. Wu, W. Zheng, Q. Yan, J. Lu and R.-W. Li, *J. Mater. Chem.*, 2012, **22**, 16422.
190. X.-D. Zhuang, Y. Chen, G. Liu, P.-P. Li, C.-X. Zhu, E.-T. Kang, K.-G. Neoh, B. Zhang, J.-H. Zhu and Y.-X. Li, *Adv. Mater.*, 2010, **22**, 1731.

191. G. Liu, X. Zhuang, Y. Chen, B. Zhang, J. Zhu, C.-X. Zhu, K.-G. Neoh and E.-T. Kang, *Appl. Phys. Lett.*, 2009, **95**, 253301.
192. G. L. Li, G. Liu, M. Li, D. Wan, K. G. Neoh and E. T. Kang, *J. Phys. Chem. C*, 2010, **114**, 12742.
193. B. Zhang, Y. Chen, L. Xu, L. Zeng, Y. He, E.-T. Kang and J. Zhang, *J. Polym. Sci., Part A: Polym. Chem.*, 2011, **49**, 2043.
194. B. Zhang, Y.-L. Liu, Y. Chen, K.-G. Neoh, Y.-X. Li, C.-X. Zhu, E.-S. Tok and E.-T. Kang, *Chem.–Eur. J.*, 2011, **17**, 10304.
195. B. Zhang, Y. Chen, G. Liu, L.-Q. Xu, J. Chen, C.-X. Zhu, K.-G. Neoh and E.-T. Kang, *J. Polym. Sci., Part A: Polym. Chem.*, 2012, **50**, 378.
196. B. Zhang, Y. Chen, Y. Ren, L.-Q. Xu, G. Liu, E.-T. Kang, C. Wang, C.-X. Zhu and K.-G. Neoh, *Chem.–Eur. J.*, 2013, **19**, 6265.
197. X. Zhuang, Y. Chen, L. Wang, K.-G. Neoh, E.-T. Kang and C. Wang, *Polym. Chem.*, 2014, **5**, 2010.
198. G. Liu, Y. Chen, R.-W. Li, B. Zhang, E.-T. Kang, C. Wang and X. Zhuang, *ChemElectroChem*, 2014, **1**, 514.
199. D. I. Son, T. W. Kim, J. H. Shim, J. H. Jung, D. U. Lee, J. M. Lee, W. Il Park and W. K. Choi, *Nano Lett.*, 2010, **10**, 2441.
200. D. I. Son, J. H. Shim, D. H. Park, J. H. Jung, J. M. Lee, W. Il Park, T. W. Kim and W. K. Choi, *Nanotechnology*, 2011, **22**, 295203.
201. C. Wu, F. Li, T. Guo and T. W. Kim, *Org. Electron.*, 2012, **13**, 178.
202. C. Wu, F. Li, Y. Zhang, T. Guo and T. Chen, *Appl. Phys. Lett.*, 2011, **99**, 042108.
203. G. Khurana, P. Misra and R. S. Katiyar, *Carbon*, 2014, **76**, 341.
204. Y. Yonekuta, K. Susuki, K. Oyaizu, K. Honda and H. Nishide, *J. Am. Chem. Soc.*, 2007, **129**, 14128.
205. T. Suga, S. Takeuchi and H. Nishide, *Adv. Mater.*, 2011, **23**, 5545.

CHAPTER 7

Conjugated Polymers for Memory Device Applications

CHENG-LIANG LIU*[a] AND WEN-CHANG CHEN*[b]

[a]Department of Chemical and Materials Engineering, National Central University, Taoyuan 32001, Taiwan; [b]Department of Chemical Engineering, National Taiwan University, Taipei 10617, Taiwan
*E-mail: clliu@ncu.edu.tw, chenwc@ntu.edu.tw

7.1 Introduction

Conjugated polymers have been considered as active memory materials and have been reported to reveal electrically volatile and non-volatile memory characteristics. In general, two constituting components, the main conjugated backbone and side chain substituents can be assembled into conjugated polymeric materials. The conjugated backbone is the most important component since it contributes to most of the memory related electrical properties. On the other hand, side chain substituents of conjugated polymers not only improve the processability but are also used to fine-tune the physical properties, particularly the electronic properties. Therefore, rational design of the repeating units of the conjugated backbone and side chain is of utmost importance for further development of polymer resistor memory devices. In the following section, we focus on the advances in performance and molecular design of conjugated polymers for electrical memory device applications reported in the past few years.

7.2 Fluorene-Based Conjugated Homopolymers and Copolymers

Conjugated polymers, especially polyfluorene and its derivatives, have been explored to fabricate single component polymer memory devices which give more reliable performance and facile processability. Figure 7.1 shows the chemical structures of some fluorene-based conjugated homopolymers and copolymers for electrical memory devices, and Table 7.1 summarizes their principal memory performance parameters. Non-volatile resistive memory devices based on a polyfluorene (**P1**) layer sandwiched between two electrodes have been statically characterized.[1,2] The switching probability of the devices was linked to the required switching voltage. The working mechanism was attributed to metallic filaments with a write pulse of 4 V and an erase pulse of between 8 and 10 V. Lee and coworkers used two different polyfluorenes with and without oxidation (**P2** and **P1**) and two Al and Au electrodes for preparing memory devices.[3] Bistability was induced in all devices when Al

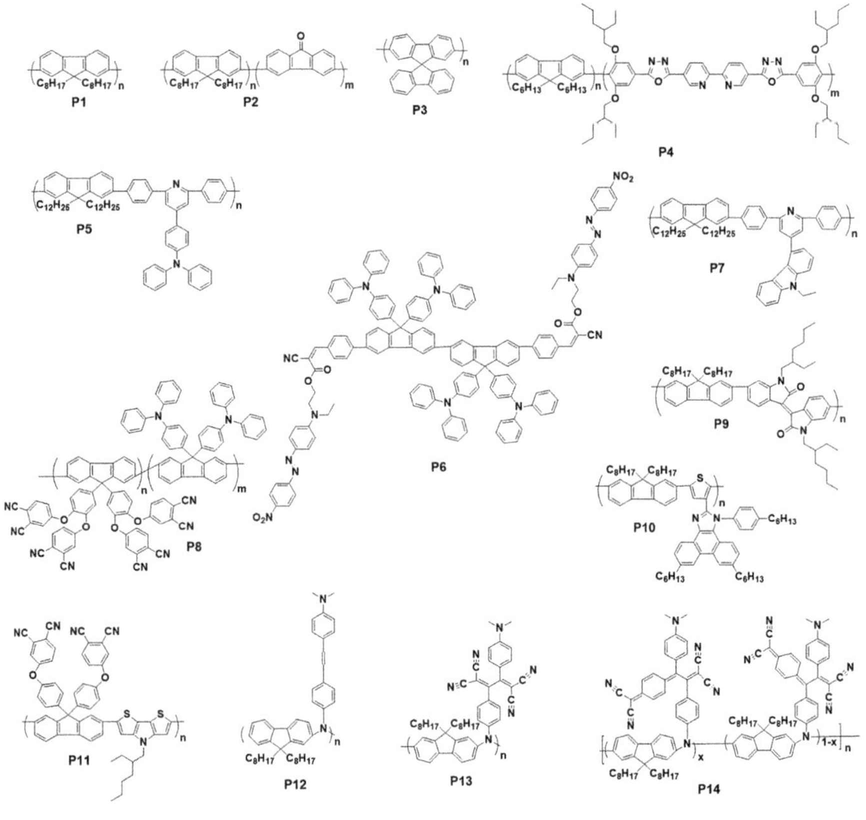

Figure 7.1 Chemical structures of fluorene-based conjugated polymers for memory device applications.

Table 7.1 Summary of the electrical memory properties of devices based on fluorene-based conjugated polymers.

Device structure	Memory type	V_{th} (V)	ON/OFF ratio	References
Polyfluorenes				
ITO/**P1**/Al	NDR	4.0	$>10^2$	1 and 2
ITO/**P2**/Al(or Au)	NDR	3.0	$>10^2$	3
Al/**P3**/Ba/Al	NDR	3–5	$\sim 10^6$	4 and 5
Fluorene-acceptor copolymers				
ITO/**P4**/Al	DRAM	−2.8	10^6	6
ITO/**P5**/Al	WORM	−2.3	10	7
ITO/**P6**/Al	Flash	−0.8 to −1.1	10^3	8
ITO/**P7**/Al	WORM (tristable)	1.8, 2.4	100, 20	9
ITO/**P8**/Al	WORM	—	6.1×10^3	10
ITO/**P9**/Al	Flash	−3.6	10^2–10^4	11
Al/**P10**/Al	Flash	−2	10^4	12
Pt/**P11**/Pt	Flash	1.5	10	13
Al/**P12**/Al	WORM	±3	10^4–10^5	14
Al/**P13**/Al	DRAM/WORM	±5/−2.7 or 2.0	10^4–10^5	14
Al/**P14**/Al	DRAM/WORM	±3.2/±1.2	10^4–10^5	14

was deposited as electrode. When Au was deposited, bistable switching was observed only in the oxidized polyfluorene (**P2**) device. Therefore, both the formation of internal trap sites and an organic/metal interface were responsible for the bistability of these fluorene-based memory devices. Gomes et al. investigated the resistive switching in poly(spirofluorene) (PSF; **P3**)-based devices using small signal impedance measurements.[4,5] Hole injection filled the states in the aluminum oxide creating a dipole layer at the oxide/polymer interface. The device remained highly resistive but the low frequency capacitance increased by several orders of magnitude. Higher external applied voltages led to an increased electrical stress across the oxide, which reduced the resistance and caused the switching.

Memory devices based on fluorene acceptor charge transfer type conjugated copolymers were demonstrated by Kang et al.[6–9] and Ma et al.[10] The incorporated electron acceptors had an important influence on the memory performance and could be viewed as trapping sites or could provide a charge transfer conducting channel for the electrical switching behavior. **P4** containing oxadiazole and bipyridine as acceptor units was synthesized by Suzuki coupling and fulfils the functionality of dynamic random access memory (DRAM) with an ON/OFF ratio of more than 10^6 (Figure 7.2(a)).[6] The memory effect was volatile due to the short retention ability of the ON state current. The ON state current could be electrically sustained by a refreshing voltage pulse every 10 s. The memory behavior was elucidated to be due to space charge and traps. The non-volatile nature of the memory effect was investigated in other fluorene-acceptor push–pull polymeric systems.[6–13] **P5** consisting of 9,9-didodeculfluorene and pendent triphenylamine donors and pyridine acceptors exhibited write-once-read-many times (WORM) memory behavior (Figure 7.2(b)).[7] Hole injection from ITO into the HOMO of **P5** was

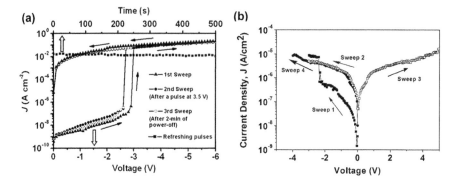

Figure 7.2 Current–voltage characteristics of the (a) **P4** and (b) **P5** memory device. Reproduced with permission from ref. 6 and 7. Copyright 2006 WILEY-VCH Verlag GmbH & Co. KGaA, Weinheim and 2007, AIP Publishing LLC.

a favored process due to the low energy barrier between the work function of ITO and the calculated HOMO level of the basic units (BUs) of **P5**. The injected hole migrated through the continuous positive electrostatic potential channel along the conjugated polymer chain and became trapped by the electron acceptor group (nitrogen atom in the pyridine ring). The filled traps could not be detrapped by a reverse voltage bias, leading to the WORM type memory behavior of **P5**. Polymer memory devices based on π-conjugated poly[9,9-bis(4-diphenylaminophenyl)-2,7-fluorene] donors covalently bridged with Disperse Red 1 (**P6**) exhibited accessible bistable conductivity states and rewritable memory with an ON/OFF current ratio of 10^3.[8] At the threshold voltage, one electron transitioned from the HOMO to the excited state; consequently the excitation of the electron donor promoted charge transfer to the conjugated channel, and finally the electron in LUMO1 could be further transferred to LUMO2 and give rise to a conductive charge separated state. The active area- or temperature-independent current density in both states indicates the absence of sample degradation or breakdown and excludes the metallic filamentary conduction effect. Tristable and nonvolatile WORM memory effects are demonstrated in poly(2,6-diphenyl-4-((9-ethyl)-9*H*-carbazole)-pyridinyl-*alt*-2,7-(9,9-didodecyl)-9*H*-fluorenyl) (**P7**).[9] The device switched from the initial low-conductivity (OFF) state to the first high-conductivity (ON-1) state at a switch-on voltage of 1.8 V, and subsequently to the second high-conductivity (ON-2) state at a higher switch-on voltage of 2.4 V. Under an applied field, the process of conformational ordering, arising from charge carrier delocalization-induced donor–acceptor interaction to form a partial or fully face-to-face conformation of the fluorene and/or carbazole units, could propagate throughout the polymer layer. An effective charge transport channel for hopping between the ordered structures switched the device from the OFF to ON-1 state. The coordinating ability of the nitrogen hetero-atoms in the carbazole pendant moieties and the indium atoms from the non-stoichiometric ITO also promoted the charge transfer interaction at the polymer/ITO interface. The enhanced absorption in the UV-vis absorption spectrum of the polymer thin film was consistent with the interfacial charge transfer interaction that elucidated the switching transition from the ON-1 to ON-2 state. A material based on the polyfluorene-based copolymer containing electron-rich triphenylamine and electron-poor 9,9-bis[3,4-bis(3,4-dicyanophenoxy)phenyl] side chains in the C-9 position of the fluorene unit (**P8**) was applied for stable non-volatile WORM memory devices with a long retention time of more than 3000 s without any degradation and a high ON/OFF ratio of 6.1×10^3.[10] The switching current of the as-fabricated device started with a high current state (ON state), decreased abruptly at a critical bias and was retained permanently in a low OFF state during the subsequent voltage sweeps. Under the low positive voltage sweep, favorable hole injection and migration resulted in the formation of the high current state. While the hole injection process went ahead, the positive charges on the triphenylamine moieties were rapidly consumed by the cyano groups as a result of the irreversible switching operation.

Zou and coworkers reported polymeric memory devices based on a low band gap polyfluorene derivative with isoindigo as electron-trapping moiety (**P9**) to lower the OFF state current and proposed another switching mechanism.[11] Metal filaments acted as conducting channels and reduced the effective distance for carrier transport across the organic layer, leading to a high-conductivity state in devices during the initial voltage sweeping process. When the voltage bias exceeded a certain value, burn out of the channels caused by Joule heating takes place. Consequently, the native properties of the conjugated polymers played an important role in the conductivity in this system. **P9** had electron-withdrawing isoindigo moieties in the polymer main chain, which served as effective electron-trapping sites. Such traps capture electrons and build up space charges, which reduced the current flowing through the polymer film. Therefore, the fabricated ITO/**P9**/Al device exhibited a switching transition with an ON/OFF ratio of 3.5×10^2. Our group also employed a conjugated poly(fluorene-thiophene) donor and a tethered phenanthro[9,10-d]imidazole acceptor (**P10**) as the active layer for a flexible bipolar resistive memory device with reliable performance in response to electric and mechanical stimuli.[12] The **P10** devices exhibited low threshold voltages (± 2 V), low switching powers (~100 µW cm^{-2}), large ON/OFF memory windows (10^4) and good retention (>10^4 s). In the backbone of **P10**, fluorene and thiophene donors act as hole transporters/trapping centers (nucleophilic sites). On the other hand, the phenanthro[9,10-d]-imidazole acceptor was more likely to serve as an electron transporter/trapping center (electrophilic site). Both donor and acceptor acted as the trapping sites that depended on the polarity of the electric field and charge association. As the applied voltage approached the threshold voltage, the majority of trapped charges were filled to create a trap-free environment. The captured charges and the charged state could be maintained due to the energy barrier for the back-transfer of charges. The deep trapping sites might not be easily recovered from the captured charge even through turning off the power. Under a reverse voltage bias, the trapped charges could be extracted and the device returned to the original OFF state, leading to bipolar flash-type memory behavior in the **P10** device. Chen and Li and coworkers designed a novel fluorene-acceptor copolymer (**P11**) for resistor memory where electron-poor 9,9-bis[4-(4-phenoxyl)phthalonitrile] side chains in the C-9 position of the fluorene unit were electron acceptors.[13] A strong dipole moment in **P11** (10.71 Debye) helped to sustain the conductive charge transfer state. The **P11**-based device exhibited a maximum memory ratio of about 10 at −2.0 V. Distinguishable from the bistable resistive switching, showing abrupt resistance or conductance jumps, the electrical transition demonstrates a smoother tuning of the conductance during the voltage sweeping processes. Ree and coworkers reported fully conjugated donor–acceptor hybrid polymers, **P12**, **P13** and **P14**, which were composed of fluorene, triphenylamine, dimethylphenylamine, alkyne, alkyne-tetracyanoethylene (TCNE) adducts, and alkyne-7,7,8,8-tetracyanoquinodimethane (TCNQ).[14] The TCNE and TCNQ units, despite their electron-acceptor characteristics, were found to enhance the π-conjugation lengths and intramolecular charge transfer of **P13** and **P14**. The TCNE and TCNQ units could diversify the digital memory modes and widen the active polymer layer thickness window. In devices with Al top and bottom electrodes, **P12** exhibited stable unipolar permanent memory behavior with high reliability. **P13** and **P14** devices showed stable unipolar permanent memory behavior over only a narrow film thickness window of 10–20 nm as well as DRAM behavior over a slightly wider thickness window of 10–30 nm at slightly higher operation voltages. The memory behavior of **P14** was shown to be driven by both hole and electron injection processes, in which the electron donor and acceptor moieties cooperatively work together as charge trapping sites. These cooperative charge injection processes indicated that the memory devices could be operated at relatively low voltages.

7.3 Thiophene-Based Conjugated Polymers

Conjugated polymer materials based on polythiophene derivatives have been also used for fabricating memory devices.[15–20] Table 7.2 summarizes the results and Figure 7.3 gives the chemical structures of thiophene-based conjugated polymers used in memory devices. Hysteresis-type behavior due to the capacitance-variant component of the displacement current was observed in an ITO/non-oriented **P15**/Al device.[15,16] Space charges were stored in the polymer layer near the metal/polymer interface that controlled charge injection and resulted in hysteresis switching. The contributions by the displacement current due to voltage variation and capacitance variation were considered to explain the hysteresis behavior. The density of stored charges was compared for different voltage amplitudes, and showed a monotonic rise during forward bias. Under reverse bias, the density saturates with an increase in the amplitude of the bias. The presence of intrinsically accumulated holes and electrons in the polymer layer near the ITO and Al electrodes

Conjugated Polymers for Memory Device Applications 239

Table 7.2 Summary of electrical memory properties of devices based on thiophene-based conjugated polymers.

Device structure	Memory type	V_{th} (V)	ON/OFF ratio	References
ITO/**P15**[a]/Al	Hysteresis without threshold	—	—	15
ITO/**P15**[b]/Al	Flash	−3.0	10^5	16
Al/**P16**/Cu	Flash	−2.1	~10	17 and 18
ITO/**P17**/Al	NDR	3.3	10^2	19
ITO/**P18**/Al	DRAM	−2.8	10^6	20

[a]Unoriented.
[b]Oriented.

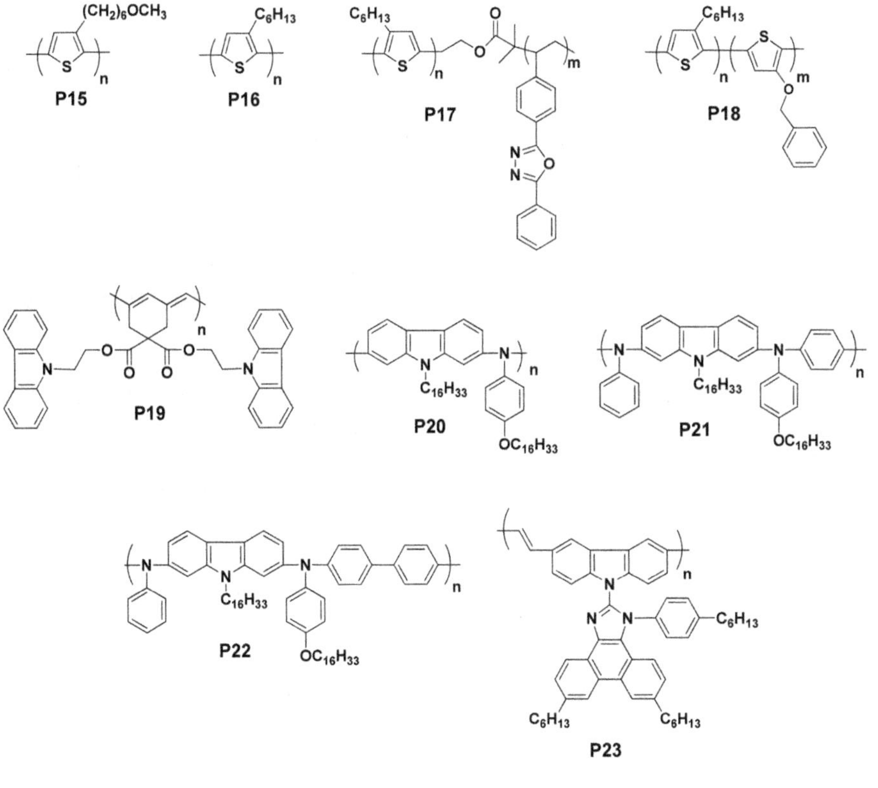

Figure 7.3 Chemical structures of thiophene/carbazole-based conjugated polymers for memory device applications.

is used to explain the asymmetric nature of the stored charges under the two bias conditions. Devices based on an oriented version of **P15** exhibited the presence of two conducting states dependent on the sweep direction of the voltage scans.[16] The origin of the ON state was explained in terms of voltage-induced tunneling of carriers, which led to a higher conductivity in the polymer layer. When the bias was swept, the charges in the quantum well from a suitable arrangement received extra energy to tunnel or hop through the chain, resulting in a high ON state device current. The current ratio between the states was as high as 10^5, due to the higher conductivity through the oriented polymers.

Joo et al. investigated poly(3-hexylthiophene) (**P16**) which reproducibly formed metal filaments by applying a forward bias for non-volatile memory applications.[17,18] The high positive voltage was believed to play an important role in ionizing the copper electrode and injecting the copper ions into the polymer layer. The polymer containing conducting π-conjugated and coordinating atom (sulfur or nitrogen) formed a strong complex with the copper ions and produced the metal filament. The formation of a conducting filament was experimentally proved by secondary ion mass spectroscopy analysis.[18]

The polymer memory performances of conjugated thiophene-based block copolymers were also reported.[19,20] Chen and coworkers prepared novel donor–acceptor rod–coil diblock copolymers of regioregular poly(3-hexylthiophene)-*block*-poly(2-phenyl-5-(4-vinylphenyl)-1,3,4-oxadiazole) (P3HT-*b*-POXD; **P17**).[19] The low-lying highest occupied molecular orbital (HOMO) energy level (−6.08 eV) of POXD was employed as a charge trap for electrical switching memory devices. The ITO/**P17**/Al memory device showed nonvolatile switching characteristics with a negative differential resistance (NDR) effect due to the charge trapped POXD block (Figure 7.4(a)). Thiophene-based all-conjugated copolymers were also explored by our group.[20] Devices with the sandwich structure of ITO/poly(3-hexylthiophene)-*block*-poly(3-phenoxymethylthiophene) (P3HT-*b*-P3PT; **P18**)/Al exhibited volatile bistable electrical switching characteristics of DRAM due to the existence of trapping sites in the P3PT domains (Figure 7.4(b)), whereas P3HT devices only showed semiconductor characteristics. This suggested the significant effect of the amorphous P3PT segments on the electrical switching behavior. The devices also exhibited a high ON/OFF current ratio of about 10^6. Both the ON and OFF states were stable under a constant voltage stress of −1.0 V and survived up to 10^8 read cycles at a constant read voltage of −1.0 V.

7.4 Carbazole-Containing Conjugated Polymers

Table 7.3 highlights the chemical structures and electric memory devices of some successful carbazole-containing conjugated polymers discovered by various groups. Ree and coworkers reported the programmable digital memory characteristics of a thin film of a conjugated polymer bearing carbazole moieties, poly[bis(9*H*-carbazole-9-ethyl) dipropargylmalonate] (**P19**).[21] The characteristics of **P19** in nanometer-scaled thin films were studied as a function of temperature and film thickness. **P19** with a thickness of 15–30 nm showed excellent unipolar DRAM behavior with a high ON/OFF ratio up to 10^8. The ON state current was dominated by ohmic conduction, and the OFF state current appeared to undergo a transition from ohmic to space charge limited conduction with a shallow-trap distribution. The ON/OFF switching

Conjugated Polymers for Memory Device Applications

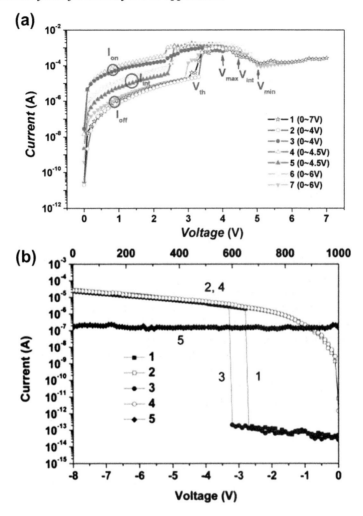

Figure 7.4 Current–voltage characteristics of the (a) **P17** and (b) **P18** memory device. Reproduced with permission from ref. 19 and 20. Copyright 2010 WILEY-VCH Verlag GmbH & Co. KGaA, Weinheim and 2011 The Royal Society of Chemistry.

Table 7.3 Summary of electrical memory properties of devices based on carbazole-containing conjugated polymers.

Device structure	Memory type	V_{th} (V)	ON/OFF ratio	References
Al/**P19**/Al	DRAM	<2.5	10^3–10^8	21
Al/**P20**/Al	DRAM	<±3	10^8–10^9	22
Al/**P21**/Al	DRAM	<±3	10^5–10^6	22
Al/**P22**/Al	DRAM	<±3	10^7–10^8	22
Al/**P23**/Al	SRAM	−1.8	>10^4	23

of the devices was mainly governed by filament formation supported by the metallic properties of the **P19** film, resulting in temperature dependence of the ON state current. Ree *et al.* also studied the electrical memory characteristics of nitrogen-linked poly(2,7-carbazole)s such as poly(9-hexadecyl-2,7-carbazole-*alt*-*N*,*N*-(4-hexadecyloxy)-aniline) (**P20**), poly(9-hexadecyl-*N*,*N*′-diphenylcarbazole-2,7-diamine-*alt*-1,3-benzene) (**P21**), and poly(9-hexadecyl-*N*,*N*′-diphenylcarbazole-2,7-diamine-*alt*-4,40-biphenyl) (**P22**). These polymers were amorphous but slightly oriented in the film plane.[22] All polymers in devices with aluminum top and bottom electrodes were found to exhibit similar DRAM behaviors without polarity. The devices were operable at low voltage (less than ±3 V) with a high ON/OFF current ratio (10^5–10^9) over the thickness range of 8–60 nm. The memory behaviors were governed by space charge limited conduction and local filament formation. These memory characteristics might originate from the electron-donating carbazole and triphenylamino units in the polymer backbones, which acted as charge-trapping sites but had weak electric polarization because of the absence of counterparts (Figure 7.5).

Chen and coworkers demonstrated a donor–acceptor conjugated poly(arylene vinylene) consisting of carbazole with pendent phenanthro[9,10-*d*]imidazole (**P23**).[23] The **P23** flexible device with the sandwich configuration of PEN/Al/**P23**/Al revealed volatile static random access memory (SRAM) characteristics. It could operate at low voltages (less than 2.5 V) with high ON/OFF current ratios (over 10^4) and exhibit excellent durability upon repeated bending tests. The charge trapping/detrapping environment and strength of charge transfer interaction in the prepared D–A conjugated polymers significantly affected the memory behavior. The carbazole donor possesses high steric hindrance with phenanthro[9,10-*d*]imidazole side chains in the polymer backbone, leading to a weak electric charge separated state. Thus, the unstable **P23** device was detrapped out of trapping sites and the separated charges easily recombined as the electrical power was turned off, resulting in volatile memory characteristics.

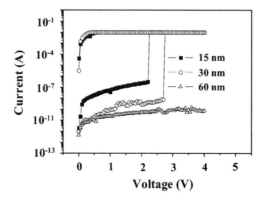

Figure 7.5 Current–voltage characteristics of the **P19** memory device with various thicknesses. Reproduced with permission from ref. 21. Copyright 2010 American Chemical Society.

7.5 Conjugated Poly(azomethine)s

Poly(azomethine)s are a family of conjugated polymers constructed from imine groups (C=N) and aromatic rings alternatively on the backbone, and exhibit good thermal stability and mechanical properties. Table 7.4 and Figure 7.6 summarize the results obtained with conjugated poly(azomethine)-based memory materials. By controlling molecular doping in conjugated poly(azomethine)s, Li and coworkers fabricated protonic acid-doped poly(azomethine)s (**P24**) that demonstrated excellent operative uniformity and multilevel stability (Figure 7.7(a)).[24] The mechanism of the resistance switching in the presented **P24** non-volatile memory devices was revealed to be through charge transfer and the electric-field induced doping/dedoping effect. Under external electric fields, electromigration of the dopants toward the imine groups of the poly(azomethine) backbone gave rise to an enhanced doping level. As a consequence, the gradually increased concentration of charge carriers led to an increase in the device's conductivity. As the field-induced doping/dedoping process of the **P24** system occurred throughout the entire thin film, very uniform cyclic operation over 700 consecutive cycles was thus expected with delocalized tuning of the effective charge carrier concentration (Figure 7.7(b)). In comparison, localized charge transfer due to intra- and/or interchain donor–acceptor interactions in undoped poly(azomethine)s was less controllable, resulting in random formation of conductive pathways and severe device-to-device variation. Li *et al.* also synthesized two donor–acceptor type poly(azomethine)s, incorporating an oxadiazole entity either acting as an electron acceptor to form donor–acceptor structured **P25** with the triphenylamine donor, or acting as a donor to form donor–acceptor structured **P26** with the 3,3′-dinitro-diphenylsulfone acceptor.[25] The variation in the role of the oxadiazole moiety in the donor–acceptor polymers leads to various resistive switching behaviors of the present poly(azomethine)s. In Pt electrode sandwich devices, the **P25**-based device showed rewriteable memory behavior with poor endurance of less than 20 cycles, while the **P26** device exhibited WORM memory characteristics. The different memory properties of the Pt electrode devices originated from the different HOMO–LUMO band gaps of the poly(azomethine)s, which influenced the degree of intra- and intermolecular charge transfer interaction between the electron donor–acceptor pairs. Due to the stronger electron

Table 7.4 Summary of electrical memory properties of devices based on conjugated poly(azomethine)s.

Device structure	Memory type	V_{th} (V)	ON/OFF ratio	References
Pt/**P24**/Pt	Flash	2.65	600	24
Pt/**P25**/Pt	Flash	−3.3	10^7	25
Pt/**P26**/Pt	WORM	−4.6	10^7	25
Pt/**P27**/Pt	Flash	1.40	>10^2	26
Pt/**P28**/Pt	Flash	1.73	>10^2	26

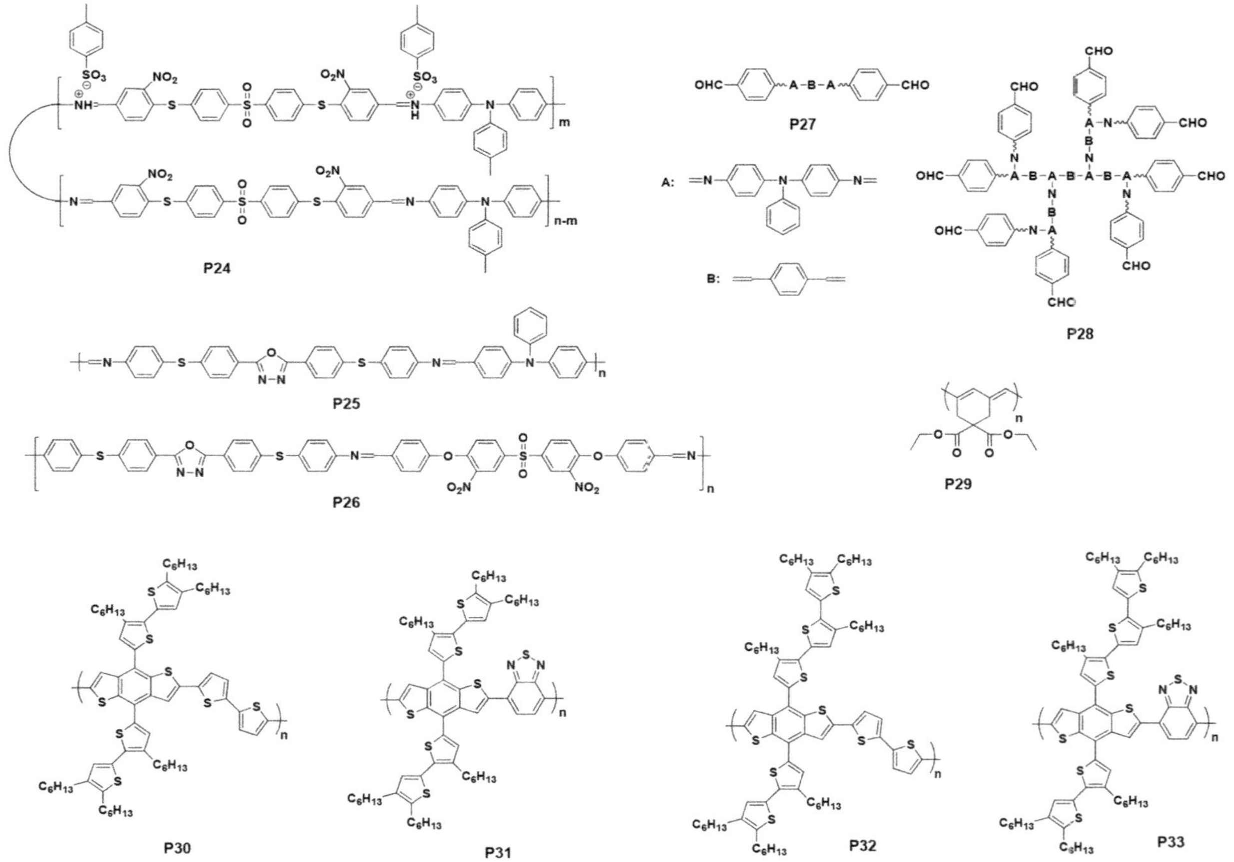

Figure 7.6 Chemical structures of polyazomethines and other intrinsic conjugated polymers for memory device applications.

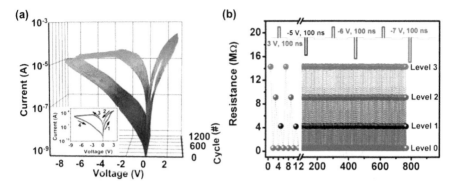

Figure 7.7 (a) Current–voltage characteristics of the **P24** memory device with various thicknesses. (b) Multilevel operation of **P24** device in pulse mode. Reproduced with permission from ref. 24. Copyright 2012 American Chemical Society.

push–pull interaction in the **P25** structure, the charge transfer effect occurred earlier and more easily in the thin film of **P26**, resulting in a lower switching threshold voltage than that of the **P26** device. Meanwhile, the moderate electron withdrawing ability of the oxadiazole acceptor made the charge transfer interaction between the triphenylamine–oxadiazole pairs reversible, and a sufficient reverse electric field initiated the back transfer of the separated charge carriers. As a result, the **P25** device recovered to its initial neutral state and the device returned to its pristine high resistance state with rewritable memory behavior. However, the strong electron withdrawing ability of the 3,3′-dinitro-diphenylsulfone moieties caused an irreversible charge transfer interaction between the oxadiazole–3,3′-dinitro-diphenylsulfone pairs, and the **P26**-based devices behaved as WORM memory.

Recently, Li and coworkers tuned the resistive switching behavior of triphenylamine-based poly(azomethine)s through structural effects. Linear (**P27**) and hyperbranched poly(azomethine)s (**P28**) with identical chemical structures but different molecular geometries and crystalline qualities were used for comparison.[26] Both **P27** and **P28** exhibited small switching voltages, ON/OFF ratios over 100, endurance capability of more than 5000 cycles and retention times exceeding 10^4 s. In comparison, the hyperbranched **P28** thin film with an isotropic architecture and semi-crystalline nature demonstrated enhanced memory behavior with more uniform distribution of the high resistance states and low resistance states since the charge transport became more efficient and stable in **P28**.

7.6 Other Intrinsic Conjugated Polymers

Table 7.5 and Figure 7.6 compare the results for optimized systems using other intrinsic conjugated polymers as resistor memory materials. Ree *et al.* reported the memory characteristics of nanoscale thin films of a

Table 7.5 Summary of electrical memory properties of devices based on other intrinsic conjugated polymers.

Device structure	Memory type	V_{th} (V)	ON/OFF ratio	References
Al(Au)/**P29**/Au(Al)	WORM/DRAM	~1.8	$10^3/10^8$	27
ITO/**P30**/Al	WORM	−2.0 or 2.5	10^3–10^5	28
ITO/**P31**/Al	WORM	−3.0 or 3.0	10^3–10^5	28
ITO/**P32**/Al	WORM	−2.0 or 2.5	10^3–10^5	28
ITO/**P33**/Al	WORM	−2.0 or 2.5	10^3–10^5	28
Ta/**P34**/Pt	Flash	−2.0	5×10^8	29

fully π-conjugated polymer, poly(diethyl dipropargylmalonate) (**P29**) in the absence of doping.[27] The good solubility of **P29** in organic solvents allowed easy processability to form nanoscale thin films through the use of conventional solution spin-, roll-, or dip-coating and subsequent drying. The memory characteristics depend on the film thickness. Films with a thickness of 30 nm were found to exhibit very stable WORM memory characteristics without polarity and an ON/OFF current ratio of 10^6, whereas films with a thickness of 62–120 nm showed excellent DRAM characteristics without polarity and an ON/OFF current ratio as high as 10^8. Both the ester units and the conjugated double bonds of the polymer backbone acted as charge trapping sites. Wang's and Li's groups prepared donor–acceptor/donor–donor conjugated polymers (**P30**–**P33**)[28] and poly(triphenylamine) (**P34**)[29] for WORM-type and flash memory, respectively, where the memory behaviors were dominated by charge transfer processes as evidenced by density functional theory (DFT) calculations.

7.7 Conjugated Polymers Containing Metal Complexes

Tremendous effort has been made to develop conjugated building blocks with metal complexes (see structures in Figure 7.8) for specific memory device applications. The performance of memory devices based on such conjugated polymers containing metal complexes is summarized in Table 7.6. A series of conjugated poly(9,9′-dialkylfluorene)s with Eu(III) complex units (**P35**–**P38**) were prepared for WORM memory applications based on ITO/polymer/Al device structures.[30–35] The polyfluorene backbone served as the electron donor and the Eu(III) complex served as the electron acceptor. The electron-accepting Eu(III) complex in the chelated form of the copolymer for electrical memory applications was produced as a solution-processable high quality film without phase separation. **P35** and **P36** with dibenzoylmethanate as ligand both showed WORM memory type behavior with a high ON/OFF current ratio up to 10^6–10^7, stable ON and OFF states with read pulse cycles up to 10^8 and excellent stability of up to 10 years at a constant stress

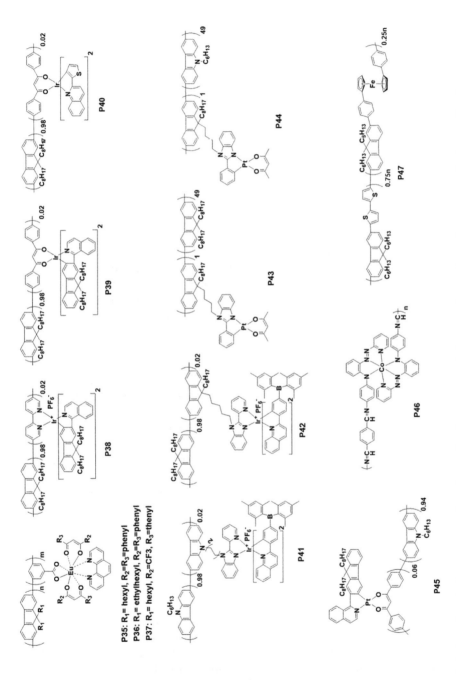

Figure 7.8 Chemical structures of conjugated polymers containing metal complexes for memory device applications.

Table 7.6 Summary of electrical memory properties of devices based on conjugated polymers containing metal complexes.

Device structure	Memory type	V_{th} (V)	ON/OFF ratio	References
ITO/**P35**/Al	WORM	3.0	10^7	30 and 31
ITO/**P36**/Al	WORM	3.0	10^6	32 and 33
ITO(or PPy)/**P37**/Al	WORM	1.0 (4.0)	80 (200)	34
ITO/PAN(+CNT)/**P37**/Al	WORM	3.8 (2.0)	10^3 (10^6)	35
ITO/**P38**/Al	Flash	−1.6	>10^5	36
ITO/**P39**/Al	Flash	−0.9	>10^5	36
ITO/**P40**/Al	Flash	−1.1	>10^5	36
ITO/**P41**/Al	Flash	−1.2	10^3	37
ITO/**P42**/Al	Flash	−1.4	10^3	37
ITO/**P43**/Al	Flash	−2.6	10^3	38
ITO/**P44**/Al	Flash	−1.6	10^3	38
ITO/**P45**/Al	Flash	−1.5	10^3	39
ITO/**P46**/LiF/Al	Flash	−1.9	10^3	40
Au/**P47**/Au	Flash	−3.0 to −5.0	10^2–10^3	42

of 1 V even with different substituted alkyl chains or composition ratios.[30–33] From the current–voltage characteristics of the device, a transition from a Schottky emission controlled current to a space charge limited current upon induction by a threshold field was demonstrated. When the electric field exceeds the energy barrier, electrons were injected into the LUMO of the Eu complexes and holes were injected into the HOMO of the fluorene moieties. The charged LUMO (radical anion) of the Eu complex and the charged HOMO (radical cation) of the fluorene moiety formed a channel for charge carriers through non-radiative intersystem transitions. The polymer became p-doped under the induction of the electric field and switched to the ON state. A sufficient degree of spin delocalization of the conjugated fluorene segments could stabilize the radical cations. The ON current of the device was stable and the high-conductivity state was maintained. However, for **P37** with 2-thenoyltrifluoro-acetone as ligand, the device performance worsened significantly with a low ON/OFF current ratio of 10^3 at a conductance switching voltage of 3.8 V.[34,35] A flexible WORM memory device based on **P37** was demonstrated with a conductive polypyrrole film as the bottom electrode and gold as the top electrode.[34] This flexible polymer memory device could meet the demand for data storage in memory devices of unique spatial construction or architecture.

Conjugated polyfluorene with cationic Ir(III) complexes (**P38**–**P42**) was selected as an active memory material that fulfils the functionality of a flash memory device.[36,37] The **P38** (polyfluorene with on-chain Ir(III) complex) device exhibited low reading, writing, and erasing voltages and a high ON/OFF current ratio of more than 10^5 (Figure 7.9(a)). Both ON and OFF states were stable under a constant voltage stress of −1.0 V and survived up to 10^8 read cycles at a read voltage of −1.0 V (Figure 7.9(b)). The flash type memory behavior was attributed to the polarized charge transfer state between the fluorene moieties (electron donors) and the cationic Ir(III) complex moieties

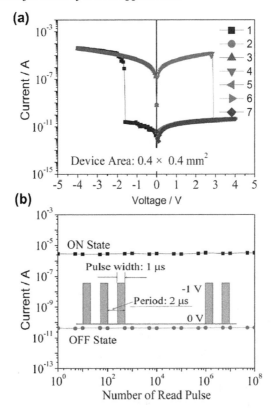

Figure 7.9 (a) Current–voltage characteristics of the **P38** memory device. (b) Effect of read cycles on ON and OFF states. Reproduced with permission from ref. 36. Copyright 2010 WILEY-VCH Verlag GmbH & Co. KGaA, Weinheim.

(electron acceptors) under an applied field. Furthermore, through the modification of the ligand structures of the Ir(III) complex units, the resulting polymers (**P39** and **P40**) also showed excellent memory behavior. Alteration of the ligands changed the threshold voltage and current of the memory device.[36] In addition, polycarbazole (**P41**) and polyfluorene (**P42**) containing Ir(III) complexes in their side chains were designed and synthesized.[37] Both polymers showed conductance switching behavior and non-volatile flash memory devices based on them were successfully realized through the formation and dissociation of space charge transfer states from the conjugated main chains to the Ir(III) complex side chains under different voltages. The devices exhibited low reading, writing, and erasing voltages and high ON/OFF current ratios. Both ON and OFF states are stable for up to 10^7 read cycles at a read voltage of 1.0 V. The threshold voltage from the OFF to ON state of the **P41** device was obviously lower than that of the **P42** device because of the low energy barrier between the work function of the ITO anode and the HOMO level of the **P41**. Different from conjugated polymers containing

Ir(III) complexes in the main chain, the acceptor Ir(III) complexes and donor polymer main chains were linked in a non-conjugated way, which avoided the electronic interaction between the donor and acceptor in the ground state. Thus, this class of polymers was more suitable for rational molecular design by separate optimization of the donor and acceptor.

Conjugated polymers containing Pt(II) complexes have also been realized for applications in resistor memory devices.[38,39] Conjugated polyfluorene and polycarbazole with Pt(II) complexes in the side chain (**P43** and **P44**, respectively) exhibited excellent flash memory behaviors in devices of structure ITO/polymer/Al with a high ON/OFF current ratio (>10^3), excellent stability and a high number of read cycles (10^7).[38] Through the study of the redox properties and theoretical calculation results for **P43** and **P44**, the memory mechanism could be attributed to the formation and dissociation of a charge transfer state induced by negative and positive voltages, respectively. The application of an electric field above the threshold value led to charge transfer from the polymer main chain as the charge donor to the Pt(II) complex units on the side chain as the charge acceptor, which was responsible for the ON state. Due to the stability of the charge transfer complex, the ON state still remained even after the driving power was turned off. A reverse bias voltage applied to the polymer, however, dissociated the charge transfer state. Thus, the device returned to the original OFF state. In addition, the main chain structures had significant influence on the threshold voltages of the devices. The threshold voltages of the devices using **P44** as active material were lower than that of the device based on **P43** because of the lower oxidation potential of polycarbazole (0.43 V) than polyfluorene (0.96 V). This resulted in a lower energy barrier between the work function of the ITO anode and the HOMO level of **P44** compared to that of **P43** and easier charge transfer for **P44** than **P43**. In addition, a conjugated polycarbazole containing a Pt(II) complex on the main chain (**P45**) was also investigated for application in a memory device.[39] The device with the sandwich structure of ITO/polymer/Al exhibited the functionality of flash memory with a low threshold voltage of −1.5 V, excellent stability, and a high ON/OFF current ratio. Compared to the side chain Pt(II) complex in **P43** and **P44**, the memory device based on **P45** exhibited a little lower threshold voltage, which indicated that the memory device is easier to write.

Choi *et al.* designed a conjugated polymer containing ferrocene (Fc) based on a poly(9,9″-dihexylfluorine-*co*-bithiophene) main chain backbone (**P46**).[40] Initially the device was in the OFF state and then the current abruptly increased at −1.9 V, implying a transition from the OFF state to the ON state, presumably because of the oxidation of the ferrocene groups. This observation could be viewed as ferrocene acting like a voltage-dependent *in-situ* dopant resulting in the enhancement of the conductivity of the polymer film. This ON state persisted until the device was switched back to the OFF state at a positive bias of 1.4 V. Hence, the redox process of ferrocene was crucial for the memory behavior. In addition, the device exhibited excellent memory performance with low driving voltage and a high ON/OFF ratio. The low driving voltage arose from the low redox potential of the ferrocene moiety. Due to the unique redox stability of ferrocene, in which ferrocenium maintained its oxidized state, the memory device retained the ON/OFF states for more than 7 h. Higuchi's group[41] tuned the non-volatile bipolar switching behavior *via* Co(III)-containing conjugated polymers with an extended azo aromatic ligand (**P47**) in lateral and sandwich device structures. Redox switching of an organic–metallic hybrid polymer generated bistable states with an ON/OFF ratio of 10^3 that supported random flip-flops for several hours. The results of cyclic voltammograms and the direct switching response of the ligand-based device together suggested that the switching phenomenon shown by both polymers is due to the reduction of Co(III) to Co(II). All the above results suggested that excellent polymer memory materials can be made by introducing metal ions into conjugated polymers.

7.8 Conjugated Polyelectrolytes

Recently, single layers of conjugated polyelectrolytes (see structures in Figure 7.10) have also been investigated for memory switching behavior. In this section, typical examples of conjugated polyelectrolytes are given and their performances in memory devices are summarized in Table 7.7. Lee *et al.* used fluorene-based polyelectrolytes as active memory materials for fabricating rewritable memory devices and a filamentary switching mechanism was found to be responsible for the observed memory behavior.[42–48] The structural constituents of the side chain in a series of fluorene-based polyelectrolytes had important effects on the characteristics of organic resistance memory devices.[42] Conjugated polyfluorene electrolytes with a hexyl side chain (**P48**) and side chains containing 2, 4, and 6 oxygen atoms (**P49**, **P50** and **P51**) were prepared and all of these conjugated polymers exhibited reversible, non-volatile memory characteristics and excellent device-to-device uniformity. Conjugated polyelectrolytes with ethylene oxide side chains (**P50** and **P51**) showed more stable memory characteristics (such as higher ON/OFF ratios, lower threshold voltage values and longer retention times than those with alkyl side chains (**P48**). It was found that the longer the ethylene oxide side chain on the fluorene-based polyelectrolyte, the better the memory characteristics exhibited. The excellent switching properties of **P50** and **P51** were attributed to the creation and rupture of redox-controlled Ag metallic bridges, thus forming highly localized current pathways and a lower density of traps than **P48** and **P49**. The effect of metal ions on the switching performance of polyfluorene electrolytes was also compared.[43] Although the basic memory behavior of **P52** (with Ca^{2+}) and **P53** (with Na^+) was not significantly affected by the metal ions, both of them show area dependence in the high resistance state, implying that metal ions assist localized current flow. The threshold voltage of **P53** was lower than that of **P52** because of more efficient Na^+ movement in **P53**. In addition, the response time of **P52** and **P53** was faster than that of **P50**, suggesting possible modulation of the memory performance by the addition of metal ions in the polymer layer.

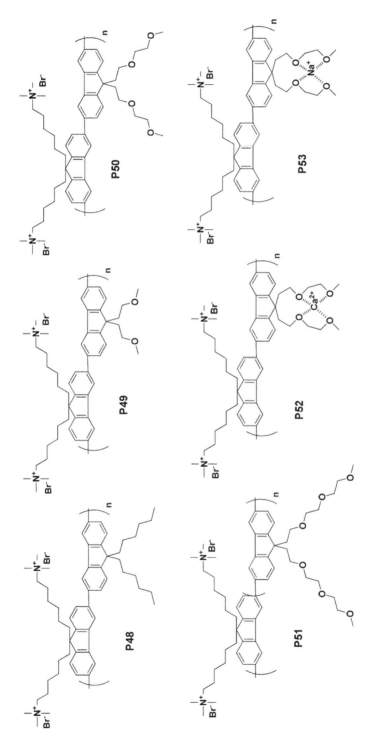

Figure 7.10 Chemical structures of conjugated polyelectrolytes for memory device applications.

Table 7.7 Summary of electrical memory properties of devices based on conjugated polyelectrolytes.

Device structure	Memory type	V_{th} (V)	ON/OFF ratio	References
Si/**P48**/Ag	Flash	4.7	10^2	43
Si/**P49**/Ag	Flash	4.0	10^3	43
Si/**P50**/Ag	Flash	3.4	10^5	43
Si/**P51**/Ag	Flash	3.5	10^5	43
Si/**P52**/Ag	Flash	3.6	10^4	44
Si/**P53**/Ag	Flash	3.0	10^4	44

7.9 Conclusion

This chapter summarizes the advancements in conjugated polymer-based memory materials in terms of the molecular structures, device structures and switching mechanism for electrical memory properties. The performances of a great variety of conjugated polymers for resistor memory devices were analysed and discussed. The characteristics of the conjugated polymers could be controlled fairly well through careful design and led to materials with desirable electrical memory properties although various switching mechanisms, including charge transfer, charge trapping and filament formation, were proposed. The ability to fine-tune the electrical conditions *via* synthetic design gives conjugated polymer memory materials long-term stable and reproducible memory performance as well as allowing for optimization of future device scaling.

References

1. B. Lei, W. L. Kwan, Y. Shao and Y. Yang, *Org. Electron.*, 2009, **10**, 1048.
2. W. L. Kwan, B. Lei, Y. Shao and Y. Yang, *Curr. Appl. Phys.*, 2010, **10**, E50.
3. C. W. Joo, S. O. Jeon, K. S. Yook and J. Y. Lee, *Synth. Met.*, 2009, **159**, 1809.
4. H. L. Gomes, A. R. V. Benvenho, D. M. de Leeuw, M. Colle, P. Stallinga, F. Verbakel and D. M. Taylor, *Org. Electron.*, 2008, **9**, 119.
5. P. R. F. Rocha, A. Kiazadeh, D. M. De Leeuw, S. C. J. Meskers, F. Verbakel, D. M. Taylor and H. L. Gomes, *J. Appl. Phys.*, 2013, **113**, 134504.
6. Q.-D. Ling, Y. Song, S.-L. Lim, E. Y.-H. Teo, Y.-P. Tan, C. Zhu, D. S. H. Chan, D.-L. Kwong, E.-T. Kang and K.-G. Neoh, *Angew. Chem., Int. Ed.*, 2006, **45**, 2947.
7. G. Liu, Q.-D. Ling, E.-T. Kang, K.-G. Neoh, D.-J. Liaw, F.-C. Chang, C.-X. Zhu and D. S.-H. Chan, *J. Appl. Phys.*, 2007, **102**, 024502.
8. Q.-D. Ling, E.-T. Kang, K.-G. Neoh, Y. Chen, X.-D. Zhuang, C. Zhu and D. S. H. Chan, *Appl. Phys. Lett.*, 2008, **92**, 143302.
9. G. Liu, D.-J. Liaw, W.-Y. Lee, Q.-D. Ling, C.-X. Zhu, D. S.-H. Chan, E.-T. Kang and K.-G. Neoh, *Philos. Trans. R. Soc., A*, 2009, **367**, 4203.
10. X.-D. Zhuang, Y. Chen, B.-X. Li, D.-G. Ma, B. Zhang and Y. Li, *Chem. Mater.*, 2010, **22**, 4455.

11. X. Xu, L. Li, B. Liu and Y. Zou, *Appl. Phys. Lett.*, 2011, **98**, 063303.
12. H.-C. Wu, A.-D. Yu, W.-Y. Lee, C.-L. Liu and W.-C. Chen, *Chem. Commun.*, 2012, **48**, 9135.
13. C. Wang, G. Liu, Y. Chen, S. Liu, Q. Chen, R. Li and B. Zhang, *ChemPlusChem*, 2014, **79**, 1263.
14. Y.-G. Ko, D. M. Kim, K. Kim, S. Jung, D. Wi, T. Michinobu and M. Ree, *ACS Appl. Mater. Interfaces*, 2014, **6**, 8415.
15. H. S. Majumdar, A. Bandyopadhyay, A. Bolognesi and A. J. Pal, *J. Appl. Phys.*, 2002, **91**, 2433.
16. H. S. Majumdar, A. Bolognesi and A. J. Pal, *Synth. Met.*, 2004, **140**, 203.
17. W.-J. Joo, T.-L. Choi, J. Lee, S. K. Lee, M.-S. Jung, N. Kim and J. M. Kim, *J. Phys. Chem. B*, 2006, **110**, 23812.
18. W.-J. Joo, T.-L. Choi, K.-H. Lee and Y. Chung, *J. Phys. Chem. B*, 2007, **111**, 7756.
19. Y.-K. Fang, C.-L. Liu, C. Li, C.-J. Lin, R. Mezzenga and W.-C. Chen, *Adv. Funct. Mater.*, 2010, **20**, 3012.
20. Y.-C. Lai, K. Ohshimizu, W.-Y. Lee, J.-C. Hsu, T. Higashihara, M. Ueda and W.-C. Chen, *J. Mater. Chem.*, 2011, **21**, 14502.
21. S. Park, T. J. Lee, D. M. Kim, J. C. Kim, K. Kim, W. Kwon, Y.-G. Ko, H. Choi, T. Chang and M. Ree, *J. Phys. Chem. B*, 2010, **114**, 10294.
22. S. G. Hahm, T. J. Lee, D. M. Kim, W. Kwon, Y.-G. Ko, T. Michinobu and M. Ree, *J. Phys. Chem. C*, 2011, **115**, 21954.
23. H.-C. Wu, C.-L. Liu and W.-C. Chen, *Polym. Chem.*, 2013, **4**, 5261.
24. B. Hu, X. Zhu, X. Chen, L. Pan, S. Peng, Y. Wu, J. Shang, G. Liu, Q. Yan and R.-W. Li, *J. Am. Chem. Soc.*, 2012, **134**, 17408.
25. L. Pan, B. Hu, X. Zhu, X. Chen, J. Shang, H. Tan, W. Xue, Y. Zhu, G. Liu and R.-W. Li, *J. Mater. Chem. C*, 2013, **1**, 4556.
26. W. Zhang, C. Wang, G. Liu, J. Wang, Y. Chen and R.-W. Li, *Chem. Commun.*, 2014, **50**, 11496.
27. T. J. Lee, S. Park, S. G. Hahm, D. M. Kim, K. Kim, J. Kim, W. Kwon, Y. Kim, T. Chang and M. Ree, *J. Phys. Chem. C*, 2009, **113**, 3855.
28. H.-J. Yen, H. Tsai, C.-Y. Kuo, W. Nie, A. D. Mohite, G. Gupta, J. Wang, J.-H. Wu, G.-S. Liou and H.-L. Wang, *J. Mater. Chem. C*, 2014, **2**, 4374.
29. W. Zhang, C. Wang, G. Liu, X. Zhu, X. Chen, L. Pan, H. Tan, W. Xue, Z. Ji, J. Wang, Y. Chen and R.-W. Li, *Chem. Commun.*, 2014, **50**, 11856.
30. Q. D. Ling, Y. Song, E. Y. H. Teo, S. L. Lim, C. X. Zhu, D. S. H. Chan, D. L. Kwong, E. T. Kang and K. G. Neoh, *Electrochem. Solid-State Lett.*, 2006, **9**, G268.
31. Y. P. Tan, Y. Song, E. Y. H. Teo, Q. D. Ling, S. L. Lim, P. G. Q. Lo, D. S. H. Chan, E. T. Kang and C. Zhu, *J. Electrochem. Soc.*, 2008, **155**, H17.
32. Y. Song, Y. P. Tan, E. Y. H. Teo, C. Zhu, D. S. H. Chan, Q. D. Ling, K. G. Neoh and E. T. Kang, *J. Appl. Phys.*, 2006, **100**, 084508.
33. Y. Song, Q. D. Ling, C. Zhu, E. T. Kang, D. S. H. Chan, Y. H. Wang and D. L. Kwong, *IEEE Electron Device Lett.*, 2006, **27**, 154.
34. L. Li, Q.-D. Ling, S.-L. Lim, Y.-P. Tan, C. Zhu, D. S. H. Chan, E.-T. Kang and K.-G. Neoh, *Org. Electron.*, 2007, **8**, 401.

35. L. Li, Q.-D. Ling, C. Zhu, D. S. H. Chan, E.-T. Kang and K.-G. Neoh, *J. Electrochem. Soc.*, 2008, **155**, H205.
36. S.-J. Liu, Z.-H. Lin, Q. Zhao, Y. Ma, H.-F. Shi, M.-D. Yi, Q.-D. Ling, Q.-L. Fan, C.-X. Zhu, E.-T. Kang and W. Huang, *Adv. Funct. Mater.*, 2011, **21**, 979.
37. S.-J. Liu, W.-P. Lin, M.-D. Yi, W.-J. Xu, C. Tang, Q. Zhao, S.-H. Ye, X.-M. Liu and W. Huang, *J. Mater. Chem.*, 2012, **22**, 22964.
38. P. Wang, S.-J. Liu, Z.-H. Lin, X.-C. Dong, Q. Zhao, W.-P. Lin, M.-D. Yi, S.-H. Ye, C.-X. Zhu and W. Huang, *J. Mater. Chem.*, 2012, **22**, 9576.
39. W. Lin, H. Sun, S. Liu, H. Yang, S. Ye, W. Xu, Q. Zhao, X. Liu and W. Huang, *Macromol. Chem. Phys.*, 2012, **213**, 2472.
40. T.-L. Choi, K.-H. Lee, W.-J. Joo, S. Lee, T.-W. Lee and M. Y. Chae, *J. Am. Chem. Soc.*, 2007, **129**, 9842.
41. A. Bandyopadhyay, S. Sahu and M. Higuchi, *J. Am. Chem. Soc.*, 2011, **133**, 1168.
42. S.-H. Lee, S.-H. Oh, Y. Ji, J. Kim, R. Kang, D. Khim, S. Lee, J.-S. Yeo, N. Lu, M. J. Kim, H. C. Ko, T.-W. Kim, Y.-Y. Noh and D.-Y. Kim, *Org. Electron.*, 2014, **15**, 1290.
43. T.-W. Kim, S.-H. Oh, J. Lee, H. Choi, G. Wang, J. Park, D.-Y. Kim, H. Hwang and T. Lee, *Org. Electron.*, 2010, **11**, 109.
44. T.-W. Kim, H. Choi, S.-H. Oh, M. Jo, G. Wang, B. Cho, D.-Y. Kim, H. Hwang and T. Lee, *Nanotechnology*, 2009, **20**, 025201.
45. T.-W. Kim, K. Lee, S.-H. Oh, G. Wang, D.-Y. Kim, G.-Y. Jung and T. Lee, *Nanotechnology*, 2008, **19**, 405201.
46. T.-W. Kim, S.-H. Oh, H. Choi, G. Wang, H. Hwang, D.-Y. Kim and T. Lee, *Appl. Phys. Lett.*, 2008, **92**, 253308.
47. T.-W. Kim, H. Choi, S.-H. Oh, G. Wang, D.-Y. Kim, H. Hwang and T. Lee, *Adv. Mater.*, 2009, **21**, 2497.
48. B. Cho, J.-M. Yun, S. Song, Y. Ji, D.-Y. Kim and T. Lee, *Adv. Funct. Mater.*, 2011, **21**, 3976.

CHAPTER 8

Non-Volatile Memory Properties of Donor–Acceptor Block Copolymers

NAM-GOO KANG[a], MYUNG-JIN KIM[b], AND JAE-SUK LEE*[b]

[a]Department of Chemistry, University of Tennessee, Knoxville, Tennessee 37909, United States; [b]School of Materials Science and Engineering Gwangju Institute of Science and Technology (GIST), 123 Chemdangwagi-ro, Gwangju 500-712, Korea
*E-mail: jslee@gist.ac.kr

8.1 Introduction

Due to the rapid development of information-based technology and industry and the increasing demand for quantitative and qualitative information technology, high-speed memory storage devices with large capacities are needed. To preserve information, volatile memory requires an active external electrical source. In contrast, non-volatile memory preserves information even if there is no external electrical source, and it is categorized as a new memory device that is able to write and erase information using electric signals.[1]

Basically, NAND flash memory based on silicon (Si) technologies still dominates the storage media market. However, the demand for scalability to nanosized devices and advanced performance from information devices is continuously increasing. When silicon-based non-volatile memory is used in various applications to write and erase information, the operating voltage ranges from 1.5–5.0 V to 17–20 V and should therefore be elevated. Additionally, there are theoretical and physical limits to downscaling Si-based technologies. If the gate length of the device is reduced to less than 100 nm, many issues appear, such as cross talk between devices, reduced capacitance voltage of floating-gates, dielectric breakdown and leakage currents *via* tunneling oxidation. This limitation of information storage has motivated the development of innovative memory technologies principally using various bistable materials with intrinsic characteristics such as ferroelectricity, magnetic polarity, phase, conformation, and conductivity.[2–6] These new technologies are categorized as ferroelectric random access memory,[2] magnetoresistive random access memory,[3,4] phase-change random access memory,[5] and resistance random access memory (RRAM).[6,7]

In particular, the Yang group[8–13] developed RRAM using organic materials, and this technology is emerging as a promising candidate for next-generation non-volatile memory devices. The structures of two-terminal crossbar memory devices were designed by introducing organic materials between two conductive electrodes. For organic non-volatile memory devices, various organic materials including small molecules and polymers are widely used because they have many advantages, such as low fabrication costs, printability, and simple device structures. Non-volatile memory has two intrinsic characteristics: (1) the resistance state can be read nondestructively and (2) electrical power is not needed for the maintenance of a given resistance state.[13–15] Additionally, for high performance memory, a fast switching speed and a high ON/OFF ratio are required.[9,16] Although it is difficult to identify the mechanisms of memory operation, the development of advanced measurement techniques and analytical tools has accelerated the detailed study of the mechanisms of charge conduction and has contributed to the understanding of the switching phenomena. In addition, to reveal the origins of the switching mechanisms, structural and electrical optimization have been successfully utilized to improve performance and advance the realization of more practical information storage devices.[16–29]

8.1.1 Categories of Electronic Memory

There are two primary categories of electronic memory depending on its volatility: (1) volatile memory, which immediately loses its data when the system turns off, requires constant electrical power to preserve the stored information. Dynamic random access memory (DRAM) stores each bit of data in a separate capacitor. If the electric power turns off, DRAM loses its data, and it can therefore be considered a type of volatile memory. In general, DRAM is utilized widely as the main memory in computers. Some polymer memory devices, as introduced in this chapter, show volatile memory characteristics.[15] (2) Non-volatile memory, which can store information without electrical power, can be classified as write-once-read-many-times (WORM) memory, hybrid, non-volatile and rewritable (flash) memory based on the representative polymer memory. Among the non-volatile memory types, WORM memory retains data permanently, can be read repeatedly and contains data that is impossible to change. WORM memory includes conventional CDR, DVD ± R, electronic labels, and RFID tags. Flash memory is another type of non-volatile memory that can repeatedly be written, read, and erased and retain its stored state. Flash memory is commonly used in PDAs, mobile PCs, video/audio players and digital cameras.[15]

8.1.2 Brief Description of Organic Memory Materials and Devices

Many organic materials, such as small molecules, polymers, and composites, demonstrate bistable switching behavior when used as the active organic layer sandwiched between bottom and top electrodes by changing the resistance in two-terminal device structures.[10,13,30–32] Extensive research has addressed the switching phenomena based on the characteristics of organic memory devices. Using theoretical simulations and experimental results on memory performances, several solid-switching mechanisms have been well described by many research groups. Detailed summaries of the organic materials, device structures, and switching behaviors of organic resistive memory devices have been provided by Scott et al.[14] and Ling et al.[15] Various organic materials, including small semiconducting organic molecules,[33,34] polymers,[13,35–38] and composites of small molecules or polymers containing nanoparticles (NPs),[20,25,39,40] show different resistances when the voltage is applied gradually[10,20] and show conductance switching as well.

Films of organic small molecules are fabricated on memory device substrates by thermal deposition and electrostatic self-assembly.[33,41] In these cases, some defects, which form in self-assembled monolayer junction devices, reduce the reliability and reproducibility of the electrical properties in molecular junctions sandwiched between electrodes. To prove the original switching mechanisms, many research groups have reported conductance and switching phenomena,[33,42–45] but several of the proposed switching mechanisms have not yet been supported. Therefore, the study of switching mechanisms is a hot issue, but it seems dubious that memory devices fabricated with small molecules will result in useful technologies in the future. In contrast, films prepared with various polymeric materials using solution fabrication methods, such as spin casting, roll-to-roll processing, and ink-jet printing, showed uniform conditions and high quality. These polymer-coated films with single-polymer-layer structures are more appropriate and attractive for realistic memory applications because of their simple device fabrication processes and their solution processability. Small molecules and polymers in composite materials, which are blended systems containing organic or inorganic nanoparticles (NPs), are generally employed as the active materials in resistive memory devices to induce resistive switching behaviors.[20,25,46–48] Conductance switching behavior has been demonstrated in memory devices based on composite materials containing core–shell hybrid NPs.[49] For example, organic memory devices fabricated with cadmium selenide (CdSe)/zinc sulfide (ZnS) NPs embedded in conducting polymers demonstrated bistable behavior based on the trapping, storage, and emission of charges in the electronic states of the CdSe NPs.[49] In these devices, changes in resistance were triggered by introducing NPs into the polymers.[20] Accordingly, in these blended systems, the concentration and aggregation of the NPs are the most important factors, and these two factors should be controlled during solution fabrication.[20,30,45,50]

The four representative structures of two-terminal organic resistive memory devices are classified based on the configuration of the active material between the electrodes: (I) a single organic layer structure based on one type of small molecule and/or polymer only, (II) a bilayer organic structure containing two different types of small molecules and/or polymers, (III) a nanowire- or nanorod-defused structure in the middle of an organic layer, and (IV) the blending of organic structures composed of randomly distributed NPs in a polymer matrix, as shown in Figure 8.1a. The volatile and non-volatile switching behaviors which are the general electrical switching characteristics of memory devices according to their abilities to retain are shown in Figure 8.1b.

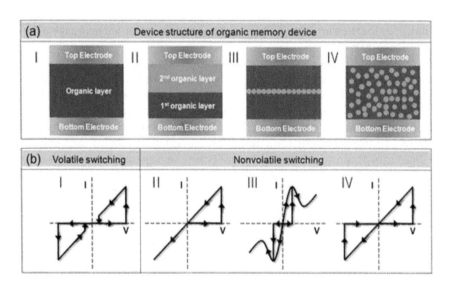

Figure 8.1 (a) Typical structures of organic memory devices; (I) a single-layer device without NPs, (II) a bilayer structure containing two types of pure polymers, (III) a structure in which the nano-traps are located in the middle of the organic layer, and (IV) spin-cast polymer–NP blends in which the nano-traps are randomly distributed throughout the entire thickness of the host matrix. (b) Typical switching characteristics of organic resistive memory devices; (I) I–V curve of DRAM-like volatile switching behavior, (II) I–V curve of WORM switching behavior, (III) I–V curve of unipolar switching, and (IV) I–V curve of bipolar switching behavior. Reproduced with permission from ref. 22, © 2011 WILEY-VCH Verlag GmbH & Co. KGaA, Weinheim.

For use as DRAM, volatile switching shows periodic refreshing, which is the loss of stored information.[51] In contrast, non-volatile organic switching memory is programmed electrically and can store data for extended periods of time. These characteristics are demonstrated in conventional flash memory. Non-volatile switching behaviors are categorized into three types based on current–voltage (I–V) curves: write-once-read-many-times (WORM),[52,53] unipolar,[54] and bipolar switching memory[55] (II, III, and IV in Figure 8.1b). WORM memory devices[52,53] demonstrate electrically irreversible switching characteristics, and the original state is never recovered. Unipolar and bipolar memory systems show electrically reversible switching characteristics. Unipolar memory devices work at the same voltage polarity for writing and erasing, but bipolar memory devices use different voltage polarities.[53]

8.1.3 Block Copolymers and Self-Assembled Nanostructures

Block copolymers (BCs) consist of two or more chemically different homopolymers as each block. These blocks are linked by covalent bonds to form more complex macromolecules, such as linear di-, tri-, or multiblock copolymers, multi-arm, star-shape, graft, and multigraft copolymers.[56] Block copolymers with different block segments exhibit different compatibilities and thermodynamic properties and form a variety of nanostructures *via* self-assembly at the molecular level.

A variety of self-assembled morphologies and phase behaviors have been extensively studied both experimentally and theoretically. In self-assembly processes using block copolymers, the relative parameters of the block segments in the block copolymers, such as the solvent characteristics, temperature, pH, *etc.*, dominate the aggregation behavior in solution as well as in the bulk. In addition, the annealing temperature has a very significant effect on the final morphology of the block copolymer aggregate. This self-assembly process of block copolymers can produce spherical, lamellar, cylindrical, and gyroid structures with periodic nanodomains. The self-assembly of block copolymers into nanostructures with novel morphologies and properties has received significant attention as a new approach for materials science, chemical synthesis, nanofabrication, biology, electrochemistry, and organic electronics.[50] For most theoretical and experimental studies on self-assembly, coil–coil block copolymers are utilized extensively; rod–coil block copolymers have also attracted significant attention because they provide an attractive strategy for the classification of a variety of highly functional rod-like polymers, such as conducting polymers with rigid π-conjugated backbones.[57] Solvatochromic behaviors have been demonstrated through π–π stacking and planarization of the conjugated segments in rod–coil block copolymers induced by aggregation.[58]

Coil–coil and coil–rod block copolymers containing electron-donating and -accepting polymer segments, such as electron-donating polymers, including polycarbazole, polyfluorene, polytriphenylamine, and polythiophene, and electron accepting polymers, including polyoxadiazole, polypyridine, and polytriazole, are important in the field of organic electronics because of their valuable characteristics, such as their simple processability, low cost, possible large area fabrication, and flexibility. In organic electronics, the high performances and efficiencies of these devices, including organic light emitting devices, photovoltaic cells, and organic thin field transistors, are attributed to charge separation and transport in devices based on self-assembled nanostructures of block copolymers. These types of block copolymers containing electron donors and acceptors with highly ordered nanostructures could have an enhanced interfacial area for charge separation and an efficient pathway for charge transport.[57,59]

For the successful design and synthesis of well-defined block copolymers and the formation of their different self-assembled nanostructures, a synthetic method to precisely control the molecular weight and distribution is required. Therefore, the living radical and anionic polymerization techniques among living polymerizations have been predominantly used for the preparation of these block copolymers. Accordingly, using these techniques, block copolymers consisting of electron-donating and accepting units based on heteroatoms of nitrogen, oxygen, sulfur, *etc.* have been successfully synthesized. However, careful attention must be paid to prevent complicated side reactions of crosslinking and branching caused by initiators, living polymer ends, and heteroatom-containing units.[60,61]

8.2 Non-Volatile Memory Based on Well-Defined Polymer Structures

Typical memory devices show biswitching behaviors with high and low conductance responses in current–voltage electric field using simple sandwich structures that consist of active polymers between two electrodes composed of the same or different metals. In general, carbazole-, fluorine-, thiophene-, and triphenylamine-containing polymers are representative electron-donating materials because of their electron rich structures. Therefore, they have been introduced as hole transfer layers in organic electronic devices. In contrast, oxadiazole-, triazole-, tetrazole-, metal complex-, and imide-containing polymers are used widely as electron transfer layers because they are electron-accepting materials due to their electron-poor molecular structures.[61,62] Using their basic electron-donating and accepting properties, many research groups have reported different types of non-volatile memory with biswitching behavior.

8.2.1 Rod–Coil block copolymers

In general, rod–coil block copolymers are synthesized *via* two steps. Rigid polymers such as poly(3-hexylthiophene) and poly[2,7-(9,9-dihexylfluorene)], as rod blocks in rod–coil block copolymers, are synthesized first and end-capped with functional groups as initiating sites. Next, flexible polymers such as poly(2-phenyl-5-(4-vinylphenyl)-1,3,4-oxadiazole) (POXD), poly(2-vinylpyridine) (P2VP), polystyrene (PS), poly(9-(4-vinylbenzyl)-9H-carbazole) (P(VCz)), as coil blocks, are synthesized *via* living polymerization such as living anionic or radical polymerization with initiation sites at the chain ends of rigid polymers. In particular, electron-donating polymers were used as rods and electron-accepting or insulating materials as coils.[63–65] Among the rod–coil block copolymers, donor–acceptor rod–coil block copolymers have been introduced into memory devices directly. On the other hand, donor–insulator rod–coil block copolymers were utilized with additives such as [6,6]-phenyl-C_{61}-butyric acid methyl ester (PCBM),[64] and graphene oxide (GO).[65] Regarding memory devices based on rod–coil block copolymers, there are some reports in this section.

Chen *et al.* synthesized new rod–coil block copolymers of poly(3-hexylthiophene) (P3HT)-*b*-poly(2-phenyl-5-(4-vinylphenyl)-1,3,4-oxadiazole) (POXD) (Figure 8.2a).[63] The effects of the block ratios of P3HT (electron donor) to

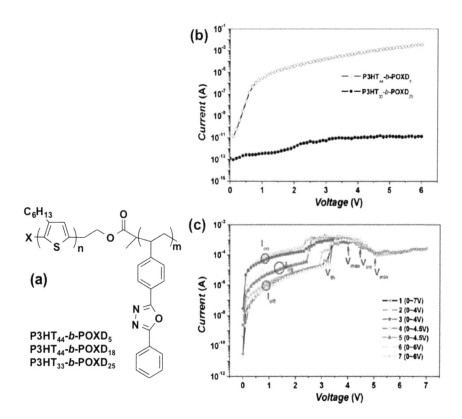

Figure 8.2 (a) Chemical structure of poly(3-hexylthiophene)-*b*-poly(2-phenyl-5-(4-vinylphenyl)-1,3,4-oxadiazole), (P3HT$_x$-POXD$_y$). *I*–*V* characteristics of (b) P3HT$_{44}$-*b*-POXD$_5$, P3HT$_{33}$-*b*-POXD$_{25}$, and (c) P3HT$_{44}$-*b*-POXD$_{18}$ memory devices. Reproduced with permission from ref. 63, © 2010 WILEY-VCH Verlag GmbH & Co. KGaA, Weinheim.

POXD pendant (electron acceptor) blocks on the memory device characteristics were explored. Typical I–V curves of ITO/polymer/Al sandwich devices with P3HT$_{44}$-b-POXD$_5$, P3HT$_{33}$-b-POXD$_{25}$ (Figure 8.2b), and P3HT$_{44}$-b-POXD$_{18}$ (Figure 8.2c) were observed. The device based on P3HT$_{44}$-b-POXD$_5$ demonstrated hole transport diode behavior due to efficient P3HT interchain hopping via a conductive percolation pathway without any available trapped POXD blocks in the active layer (Figure 8.2b). P3HT$_{44}$-b-POXD$_{18}$ exhibited a similar non-volatile nature with NDR behavior in an ITO/polymer/Al sandwich device (Figure 8.2c). The memory devices could be repeatedly written, read and erased with an ON/OFF current ratio of approximately 10^2. Additionally, under a constant stress of 0.5 V, no obvious degradation in current was observed for the ON, intermediate, and OFF states for at least 10^4 s during the readout test. The ON- and OFF-currents were maintained over 200 cycles with good electrical switching reliability. The electrical switching between the top and bottom electrodes depends on the amount of trapped POXD blocks. In the case of the P3HT$_{44}$-b-POXD$_{18}$ device, holes are injected into the polymer films and trapped by the POXD blocks, suggesting that the switching behavior triggered by the electric field can be determined by structural blocks with electric functionality.[63]

Conjugated rod–coil poly[2,7-(9,9-dihexylfluorene)]-b-poly(2-vinylpyridine) diblock copolymers (PF-b-P2VP) were synthesized by combining typical coupling reactions and living anionic polymerization (Figure 8.3a). PF$_{10}$-b-P2VP$_{37}$ and PF$_{10}$-b-P2VP$_{68}$-based devices exhibited volatile static random access memory (SRAM) characteristics with ON/OFF current ratios of up to 1×10^7, which were explained by the trapping/back transferring of charge carriers (Figure 8.3b and c). PF$_{10}$-b-P2VP$_{68}$ had a longer holding time in the ON state than PF$_{10}$-b-P2VP$_{37}$ because of the delayed back transfer of trapping carriers originally from the longer P2VP blocks. Hybrid composite [6,6]-phenyl-C$_{61}$-butyric acid methyl ester (PCBM)/PF-b-P2VP devices were investigated as a function of PCBM composition in the block copolymer matrix. As the PCBM composition was increased, the device behavior changed from non-volatile WORM memory to conductive. The charge transfer interactions in the block copolymer composites are responsible for the WORM memory effect. The PCBM molecules (electron donors) form corresponding negatively charged counterions via noncovalent interaction and by exchanging electrons with electron donors, leading to a stable charge transfer state, which could retain the ON state for a long time, as observed in the WORM memory device, and could prevent recombination even under a reverse bias.[64]

Polymer–graphene composites have been applied in memory devices as the active materials in Au/PMrGO nanohybrid/Al memory device cells (Figure 8.4a) via the modification of graphene oxide (GO) with a conjugated block copolymer consisting of rod–coil poly-(paraphenylene) (PPP) and polystyrene (PS), PS-b-PPP (5k/1k), as a polymeric dispersant. The conjugated PPP block is expected to strongly adhere to the GO surface because its molecular skeleton is isomorphic with the honeycomb structure present on the grapheme surface, while the PS block offers good solubility in various

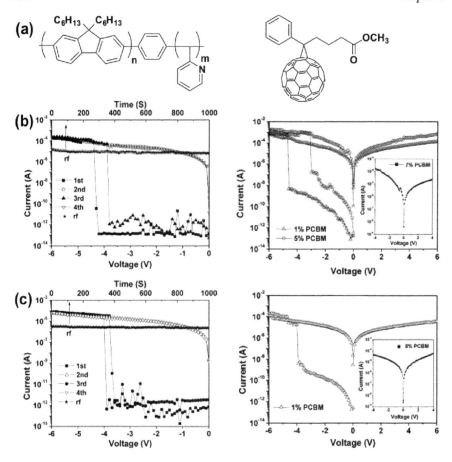

Figure 8.3 (a) Polymer structures of poly[2,7-(9,9-dihexylfluorene)]-*b*-poly-(2-vinylpyridine) (PF-*b*-P2VP) and [6,6]-phenyl-C_{61}-butyric acid methyl ester (PCBM). *I–V* characteristics of (b) ITO/PF$_{10}$-*b*-P2VP$_{37}$/Al (left) and ITO/PCBM:PF$_{10}$-*b*-P2VP$_{37}$/Al device (right), (c) ITO/PF$_{10}$-*b*-P2VP$_{68}$/Al (left) and ITO/PCBM:PF$_{10}$-*b*-P2VP$_{68}$/Al devices (right). Reproduced with permission from ref. 64, © 2011 American Chemical Society.

nonpolar solvents. Different memory behaviors with different contents of GO were achieved in ITO/PS-*b*-PPP, graphene/Al sandwich devices. Typical insulating behavior was displayed for nanohybrids containing zero (Figure 8.4b, red) and very low GO contents. A nanohybrid with 1 wt% PMGO also exhibited a high resistive state (HRS) with its electrical resistance slightly lower than that of the neat PS-*b*-PPP without any bistable switching behavior. In a device with 2 wt% GO with respect to PS-*b*-PPP, non-volatile memory behavior was observed with excellent data retention and endurance cycling properties as shown in Figure 8.4b (green), c and d which can be attributed to the PS-*b*-PPP and GO interface formed *via* π–π interactions between the PPP block and GO.[65]

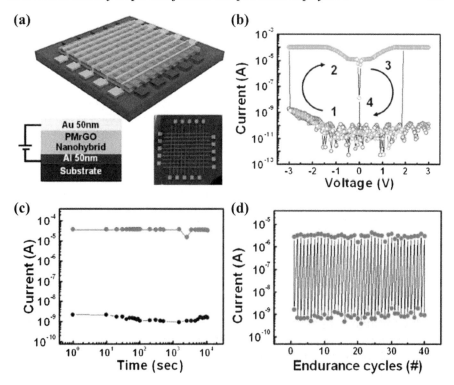

Figure 8.4 (a) Schematic illustration of the cross-bar type arrays of Au/PMGO nanohybrid/Al memory device cells. A photograph of the 10 × 10 arrays is shown in the bottom right of (a). (b) *I–V* characteristics of the memory cells with a pure PS-*b*-PPP layer (red color) and a nanohybrid film of PS-*b*-PPP and GO with 2 wt% GO (green color). (c) Data retention and (d) write–erase cycle endurance for a memory cell with a nanohybrid film of PS-*b*-PPP and GO with 2 wt% GO. Reproduced with permission from ref. 65, © 2012 Royal Society of Chemistry.

High temperature stability for reliable memory performance has been considered from an industrial point of view. Park *et al.*[66] reported a high-performance printable non-volatile polymer memory device containing single-wall carbon nanotubes (SWNTs) by simple solution blending for high temperature operation (Figure 8.5a). To meet demands, polymer memory devices based on nanocomposite films of the conjugated block copolymer, poly(styrene-*block-para*phenylene) (PS-*b*-PPP), and single-wall carbon nanotubes (SWNTs) were fabricated. Through π–π interactions, the conjugated PPP block could strongly adhere to the SWNT surface due to its isomorphic molecular skeleton and the honeycomb structure present on the nanotube surface. Additionally, the PS block offers facile film formation with sufficient mechanical flexibility.[67] Devices composed of PS-*b*-PPP/SWNT nanocomposite films exhibited bipolar non-volatile memory characteristics with a high ON/OFF ratio (Figure 8.5b) and good data retention times. The devices were

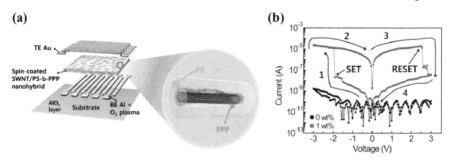

Figure 8.5 (a) A schematic illustration of the fabrication process for arrays of crossbar-type non-volatile conjugated block polymer memory devices with SWNTs. The zoomed-in schematic in the top right contains a snap shot of the simulated assembly of SWNTs and PS-*b*-PPP which has 50 and 14 monomeric units of PS and PPP, respectively. (b) *I–V* characteristics of the conjugated block copolymer memory cells with 0 wt% and 1 wt% SWNTs with respect to the polymer. Reproduced with permission from ref. 66, © 2013 WILEY-VCH Verlag GmbH & Co. KGaA, Weinheim.

suitable for high temperature operation at up to approximately 100 °C. The excellent thermal stability and reliability of our devices is due to the strong adhesion of the high T_g PPP blocks to the carbon nanotube surfaces due to π–π interactions and the adsorption of the PPP blocks onto the curved surfaces of the SWNTs. An alternately tilted conformation of PPP is not only advantageous for tight contact with the curved surfaces of the SWNTs but also decreases the PPP–PPP self-attraction, which may result in strong adsorption of PPP onto the SWNT surfaces. Therefore, nanocomposites of block copolymers and SWNTs are promising materials for thermally stable, low-cost, printable non-volatile polymer memory devices.[66]

As random copolymers with 2-(9*H*-carbazol-9-yl)ethyl methacrylate and an iridium complex (PCzMA–Ir) demonstrated ternary memory characteristics *via* a combination of conformational change and the CT mechanism,[59] the random copolymer of poly(9-(4-vinylbenzyl)-9*H*-carbazole)-random-poly(1-(4-nitro-azo-phenyl)-pyrrole-2,5-dione) (P(VCz)$_x$(MIDO3)$_y$) with five different chemical compositions, $x:y$ = 45:5, 41:9, 18:7, 33:16, and 29:21 (Figure 8.6a), was designed and synthesized by introducing electron-donating carbazole and accepting nitroazobenzene entities into the lateral chains for the study of polymer-based ternary memory behavior explained by two or more mechanisms, resulting in more than two electronic states. Even though these polymers of P(VCz)$_x$(MIDO3)$_y$ are not block copolymers, these polymers consisted of rods of MIDO3 and coils of P(VCz) randomly so we introduced them here. Five random copolymer-based devices (ITO/P(VCz)$_x$(MIDO3)$_y$/Al) demonstrated memory characteristics with switching behavior and different OFF and ON switching voltages depending on the different donor/acceptor contents in the random copolymers (Figure 8.6b). Among the five random copolymers, the fabricated devices with ITO/P(VCz)$_{29}$(MIDO3)$_{21}$/Al, showing typical bistable switching behavior

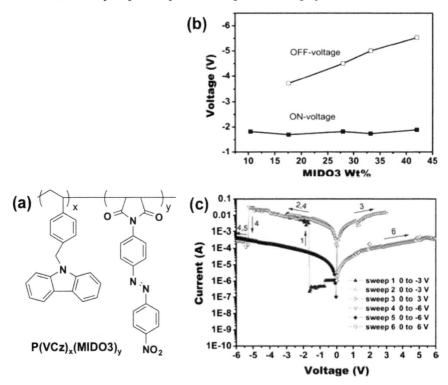

Figure 8.6 (a) Chemical structure of random copolymer; (b) typical switch-ON and switch-OFF voltages for P(VCz)$_x$(MIDO3)$_y$ copolymers with different MIDO3 content: no switch-OFF phenomenon was found for P(VCz)$_{45}$(MIDO3)$_5$. (c) Typical I–V curves of the ITO/P(VCz)$_{29}$(MIDO3)$_{21}$/Al memory device with current compliance at 0.05 A. Reproduced with permission from ref. 68, © 2012 American Chemical Society.

in their I–V characteristics (Figure 8.6c), exhibited three distinct conductivity states when a negative bias was applied, which can be encoded as "0", "1" and "2". The threshold voltages for the OFF–ON transition are almost constant, which could be attributed to the intramolecular and intermolecular charge transfer between the donor/acceptor moieties. The transition from the ON- to intermediate-state under higher bias would rupture the originally formed conductive filaments because excessive current generated too much Joule heating. The filamentary mechanism for the second transition is explained by the formation or rupturing of aluminum filaments, resulting from the penetration of the aluminum electrodes and their interaction with the strongly coordinating nitrogen. This result was corroborated by the decrease in the switch-OFF voltages for the ON–intermediate state transition and the decreasing MIDO3 content within the polymers. This approach to achieving devices with multi-stable states through a combination of different switching mechanisms in one device may be a strategy for future ternary permanent data storage.[68]

8.2.2 Well-Defined Homopolymers

To synthesize well-defined block copolymers containing various functional groups, well-defined homopolymers could be prepared successfully *via* living anionic polymerization or living radical polymerization (RAFT and ATRP) as promising techniques for the preparation of well-controlled polymers. However, due to the various possible side reactions, careful handling is required.[50] Despite this, some well-defined homopolymers containing fluorene, carbazole, or oxadiazole have been synthesized successfully and utilized in non-volatile memory applications.

Hirao and Chen *et al.*[69] reported the influence of both the fluorene chain length and polymer morphology on the switching behaviors of device performance. At first, this group synthesized three *para*-substituted polystyrene (PS) derivatives with mono-, di-, and tri(9,9-dihexylfluorene) units, (poly(St-Fl), poly(St-Fl2), and poly(St-Fl3)), *via* living anionic polymerization (Figure 8.7a), and devices with structures of Al/polymer/ITO were fabricated by spin-coating these poly(St-Fl), poly(St-Fl2), and poly(St-Fl3) polymers onto ITO (Figure 8.7b). These devices exhibited non-volatile bistable switching behaviors with high ON and OFF states as shown in Figure 8.7. Depending on whether there was one, two, or three different fluorene groups, each device based on poly(St-Fl), poly(St-Fl2), and poly(St-Fl3) showed the same

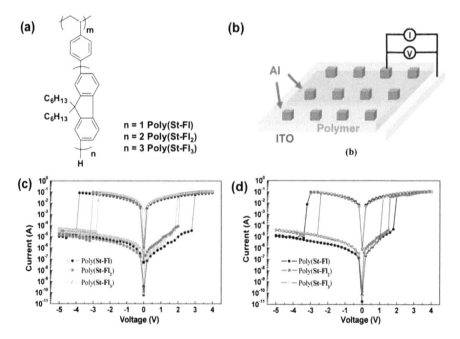

Figure 8.7 (a) Chemical structures of poly(St-Fl), poly(St-Fl2), and poly(St-Fl3). (b) Geometry of the ITO/polymer/Al memory device. Typical *I–V* curves of (c) poly(St-Fl), poly(St-Fl2), and poly(St-Fl3) thin films (prepared from a CB solution) and (d) thin film (prepared from a CB–DMF mixed solvent) memory devices. Reproduced with permission from ref. 69, © 2009 American Chemical Society.

non-volatile memory characteristics but changes in the applied writing voltage. The mechanism of the flash-type switching behavior was attributed to the space-charge-limited-current (SCLC) theory with metallic filamentary conduction. In addition, by investigating the hole injection/transporting barrier from the anode, it was verified that longer pendant chains from one to three enhanced the HOMO level and reduced the turn-on threshold voltage. Interestingly, this group also reported the effect of different polymer morphologies formed using an optimal CB(chlorobenzene):DMF solvent ratio of 9:1 on the switching behavior and threshold voltage; the addition of a poor solvent into the pristine polymer solution changed the turn-on threshold voltage. These results indicated that the switching threshold voltage could be controlled by a change in the solvent type and ratio (Figure 8.7). Therefore, the turn-on voltage of the memory devices was reduced more by increasing the aggregation domain size based on the mixed solvent approach.[69]

Shang and Li et al.[70] successfully prepared well-defined poly(9-(2-(4-vinyl(benzyloxy)ethyl)-9H-carbazole)) (PVBEC) brushes on silicon surfaces via surface-initiated atom transfer radical polymerization (ATRP) (Figure 8.8a). Devices based on the PVBEC brushes grafted on surfaces showed conducting switching behaviors in the I–V characteristics with high ON/OFF current ratios (up to 10^5) that lasted 10^6 read cycles. Additionally, compared with conventional Si/PVBEC/Al devices fabricated by spin-coating, the switching voltages of the PVBEC brush-grafting memory devices were lower and the ON/OFF current ratios of the devices were higher (Figure 8.8b and c). The non-volatile PVBEC WORM memory is very similar to that of PVK because, similar to PVK, PVBEC brush grafting polymers also have the same carbazole pendant groups and a nonconjugated main chain. In general, carbazole moieties act as electron-donors and hole-transfer layers in organic electronics,[61] and form face-to-face conformations with neighboring carbazole moieties to provide extended electron delocalization.[71] Through the extended electron delocalization, charge carrier hopping is easier via the carbazole moieties in the electric field. Therefore, the switching mechanism of the PVBEC memory devices was most likely attributed to a conformation change of the polymers via rotation of the carbazole groups, which provide a more regioregular arrangement.[70]

In general, the incorporation of fullerenes (C_{60}s) into polymers has been limited to 20 wt% with respect to the polymer weight due to immiscibility issues. Although blending <20 wt% C_{60}s with polymers has been achieved, the aggregation of fullerene molecules is severe, which results in unusually high instability in the associated devices.

Recently, Hirao and Chen et al.[72] reported that poly(2-vinylpyridine)-based block copolymer blends with fullerenes showed stable WORM memory behavior due to favorable interactions between the pyridine moieties and fullerene molecules. However, their C_{60} loading was still limited to less than 10 wt% because of fullerene aggregation due to strong interactions between the C_{60}s. Furthermore, a polymer system with covalently bonded fullerene has also been tested, but the fullerene loading was limited to only 3 wt%.[73] These types of devices seem to be unstable and exhibit only volatile memory behavior rather than non-volatile memory characteristics.

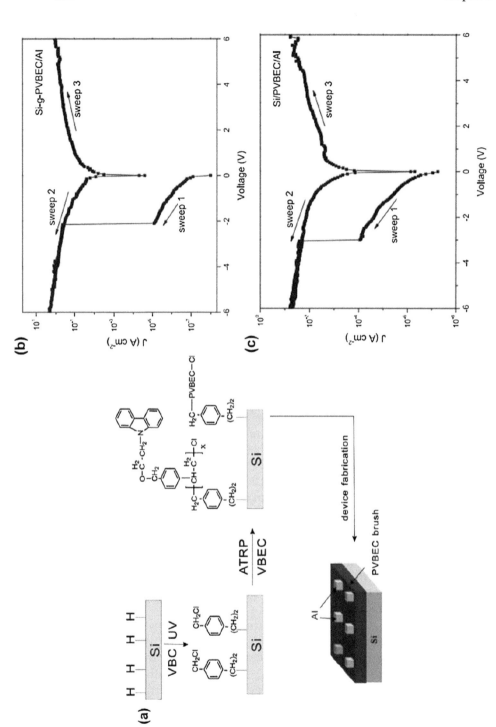

Figure 8.8 (a) Schematic diagram illustrating the preparation of the polymer device. Current density-voltage curves of the (b) Si-g-PVBEC/Al device and (c) conventional Si/PVBEC/Al device. Reproduced with permission from ref. 70, © 2011 Elsevier Ltd.

Therefore, the development of fullerene-based, high-performance polymers for non-volatile memory devices remains in its early stages.

Lee and Ree et al.[74] designed and synthesized a new poly(2-(N-carbazolyl) ethyl methacrylate) end-capped with fullerene (PCzMA-C_{60}) via living anionic polymerization and successfully controlled the location of the fullerene at one end of the polymer chain (Figure 8.9a). The PCzMA-C_{60} polymer contained 21.3 wt% C_{60}, which is the highest C_{60} loading achieved without immiscibility problems. Moreover, only one chain end of the polymer was capped with a C_{60} moiety. This well-defined PCzMA-C_{60} polymer contains carbazole moieties as electron donors and C_{60}s as electron acceptors, and this structure results in synergistic effects, acting as charge trap sites in memory devices. Nanoscale films of the PCzMA-C_{60} polymer, fabricated by conventional solution spin-coating and the evaporation of metal electrodes, exhibit excellent write–read–erase memory characteristics with high ON/OFF current ratios (Figure 8.9b–g), long retention times, high reliabilities, and low power consumption. Interestingly, films of this polymer exhibit bipolar switching behavior as well as unipolar switching behavior. Thus, the chemical structure of PCzMA in which the electron acceptor of C_{60} is attached to one chain end of one electron-donating carbazole polymer directly generated excellent non-volatile memory performance with high reliability. The devices used for the reliability test 3 years ago exhibited long-term reliability by showing bistable switching behaviors again.[74]

Kang and Chen et al.[75] grew PVK directly on the surface of graphene oxide (GO) via reversible addition fragmentation chain transfer (RAFT) radical polymerization using S-1-dodecyl-S'-(α,α'-dimethyl-α'-acetic acid)trithiocarbonate (DDAT)-covalently functionalized graphene oxide (GO) as a RAFT chain transfer agent and a "grafting from" approach (Scheme 8.1). In the field of organic electronics, GO plays the role of electron acceptor, similar to other carbon materials, such as C_{60} and CNT; therefore, PVK-GO could be a donor–acceptor polymer (DAP). This PVK-GO has good solubility in organic solvents compared with pristine GO. Because GO-based donor–acceptor materials show bistable electrical switching and memory effects, the device with the structure Al/PVK-GO/ITO exhibited non-volatile rewritable memory with bistable electrical switching behavior (Figure 8.10), resulting in a low turn-on voltage (−1.7 V) and an ON/OFF state current ratio in excess of 10^3.[75]

Oxadiazole-containing materials have higher affinities for electrons than holes. Therefore, oxadiazole-containing polymers are considered good electron transporting materials for organic electronics. Some oxadiazole-containing polymers demonstrate volatile random access memory behavior instead of non-volatile memory behavior. Interestingly, oxadiazole moieties in this type of polymer act as electron donors rather than acceptors. Based on these basic concepts, a new oxadiazole-containing brush polymer, poly-(5-phenyl-1,3,4-oxadiazol-2-yl-[1,1'-biphenyl] carboxyloxynonyl acrylate) (PPOXBPA), was synthesized via nitroxide-mediated radical polymerization by Ree and Kim et al.[76] (Figure 8.11a). Thin films of PPOXBPA molecules showed self-assembly abilities because the oxadiazole moieties are linked

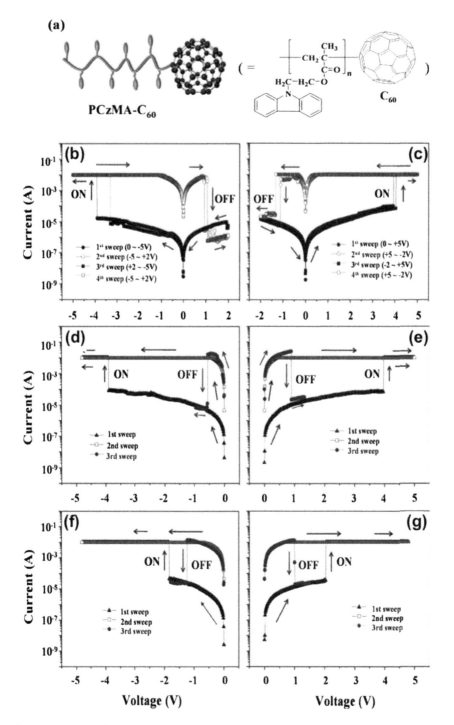

Figure 8.9 (a) Chemical structures of PCzMA-C$_{60}$ and PCzMA. Typical bipolar I–V curves of the Al/PCzMA-C$_{60}$ (60 nm thick)/ITO devices: (b) negative voltage sweeps and reverse voltage sweeps; (c) positive voltage sweeps and reverse voltage sweeps. Typical unipolar I–V curves of (d and e) Al/PCzMA-C$_{60}$ (60 nm thick)/ITO and (f and g) Al/PCzMA (60 nm thick)/ITO devices: (d, f) negative voltage sweeps; (e, g) positive voltage sweeps. The electrode contact area was 0.6 × 0.6 mm^2. Reproduced with permission from ref. 74, © 2012 WILEY-VCH Verlag GmbH & Co. KGaA, Weinheim.

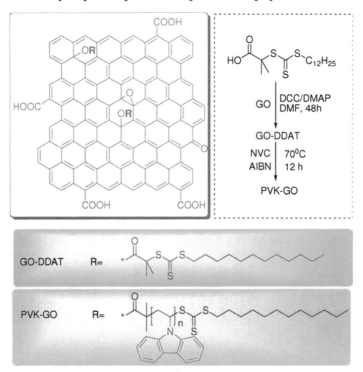

Scheme 8.1 Syntheses of GO-DDAT and PVK-GO. Reproduced with permission from ref. 75, © 2011 Wiley Periodicals, Inc.

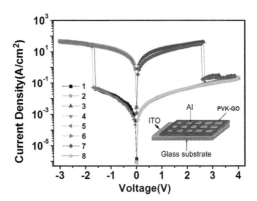

Figure 8.10 Current density–voltage (J–V) characteristics of a 0.16 mm^2 Al/PVK-GO/ITO device. Inset: the structure of the device. Reproduced with permission from ref. 75, © 2011 Wiley Periodicals, Inc.

to the polymer main chains *via* long alkyl chains between the arylate main chain and the oxadiazole moieties as mesogens in the bristles, and these oxadiazole moieties as mesogens formed a molecular multibilayer structure *via* π–π stacking (Figure 8.11b and c). In addition, the ordering and orientation of these multibilayer structures were improved by a thermal

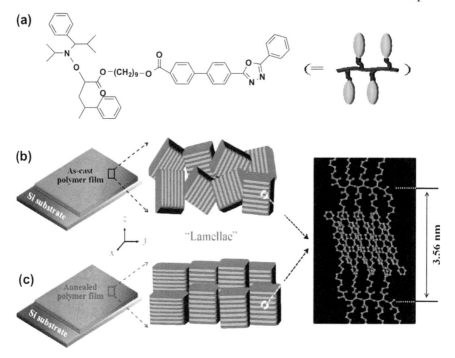

Figure 8.11 (a) Chemical structures of poly-(5-phenyl-1,3,4-oxadiazol-2-yl-[1,1′-biphenyl]carboxyloxynonyl acrylate) (PPOXBPA). Schematic representation of the multibilayer structure of PPOXBPA molecules: (b) as-cast film; (c) thermally annealed film. The atoms are colored: C, gray; O, red; N, blue; hydrogen atoms are omitted for clarity. In the multibilayer structure the bristles are partially interdigitated via π–π stacking of the mesogenic part containing the oxadiazole unit. Reproduced with permission from ref. 76, © 2011 American Chemical Society.

annealing process. These types of changes influenced the electrical memory characteristics of the brush polymer film. In particular, the as-cast films without thermal annealing revealed excellent volatile memory behavior. In contrast, the films treated with thermal annealing exhibited excellent non-volatile memory characteristics. The devices with nanoscale polymer films sandwiched between a bottom and top electrode exhibited either SRAM or WORM memory behavior, depending on the ordering and orientation of the multibilayer structure formed in the film. In detail, the as-cast film, which showed a less ordered multibilayer structure with a preferentially out-of-plane orientation, exhibited unipolar SRAM behavior with a relatively low switching voltage (2.7 V) and a high ON/OFF current ratio of 10^4–10^8. The thermally annealed films formed highly ordered multibilayer structures with almost perfect in-plane orientation and showed unipolar WORM memory behavior with relatively low switching voltages (2.8 V) and high ON/OFF current ratios of 10^4–10^7. Both memory behaviors of SRAM and WORM were governed by shallow trap-limited space charge limited conduction and local filament formation.[76]

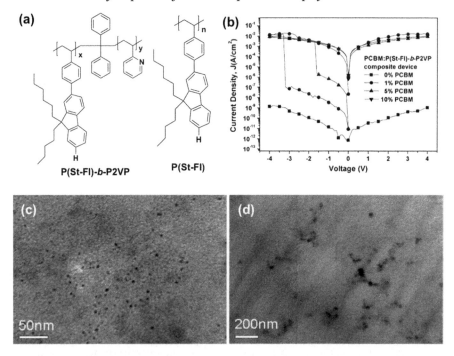

Figure 8.12 (a) Chemical structures of the studied materials. (b) Current density vs. voltage plots for ITO/PCBM:P(St-Fl)-b-P2VP/Al composite devices. TEM images of (c) 5% PCBM:P(St-Fl)-b-P2VP and (d) 5% PCBM:P(St-Fl) films. Reproduced with permission from ref. 72, © 2011 WILEY-VCH Verlag GmbH & Co. KGaA, Weinheim.

8.2.3 Well-Defined Coil–Coil Block Copolymers

Block copolymers have been extensively studied recently because of their ability to self-assemble and form nanostructured morphologies as well as their ability to easily incorporate functional moieties. Recently, control of the morphologies of polymers based on block copolymers containing electron donors and acceptors has allowed for significant manipulation of the electrical characteristics of these materials, including their conductance and switching behaviors.[73,76] However, random copolymers do not allow for control of the memory switching characteristics *via* morphology tuning because it is very difficult to achieve a systematically controlled block ratio and molecular weight distribution. As a result, different polymerization methods for the synthesis of block copolymers with electron donors and acceptors are required to study the effects of the manipulated morphologies on the memory characteristics.[76,77,78]

Homopolymers of poly(4-(9,9-dihexylfloren-2-yl)styrene) (P(St-Fl)) and block copolymers of poly(4-(9,9-dihexylfloren-2-yl)styrene)-*block*-poly(2-vinylpyridine) (P(St-Fl)-*b*-P2VP) (Figure 8.12a) were synthesized *via* living anionic polymerization.[79,80] To investigate the effects of the chemical composition of PCBM:P(St-Fl)-b-P2VP blending on the electrical switching behavior, PCBM:P(St-Fl)-b-P2VP composites containing 1, 5, and 10% PCBM in the polymer were prepared, and the morphology of the 5% PCBM:P(St-Fl)-b-P2VP film was investigated through TEM. The TEM images of 5% PCBM:P(St-Fl)-b-P2VP film showed an approximately 5 nm aggregate size for PCBM in P(St-Fl)-b-P2VP. The 5 nm size is much smaller than the 30 nm aggregate size of PCBM in the P(St-Fl) homopolymer, as shown in Figure 8.12c and d. This TEM result indicated that the P2VP block and PCBM interact *via* charge transfer[81] and the microphase separation of P(St-Fl)-b-P2VP reduces the aggregate size of PCBM. The devices based on the 1, 5, or 10% PCBM:P(St-Fl)-b-P2VP composite and P(St-Fl)-b-P2VP were fabricated with the sandwich structure of ITO/polymer/Al *via* spin-coating the polymers and thermal evaporation of Al. The electrical characteristics were found to depend on the doping concentration of PCBM in P(St-Fl)-b-P2VP. The electrical characteristics of the pristine P(St-Fl)-b-P2VP diblock copolymer without PCBM remained in the low conductivity state in the *I–V* curve, showing insulating characteristics. In contrast, the PCBM:P(St-Fl)-b-P2VP composite-based devices exhibited non-volatile memory with biswitching behavior and conductivity at the applied writing voltage (Figure 8.12b). Both devices based on 1 and 5% PCBM:P(St-Fl)-b-P2VP films showed non-volatile WORM memory behavior with high ON/OFF ratios from at least 10 memory cells. However, only the high-conductivity state was observed in the device based on 10% PCBM:P(St-Fl)-b-P2VP because of the high conductivity due to the PCBM concentration (10%). The device based on the 5% PCBM:P(St-Fl) composite demonstrated a single conducting state without any memory effect. These two results verified that the aggregate size of the PCBM dominated the memory characteristics. The switching mechanism could be described by charge injection-dominated thermionic emission in the OFF state and the charge transfer-dominated Poole–Frenkel (PF) emission model in the ON state. This method was suggested to be a novel system for tuning the electronic characteristics of polymer devices from insulating to conducting due to bistable switching behavior *via* a supramolecular materials approach, such as the morphology-controllable and structural design of the electronic functionality of PCBM:block copolymer composites.[72]

The factors that determine the morphological structures of block copolymer films and the relationship between the structure and the electrical memory device performance were investigated by Ree *et al.*[82] An amphiphilic brush–linear diblock copolymer, poly(4-di(9,9-dihexylfluoren-2-yl) styrene)-b-poly(2-vinylpyridine) PStFl2$_7$-b-P2VP$_{121}$ (volume fraction ratio = 29:71 v/v, Figure 8.13), was prepared and formed a variety of phase-separated nanostructures in nanoscale thin films, depending on the film fabrication conditions. The as-cast and thermally annealed films showed random two-phase structures (*i.e.*, sea and island structures). In contrast, lamellar structures with wide and narrow distributions of orientations were observed in films annealed at room temperature under chlorobenzene (CB) and carbon disulfide (CS$_2$), respectively. The distinct electrical memory properties were influenced by a variety of morphologies, as shown in Figure 8.14. The device with the as-cast film exhibited dielectric characteristics because

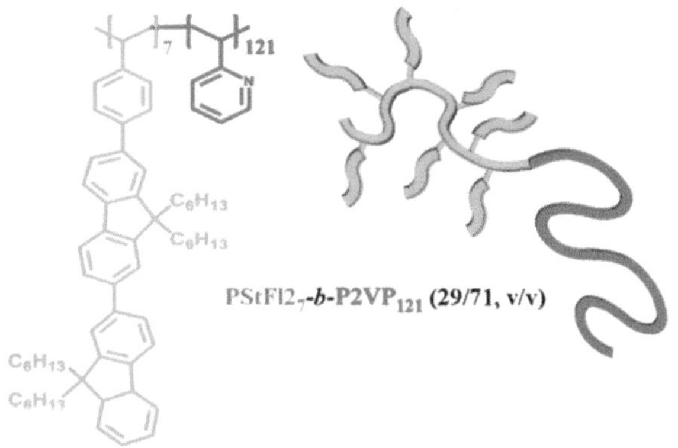

Figure 8.13 Chemical structure of the PStFl2$_7$-b-P2VP$_{121}$ copolymer. Reproduced with permission from ref. 82, © 2013 American Chemical Society.

the charges on the small, individual PStFl2 block chains and/or aggregates are not transferrable to neighboring domains through the dielectric-like P2VP phase. In contrast, the WORM memory behavior of the CB-annealed film resulted from the contact of some fraction of the vertically oriented PStFl2 lamellae with both the bottom and top electrodes, which created physical routes for charge trapping and transport. In the case of the CS$_2$-annealed film, DRAM behavior was observed due to the destabilization of the overall charge trapping and transport of the highly oriented P2VP lamellae that exist along the film plane.[82]

3-arm AB$_2$ and 5-arm AB$_4$ asymmetric amphiphilic star polymers of poly[4-(9,9-dihexylfloren-2-yl)styrene]-*block*-poly(2-vinylpyridine) (P(St-Fl)-*b*-P2VP) (Figure 8.15a) were blended with electron-accepting phenyl-C$_{61}$-butyric acid methyl ester (PCBM) for electrical memory device applications. Poly(vinylpyridine)s are known to have affinities for electron accepting species[83] and have been shown to control fullerene aggregation domains through charge transfer interactions.[71] The morphologies of the AB$_2$ and AB$_4$ films show hexagonally perforated lamellae phases (HPL) that consist of hexagonally packed P2VP spots in the P(St-Fl) domains. When blended with PCBM, the increasing PCBM content provides a high degree of order to the liquid crystal-like morphology in the P(St-Fl) block (Figure 8.16), which was deduced from the possible interactions between the P(St-Fl) units and the PCBM. Additionally, the HPL and lamellar morphologies of the blends were similar to those of the pure asymmetric star polymers due to the macroscopic phase separation at high PCBM loads. The resulting memory characteristics of the PCBM:AB$_2$ composite films, *i.e.*, acting as an insulator, WORM, WORM, or conductor, were observed at PCBM compositions of 0%, 1%, 5%, and 10%, respectively, as shown in Figure 8.15b.

These switching behaviors are explained by the electric-field-induced charge transfer process between the electron-donating pendent fluorene moieties and the electron acceptor PCBM. The WORM memory behavior at 1% and 5% PCBM was observed because there was no macroscopic phase

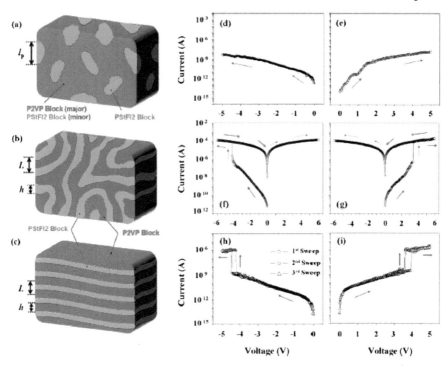

Figure 8.14 Schematic representations of the phase-separated nanostructures in the PStFl2$_7$-b-P2VP$_{121}$ films: (a) random two-phase structure (i.e., sea and island structure) developed in the as-cast film; (b) lamellar structure with a wide distribution of orientations developed in the CB-annealed film; (c) lamellar structure with a narrow distribution of orientations developed in the CS$_2$-annealed film. The size (i.e., correlation length) of domains in the random two-phase structure is denoted with l_p. Long period and lamellar thickness are denoted as L and h, respectively. Typical I–V curves of the devices fabricated with PStFl2$_7$-b-P2VP$_{121}$ films (30 nm thick), which were measured with a compliance current set at 0.01 A: (d and e) as-cast film; (f and g) CB-annealed film; (h and i) CS$_2$-annealed film. Al top and ITO bottom electrodes were used. Reproduced with permission from ref. 82, © 2013 American Chemical Society.

separation of PCBM. However, in the case of the highest PCBM concentration (10%), the large PCBM aggregate size leads to electrically conductive channels percolating through the film, resulting in short-circuited devices.[83]

Devices based on random polymers and composites incorporating nanoparticles (e.g., gold (Au),[84,85] titanium dioxide (TiO$_2$),[77] and quantum dots (CuInS$_2$/ZnS),[78]) or carbon materials (e.g., graphene,[86] and carbon nanotubes (CNTs)[30]) into carbazole polymers have exhibited bistable switching behavior. However, poor compatibility between random polymers and nanoparticles triggered irregular polymer structures and non-uniform dispersion. Therefore, the aggregation of nanoparticles and irregular phase separation could cause poor reproducibility of device performance.

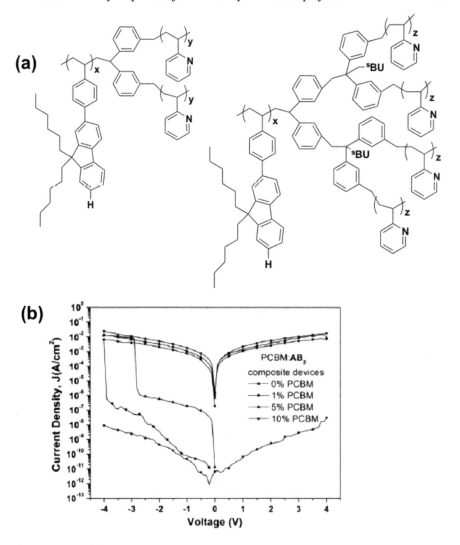

Figure 8.15 (a) Chemical structures of asymmetric amphiphilic star polymers of P(St-Fl)-b-P2VP and (b) Current density vs. voltage plots for ITO/PCBM:AB$_2$/Al composite devices. Reproduced with permission from ref. 83, © 2011 Royal Society of Chemistry.

To improve the non-uniform polymer structures and investigate the effects of various morphologies of block copolymers on the electric characteristics, a total of five different polymeric materials of three diblock copolymers of poly(9-(4-vinylphenyl)carbazole)-b-poly(2-vinylpyridine) (PVPCz-b-P2VP) with a block ratio (mol%) of PVPCz:P2VP = 17:83, 59:41, and 86:14 and two homopolymers of poly(9-(4-vinylphenyl)carbazole) (PVPCz) and poly(2-vinylpyridine) (P2VP) were synthesized by anionic polymerization (Figure 8.17b). The phase separation of these block copolymers was controlled using

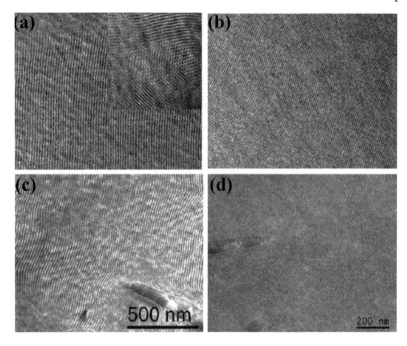

Figure 8.16 TEM images of (a) star block copolymer AB_2, (b) star block copolymer AB_2 blended with PCBM (10 wt%), (c) star block copolymer AB_4, (c) star block copolymer AB_4 blended with PCBM (10 wt%). Reproduced with permission from ref. 83, © 2011 Royal Society of Chemistry.

different block compositions. Among the PVPCz-b-P2VP polymers with block ratios PVPCz:P2VP = 17:83, 59:41, and 86:14 (PVPCz17, PVPCz59, and PVPCz86), spherical structures were observed in the PVPCz17 and PVPCz86 films, and a lamellar structure was observed in the PVPCz59 film. TEM images and schematic illustrations are shown in Figure 8.18. Devices with the sandwich structure ITO/polymer/Al exhibited reproducibly different electrical behaviors of metallic, bistable switching, and insulating, depending on the block ratio of PVPCz and P2VP (Figure 8.19a). Devices based on P2VP and PVPCz17 remained in low current states without switching behaviors, but the PVPCz17-based device showed a slightly higher current than P2VP because PVPCz17 contains 17% carbazole blocks. The 86% carbazole-based PVPCz86 and the homopolymer of PVPCz exhibited high current states. The 59% carbazole-based PVPCz59 exhibited typical unipolar switching behavior and non-volatile memory characteristics (Figure 8.19b). These properties are critical to the development of more realistic memory applications involving a one-diode and one-resistor (1D–1R) memory device architecture.[87]

In addition, Lee *et al.*[87] investigated the I–V characteristics of devices fabricated using a blended system of PVPCz and P2VP homopolymers as well as a random copolymer of PVPCz59-r-P2VP$_{41}$ containing PVPCz and P2VP. Both devices did not show any switching behaviors, which indicates that the

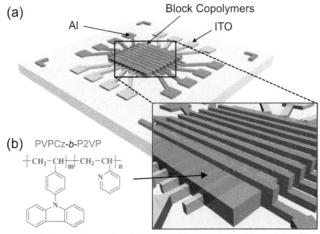

Figure 8.17 (a) A schematic diagram of organic memory devices consisting of Al/polymer/ITO junctions in an 8 × 8 array structure. (b) Structure of the block copolymer PVPCz-*b*-P2VP. Various block ratios (mol%) of PVPCz:P2VP ($m:n$) = 0:100, 17:83, 59:41, 86:14, and 100:0 were used in this study. m and n indicate the block ratio of PVPCz and P2VP in the block copolymer. Reproduced with permission from ref. 87, © 2011 WILEY-VCH Verlag GmbH & Co. KGaA, Weinheim.

arrangement and amount of each segment in the block copolymer is important for resistive switching. The possible mechanism was described by a filament model. Conductive paths mainly through carbazole segments can be formed or ruptured by an electric field, and this operation results in a reversible change in resistance. In particular, filamentary paths are favorably grown and then easily broken up into lamellar structures with conducting and insulating layers.[87]

Morphological structures are related to the development of high performance memory polymers. Two linear–brush diblock copolymers, poly(styrene-*b*-9-carbazolylethylenyloxycarbonylethylenylcarbonyloxyethylenyl methacrylate), bearing carbazole moieties in the brush block with different block volume ratios (PS245-*b*-PCzMA34 (67:33 v/v) and PS1269-*b*-PCzMA46 (89:11 v/v)) (Figure 8.20b) were used in a study on the relationship between memory behavior and morphological structure. Upon annealing with different solvents, *i.e.*, toluene or carbon disulfide (CS_2), different morphological structures were formed. During toluene-based annealing, the toluene vapor selectively mobilizes the PS blocks in PS-*b*-PCzMA, and thus, the mobilization of the PCzMA blocks was further induced. The phase separation of the block components is promoted. As a result, a vertically oriented hexagonal PCzMA cylinder structure was formed in the block copolymer film. During the subsequent CS_2-based annealing, the CS_2 vapor mobilized the PS blocks in PS-*b*-PCzMA selectively. The mobilization of the PCzMA block was induced sequentially, and a horizontally oriented lamellar structure was finally formed

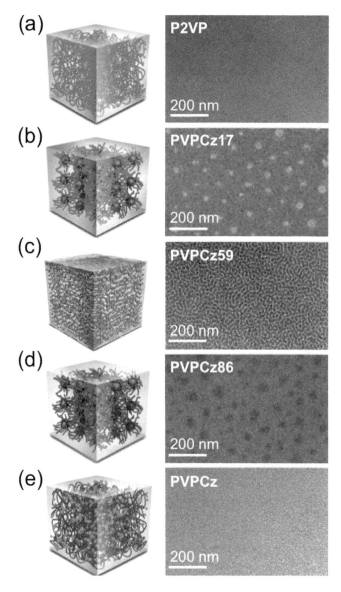

Figure 8.18 Schematic illustrations (left) and TEM images (right) of PVPCz-b-P2VP block copolymers of (a) P2VP (PVPCz:P2VP = 0:100), (b) PVPCz17 (PVPCz:P2VP = 17:83), (c) PVPCz59 (PVPCz:P2VP = 59:41), (d) PVPCz86 (PVPCz:P2VP = 86:14), and (e) PVPCz (PVPCz:P2VP = 100:0). The scale bars on the TEM images are 200 nm. Reproduced with permission from ref. 87, © 2011 WILEY-VCH Verlag GmbH & Co. KGaA, Weinheim.

Non-Volatile Memory Properties of Donor–Acceptor Block Copolymers 283

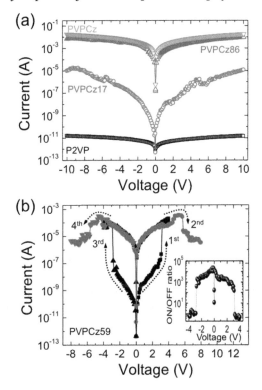

Figure 8.19 (a) *I–V* characteristics of device-P2VP, -PVPCz17, -PVPCz86, and -PVPCz fabricated with the polymer P2VP, PVPCz17, PVPCz86, and PVPCz, respectively. (b) *I–V* characteristics of device-PVPCz59 made with the PVPCz59 block copolymer. The inset shows the ON/OFF ratio as a function of voltage. Reproduced with permission from ref. 87, © 2011 WILEY-VCH Verlag GmbH & Co. KGaA, Weinheim.

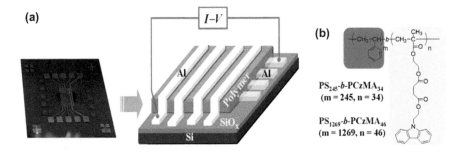

Figure 8.20 (a) An optical image and schematic diagram of memory devices fabricated with PS245-*b*-PCzMA34 and PS1269-*b*-PCzMA46 films. (b) Chemical structure of PSm-*b*-PCzMAn polymer in two different compositions. Reproduced with permission from ref. 88, © 2014 American Chemical Society.

in the block copolymer film. Illustrations of the phase-separated structures are shown in Figures 8.21a and b and Figure 8.22a–c.[88]

Devices were fabricated with polymer films deposited on Al-deposited substrates, and then, Al top electrodes were deposited using thermal deposition (Figure 8.20b). As shown in Figure 8.21c the current–voltage (I–V) characteristics of the as-cast PS245-b-PCzMA34 film indicated a high resistance state, and the current level increased gradually with increasing voltage. The as-cast PS1269-b-PCzMA46 film demonstrated DRAM behavior in Figure 8.21d, and this memory performance was much better than those of the as-cast PS245-b-PCzMA34 films because of the lower threshold voltage and higher ON/OFF current ratio (approximately 10^5).[88]

As shown in Figure 8.22d, the PS245-b-PCzMA34 film was in the OFF state initially, but the current level in the film increased abruptly, indicating that the film experienced a sharp electrical transition from the OFF state to the ON state. The PS1269-b-PCzMA46 film without an OTS layer demonstrated similar WORM memory behavior, as shown in Figure 8.22f. These results indicated two important factors for advanced memory devices. First, the carbazole units can create local hopping paths with a nanoscale cross-section for charge trapping and transport. Second, block copolymers

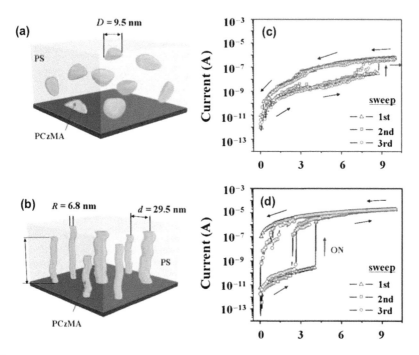

Figure 8.21 Schematic representations of the phase-separated morphologies in the as-cast copolymer films: (a) poorly developed island domain structure in the PS245-b-PCzMA34 film; (b) poorly developed vertical hexagonal cylinder structure in the PS1269-b-PCzMA46 film. Typical I–V curves of the devices: (c) PS245-b-PCzMA34 film; (d) PS1269-b-PCzMA46 film. Reproduced with permission from ref. 88, © 2014 American Chemical Society.

with orientation-controlled PCzMA cylinder structures can realize advanced memory devices with extremely high data storage capacities. Moreover, the WORM memory mode of these devices can be converted into other memory modes by changing the morphology and orientation. Interestingly, the CS_2-annealed PS245-*b*-PCzMA34 film with an OTS layer showed no switching

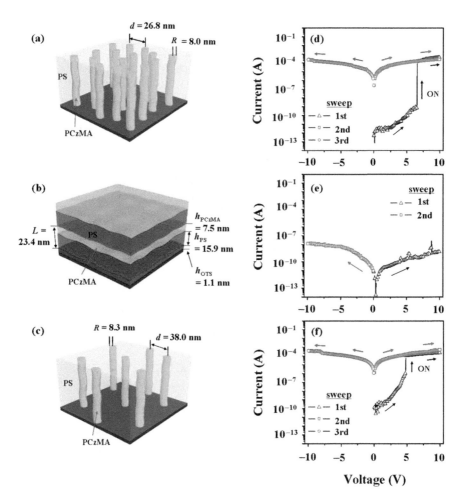

Figure 8.22 Schematic representations of the phase-separated morphologies in the solvent-annealed copolymer films: (a) well developed vertical hexagonal cylinder structure in the toluene-annealed PS245-*b*-PCzMA34 film; (b) well developed horizontal lamellar structure in the CS_2-annealed PS245-*b*-PCzMA34 film; (c) well developed vertical hexagonal cylinder structure in the toluene-annealed PS1269-*b*-PCzMA46 film. Typical *I–V* curves of the devices: (d) toluene-annealed PS245-*b*-PCzMA34 film without an OTS (octyltrichlorosilane) layer; (e) CS_2-annealed PS245-*b*-PCzMA34 film with an OTS layer; (f) toluene-annealed PS1269-*b*-PCzMA46 film without an OTS layer. Al top and bottom electrodes were used. The electrode contact area was 0.2 Å–0.2 mm^2. Reproduced with permission from ref. 88, © 2014 American Chemical Society.

behaviors and this film exhibits only dielectric-like *I–V* behavior, as shown in Figure 8.22e. The mechanisms of all of these memory behaviors were attributed to trap-limited SCLC and local hopping path formation. In detail, the ohmic conduction mechanism leads to high-performance WORM memory as well as DRAM behavior, and poor performance DRAM behavior is dominated by the trap-limited SCLC mechanism.[88]

Lee *et al.*[89] reported the synthesis of well-defined triblock copolymers with pendent triphenylamine and ethynylpyridine, poly(2-(2-(4-vinylphenyl)ethynyl)pyridine)-*b*-poly(4,4′-vinylphenyl-*N*,*N*-bis(4-tertbutylphenyl) benzenamine)-*b*-poly(2-(2-(4-vinyl-phenyl)ethynyl)pyridine) (PVPEP-*b*-PVBPTPA-*b*-PVPEP) (Figure 8.23a) in which the PVBPTPA block is an electron donor and the PVPEP block is an electron acceptor. These copolymers were formed using potassium naphthalene as a bidirectional initiator because the carbanions formed from lithium cation metal initiators result in side reactions during the living anionic polymerization of pyridine-containing monomers, such as 2-vinylpyridine,[90] 2-(4-vinylphenyl) pyridine,[91] and 2-(4′-vinylbiphenyl-4-yl)pyridine.[92] A previous report on the relationship between the self-assembly morphologies and the electrical characteristics of block copolymer-based devices elucidated that block copolymer devices with

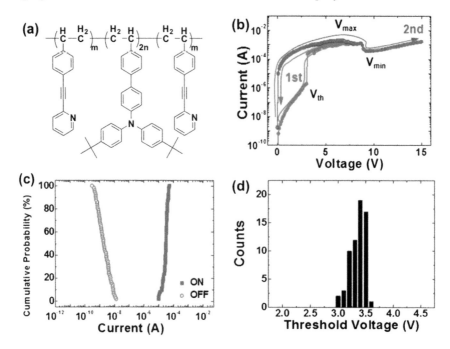

Figure 8.23 (a) Chemical structure of the triblock copolymer of PVPEP-*b*-PVBPTPA-*b*-PVPEP. (b) The representative current–voltage (*I–V*) switching characteristics of the memory device. (c) Cumulative probability of ON and OFF states. (d) Statistical distributions of the threshold voltages of the operative memory cells. Reproduced with permission from ref. 89, © 2014 Wiley Periodicals, Inc.

well-ordered microphase-separated structures exhibited metallic, resistive switching, and insulating behaviors, depending on the block ratios of the block copolymers.[87,93] Based on this previous study, triblock copolymers of PVPEP-*b*-PVBPTPA-*b*-PVPEP with both donor–acceptor structures and bicontinuous microphase-separated morphologies were fabricated on ITO substrates by spin-coating using polymer solution. The device with the configuration ITO/triblock copolymer/Al and PVPEP-*b*-PVBPTPA-*b*-PVPEP with the block ratio PVBPTPA : PVPEP = 71 : 29 showed typical bistable switching behavior in its *I–V* characteristics, as shown in Figure 8.23b. Interestingly, statistical data for the switching characteristics of all of the operative memory cells showed resistive switching memory with a 100% device yield (64 cells out of 64) as shown in Figure 8.23c. The ON and OFF states of all of the operative cells were cumulatively plotted. In addition, the statistical distributions of the threshold voltages of all operative cells demonstrated that the transition from the OFF state to the ON state appears in a very narrow voltage range from 3.0 to 3.6 V, as shown in Figure 8.23d. It was proved that a small variation in the threshold voltages could considerably decrease the failure in the operation of memory cells.[89]

Various phase-separated nanostructures in well-defined block copolymers have been related to the memory characteristics. The well-controlled block copolymer (PBAB) based on 4,4′-vinylphenyl-*N*,*N*-bis(4-*tert*-butylphenyl) benzenamine and *n*-hexyl isocyanate with block ratios of 25 : 75, 55 : 45, and 85 : 15 was denoted PBAB75, PBAB45, and PBAB15, respectively (Figure 8.24a).

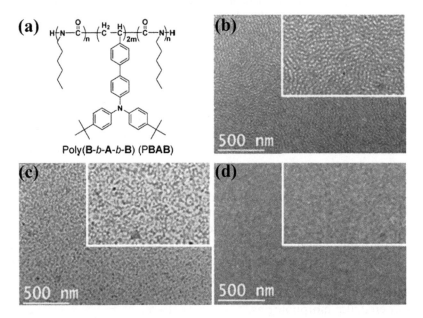

Figure 8.24 (a) Chemical structure of PBAB and TEM images of (b) PBAB75, (c) PBAB45, and (d) PBAB15. Reproduced with permission from ref. 94, © 2014 American Chemical Society.

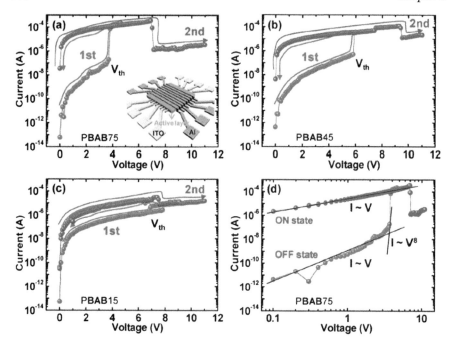

Figure 8.25 Current–voltage (I–V) switching characteristics of (a) PBAB75, (b) PBAB45, and (c) PBAB15. (d) A double-log I–V plot analysis of PBAB75. Inset of (a): schematic illustration of 8 × 8 array-type memory device consisting of ITO/PBAB/Al layers. Reproduced with permission from ref. 94, © 2014 American Chemical Society.

Compared with the poorly ordered structures in the PBAB45 and PBAB15 triblock copolymers, PBAB75 formed a well-ordered microphase-separated morphological structure (Figure 8.24b–d). These morphological behaviors of the block copolymers significantly affected the performances of the polymer resistive memory devices. Write-once read-many times (WORM) characteristics that showed negative differential resistance (NDR) were observed in memory devices consisting of ITO/PBAB/Al layers (Figure 8.25). However, depending on the morphology of the microphase-separated structure, larger threshold voltages were needed for the poorly ordered PBAB45 (~5.5 V) and PBAB15 (~6 V) structures compared with the well-ordered PBAB75 structure (~4 V). Furthermore, the PBAB75-based devices exhibited excellent memory performances with the highest ON/OFF ratios, cell-to-cell uniformities, endurance reliabilities (230 cycles), and longest retention times of 10^4 among the three types of device. These results are thought to arise from the accessibility and reproducibility of the filamentary conduction paths in the well-ordered film morphologies.[94]

Recently, Lee et al.[95] reported donor–acceptor alternating conjugated polymers for non-volatile memory applications. These polymers are not categorized in the group of coil–coil block copolymers but the basic structure of the

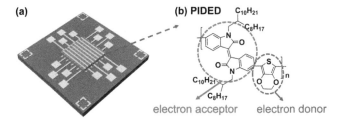

Figure 8.26 (a) Schematic of 8 × 8 crossbar array-type organic memory device consisting of Al/polymer/Al. PIDED, PEDOT/PID blends, and the PID homopolymer were used as active layers. (b) Structure of the conjugated polymer, PIDED, consisting of PEDOT and PID as the electron donor and acceptor moieties, respectively. Reproduced with permission from ref. 95, © 2015 American Chemical Society.

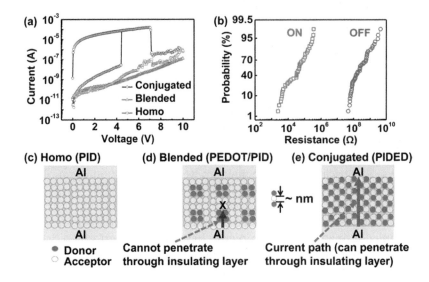

Figure 8.27 (a) *I–V* curves of MIM-type memory devices fabricated from three different active polymers, measured in ambient air: D–A conjugated polymer, PIDED, D–A blended polymer, PEDOT/PID, and acceptor homopolymer, PID. (b) Cumulative probability characteristics of the switching current for all operative memory cells (48 operative cells out of 64 fabricated cells), measured at a read voltage of 0.3 V in ambient air. Schematics of current flow in three different polymer systems: (c) acceptor homopolymer, PID, (d) D–A blended polymer, PEDOT/PID, and (e) D–A conjugated polymer, PIDED. Reproduced with permission from ref. 95, © 2015 American Chemical Society.

polymers is also based on donors and acceptors predominantly. Therefore, this polymer type is included in this section. Non-volatile resistive memory devices based on a new low bandgap donor–acceptor (D–A) conjugated polymer, poly((*E*)-6,6′-bis(2,3-dihydrothieno[3,4-*b*][1,4]dioxine-5-yl)-1,1′-bis(2-octyldodecyl)-[3,3′-biindolinyi-dene]-2,2′-dione) (PIDED), which were fabricated and operated in ambient air, were reported. The D–A conjugated polymer was synthesized from 2,3-dihydrothieno[3,4-*b*][1,4]dioxine and isoindigo as electron donor and electron acceptor, respectively, using CH-arylation polymerization (Figure 8.26). The devices show non-volatile, unipolar resistive switching behaviors with high ON/OFF current ratios ($\sim 10^4$) (Figure 8.27), excellent endurance cycles (>200 cycles), and a long retention time (>10^4 s) in ambient air. These properties remain stable in ambient air for over one year, demonstrating that the device performance is unaffected by exposure to air because the isoindigo has strong electron withdrawing character, and the PIDED exhibits a high degree of crystallinity. This study may pave the way for the use of practical non-volatile organic memory devices that operate in ambient air.[95]

8.3 Summary and Outlook

In this chapter, the electronic characteristics of non-volatile ON/OFF switching memory devices with the electrode/polymer/electrode sandwich structure that were fabricated using well-controlled donor–acceptor polymers with different chemical structures, such as rod–coil block copolymers, well-defined homopolymers, coil–coil block copolymers and coil–coil random copolymers, were demonstrated. Many different types of memory devices based on these polymers successfully exhibited switching behavior. Depending on the chemical structures and thicknesses of the polymer films, unipolar and/or bipolar memory behaviors were observed. In particular, these electrical memory behaviors were primarily dominated by the chemical natures of the functional polymers, including the donating and accepting moieties.

Broad studies on memory devices using well-defined block copolymers containing donor and acceptor moieties have not been performed due to the difficulty of controllably synthesizing donor–acceptor polymers containing various heteroatoms. In general, the living anionic polymerization of donating or accepting monomers containing heteroatoms is a significant challenge due to the strong nucleophilicity of the carbanionic initiators, which cause different side reactions in living polymers, such as cross-linking and branching. Therefore, using living radical polymerizations of NMP, ATRP, and RAFT instead of living anionic polymerizations, these types of well-defined homopolymers or block copolymers containing donors and/or acceptors have also been synthesized and utilized in the application of memory devices. Because block copolymers containing donors and acceptors were introduced in this chapter, the well-defined block copolymers have only a few types of donors, such as fluorene, carbazole, and triphenylamine groups, and strong acceptors, such as oxadiazole and pyridine derivatives. To understand the memory mechanism and to develop high performance memory, understanding the relationship between the electrical memory behavior and the chemical structure and morphology of polymers is critical.

Several groups have suggested electronic switching mechanisms due to the formation of morphologies that use these types of controlled donor–acceptor block copolymers. Accordingly, a variety of innovative functional donor–acceptor block copolymers should be designed and synthesized preferentially. Fortunately, studies on memory devices based on well-defined block copolymers have gradually increased. For the successful development of advanced memory, some factors that are needed for the preparation of various donor–acceptor block copolymers are suggested: (1) the design and synthesis of various types of donor and acceptor functional groups, (2) the establishment of optimum conditions for the synthesis of well-defined polymers, and (3) the design of ideal polymer structures with easily tunable morphologies and excellent general characteristics, such as high solubility, simple processability, and good mechanical properties. In the future, nanostructured organic memory devices with high performances will most likely be realized using advanced, well-defined donor–acceptor block copolymers.

References

1. P. Heremans, G. H. Gelinck, R. Müller, K.-J. Baeg, D.-Y. Kim and Y.-Y. Noh, *Chem. Mater.*, 2011, **23**, 341.
2. N. Setter, D. Damjanovic, L. Eng, G. Fox, S. Gevorgian, S. Hong, A. Kingon, H. Kohlstedt, N. Y. Park, G. B. Stephenson, I. Stolitchnov, A. K. Taganstev, D. V. Taylor, T. Yamada and S. Streiffer, *J. Appl. Phys.*, 2006, **100**, 051606.
3. J. D. Boeck, W. V. Roy, J. Das, V. Motsnyi, Z. Liu, L. Lagae, H. Boeve, K. Dessein and G. Borghs, *Semicond. Sci. Technol.*, 2002, **17**, 342.
4. S. Tehrani, J. M. Slaughter, E. Chen, M. Durlam, J. Shi and M. DeHerren, *IEEE Trans. Magn.*, 1999, **35**, 2814.
5. S. Hudgens and B. Johnson, *MRS Bull.*, 2004, **29**, 829.
6. R. Waser and M. Aono, *Nat. Mater.*, 2007, **6**, 833.
7. H. Y. Jeong, Y. I. Kim, J. Y. Lee and S.-Y. Choi, *Nanotechnology*, 2010, **21**, 115203.
8. L. P. Ma, J. Liu and Y. Yang, *Appl. Phys. Lett.*, 2002, **80**, 2997.
9. J. Ouyang, C.-W. Chu, C. R. Szmanda, L. Ma and Y. Yang, *Nat. Mater.*, 2004, **3**, 918.
10. C. W. Chu, J. Ouyang, J. H. Tseng and Y. Yang, *Adv. Mater.*, 2005, **17**, 1440.
11. R. J. Tseng, J. Huang, J. Ouyang, R. B. Kaner and Y. Yang, *Nano Lett.*, 2005, **5**, 1077.
12. R. J. Tseng, C. Tsai, L. Ma, J. Ouyang, C. S. Ozkan and Y. Yang, *Nat. Nanotechnol.*, 2006, **1**, 72.
13. Y. Yang, J. Ouyang, L. Ma, R. J. H. Tseng and C. W. Chu, *Adv. Funct. Mat.*, 2006, **16**, 1001.
14. J. C. Scott and L. D. Bozano, *Adv. Mater.*, 2007, **19**, 1452.
15. Q.-D. Ling, D.-J. Liaw, C. Zhu, D. S.-H. Chan, E.-T. Kang and K.-G. Neoh, *Prog. Polym. Sci.*, 2008, **33**, 917.
16. B. Mukherjee and A. J. Pal, *Org. Electron.*, 2006, **7**, 249.
17. T. Kondo, S. M. Lee, M. Malicki, B. Domercq, S. R. Marder and B. Kippelen, *Adv. Funct. Mat.*, 2008, **18**, 1112.
18. X.-D. Zhuang, Y. Chen, G. Liu, B. Zhang, K.-G. Neoh, E.-T. Kang, C.-X. Zhu, Y.-X. Li and L.-J. Niu, *Adv. Funct. Mat.*, 2010, **20**, 2916.

19. B. Cho, S. Song, Y. Ji and T. Lee, *Appl. Phys. Lett.*, 2010, **97**, 063305.
20. P. Y. Lai and J. S. Chen, *Appl. Phys. Lett.*, 2008, **93**, 153305.
21. Y. Li, D. Qiu, L. Cao, C. Shao, L. Pan, L. Pu, J. Xu and Y. Shi, *Appl. Phys. Lett.*, 2010, **96**, 133303.
22. B. Cho, S. Song, Y. Ji, T.-W. Kim and T. Lee, *Adv. Mat.*, 2011, **21**, 2806.
23. J. Chen and D. Ma, *J. Appl. Phys.*, 2006, **100**, 034512.
24. J. Lin and D. Ma, *J. Appl. Phys.*, 2008, **103**, 024507.
25. G. Liu, Q.-D. Ling, E.-T. Kang, K.-G. Neoh, D.-J. Liaw, F.-C. Chang, C.-X. Zhu and D. S.-H. Chan, *J. Appl. Phys.*, 2007, **102**, 024502.
26. J. Bo, W. Zhaoxin, H. Qiang, T. Yuan, M. Guilin and H. Xun, *J. Phys. D: Appl. Phys.*, 2010, **43**, 035101.
27. T.-W. Kim, S.-H. Oh, J. Lee, H. Choi, G. Wang, J. Park, D.-Y. Kim, H. Hwang and T. Lee, *Org. Electron.*, 2010, **11**, 109.
28. B. Lei, W. L. Kwan, Y. Shao and Y. Yang, *Org. Electron.*, 2009, **10**, 1048.
29. J. Lin and D. Ma, *Org. Electron.*, 2009, **10**, 275.
30. G. Liu, Q.-D. Ling, E. Y. H. Teo, C.-X. Zhu, D. S.-H. Chan, K.-G. Neoh and E.-T. Kang, *ACS Nano*, 2009, **3**, 1929.
31. C. N. Lau, D. R. Stewart, R. S. Williams and M. Bockrath, *Nano Lett.*, 2004, **4**, 569.
32. S. L. Lim, Ling, E. Y. H. Teo, C. X. Zhu, D. S. H. Chan, Kang and K. G. Neoh, *Chem. Mater.*, 2007, **19**, 5148.
33. Z. J. Donhauser, B. A. Mantooth, K. F. Kelly, L. A. Bumm, J. D. Monnell, J. J. Stapleton, D. W. Price, Jr, A. M. Rawlett, D. L. Allara, J. M. Tour, P. S. Weiss, *et al.*, *Science*, 2001, **292**, 2303.
34. B. C. Das and A. J. Pal, *Org. Electron.*, 2008, **9**, 39.
35. H. Carchano, R. Lacoste and Y. Segui, *Appl. Phys. Lett.*, 1971, **19**, 414.
36. H. K. Henisch and W. R. Smith, *Appl. Phys. Lett.*, 1974, **24**, 589.
37. S. L. Lim, Ling, E. Y. H. Teo, C. X. Zhu, D. S. H. Chan, Kang and K. G. Neoh, *Chem. Mater.*, 2007, **19**, 5148.
38. T.-W. Kim, S.-H. Oh, H. Choi, G. Wang, H. Hwang, D.-Y. Kim and T. Lee, *IEEE Electron Device Lett.*, 2008, **29**, 852.
39. S. Paul, A. Kanwal and M. Chhowalla, *Nanotechnology*, 2006, **17**, 145.
40. F. Verbakel, S. C. J. Meskers and R. A. J. Janssen, *Chem. Mater.*, 2006, **18**, 2707.
41. A. Bandyopadhyay and A. J. Pal, *Appl. Phys. Lett.*, 2004, **84**, 999.
42. M. Nakaya, S. Tsukamoto, Y. Kuwahara, M. Aono and T. Nakayama, *Adv. Mater.*, 2010, **22**, 1622.
43. J. Lee, H. Chang, S. Kim, G. S. Bang and H. Lee, *Angew. Chem., Int. Ed.*, 2009, **48**, 8501.
44. A. Bandyopadhyay and A. J. Pal, *Appl. Phys. Lett.*, 2003, **82**, 1215.
45. S. Y. Quek, M. Kamenetska, M. L. Steigerwald, H. J. Choi, S. G. Louie, M. S. Hybertsen, J. B. Neaton and V. Latha, *Nat. Nanotechnol.*, 2009, **4**, 230.
46. A. Laiho, H. S. Majumdar, J. K. Baral, F. Jansson, R. Osterbacka and O. Ikkala, *Appl. Phys. Lett.*, 2008, **93**, 203309.
47. F. Li, D.-I. Son, S.-M. Seo, H.-M. Cha, H.-J. Kim, B.-J. Kim, J. H. Jung and T. W. Kim, *Appl. Phys. Lett.*, 2007, **91**, 122111.

48. B. Cho, T.-W. Kim, M. Choe, G. Wang, S. Song and T. Lee, *Org. Electron.*, 2009, **10**, 473.
49. D.-I. Son, J.-H. Kim, D.-H. Park, W. K. Choi, F. Li, J. H. Ham and T. W. Kim, *Nanotechnology*, 2008, **19**, 055204.
50. N.-G. Kang, B.-G. Kang, H.-D. Koh, M. Changez and J.-S. Lee, *React. Funct. Polym.*, 2009, **69**, 470.
51. T. J. Lee, S. Park, S. G. Hahm, D. M. Kim, K. Kim, J. Kim, W. Kwon, Y. Kim, T. Chang and M. Ree, *J. Phys. Chem. C*, 2009, **113**, 3855.
52. E. Y. H. Teo, Q. D. Ling, Y. Song, Y. P. Tan, W. Wang, E. T. Kang, D. S. H. Chan and C. Zhu, *Org. Electron.*, 2006, **7**, 173.
53. S. Möller, C. Perlov, W. Jackson, C. Taussig and S. R. Forrest, *Nature*, 2003, **426**, 166.
54. V. S. Reddy, S. Karak and A. Dhar, *Appl. Phys. Lett.*, 2009, **94**, 173304.
55. J. Ouyang and Y. Yang, *Appl. Phys. Lett.*, 2010, **96**, 063506.
56. M. Lazzari and M. A. López-Quintela, *Adv. Mater.*, 2003, **15**, 1583.
57. C.-A. Dai, W.-C. Yen, Y.-H. Lee, C.-C. Ho and W.-F. Su, *J. Am. Chem. Soc.*, 2007, **129**, 11036.
58. M. LecLerc, *J. Polym. Sci., Part A: Polym. Chem.*, 2001, **39**, 2867.
59. S.-J. Liu, P. Wang, Q. Zhao, H.-Y. Yang, J. Wong, H.-B. Sun, X.-C. Dong, W.-P. Lin and W. Huang, *Adv. Mater.*, 2012, **24**, 2901.
60. A. Hirao, S. Loykulnant and T. Ishizone, *Prog. Polym. Sci.*, 2002, **27**, 1399.
61. N.-G. Kang, B.-G. Kang, Y.-G. Yu, M. Changez and J.-S. Lee, *ACS Macro Lett.*, 2012, **1**, 840.
62. L. Akcelrud, *Prog. Polym. Sci.*, 2003, **28**, 875.
63. Y. K. Fang, C. L. Liu, C. Li, C. J. Lin, R. Mezzenga and W. C. Chen, *Adv. Funct. Mater.*, 2010, **20**, 3012.
64. S. L. Lian, C. L. Liu and W. C. Chen, *ACS Appl. Mater. Interfaces*, 2011, **3**, 4504.
65. D. B. Velusamy, S. K. Hwang, R. H. Kim, G. Song, S. H. Cho, I. Bae and C. Park, *J. Mater. Chem.*, 2012, **22**, 25183.
66. S. K. Hwang, J. R. Choi, I. Bae, I. Hwang, S. M. Cho, J. Huh and C. Park, *Small*, 2013, **9**, 831.
67. J. S. Sung, J. Huh, J. H. Choi, S. J. Kang, Y. S. Choi, G. T. Lee, J. Cho, J. M. Myoung and C. Park, *Adv. Funct. Mater.*, 2010, **20**, 4305.
68. H. Zhuang, X. Xu, Y. Liu, Q. Zhou, X. Xu, H. Li, Q. Xu, N. Li, J. Lu and L. Wang, *J. Phys. Chem. C*, 2012, **116**, 25546.
69. C. L. Liu, J. C. Hsu, W. C. Chen, K. Sugiyama and A. Hirao, *ACS Appl. Mater. Interfaces*, 2009, **1**, 1974.
70. Y. Wei, D. Gao, L. Li and S. Shang, *Polymer*, 2011, **52**, 1385.
71. J. Vandendriessche, P. Palmans, S. Toppet, N. Boens, F. C. De Schryver and H. Masuhara, *J. Am. Chem. Soc.*, 1984, **106**, 8057.
72. J. C. Hsu, C. L. Liu, W. C. Chen, K. Sugiyama and A. Hirao, *Macromol. Rapid Commun.*, 2011, **32**, 528.
73. Q. D. Ling, S. L. Lim, Y. Song, C. X. Zhu, D. S. H. Chan, E. T. Chan, K. Kang and K.-G. Neoh, *Langmuir*, 2007, **23**, 312.

74. S. G. Hahm, N.-G. Kang, W. Kwon, K. Kim, Y. G. Ko, S. Ahn, B.-G. Kang, T. Chang, J.-S. Lee and M. Ree, *Adv. Mater.*, 2012, **24**, 1062.
75. B. Zhang, Y. Chen, L. Xu, L. Zeng, Ying He, En-T. Kang and J. Zhang, *J. Polym. Sci., Part A: Polym. Chem.*, 2011, **49**, 2043.
76. W. Kwon, B. Ahn, D. M. Kim, Y. G. Ko, S. G. Hahm, Y. Kim, H. Kim and M. Ree, *J. Phys. Chem. C*, 2011, **115**, 19355.
77. B. Cho, T. W. Kim, M. Choe, G. Wang, S. Song and T. Lee, *Org. Electron.*, 2009, **10**, 473.
78. J. H. Shim, J. H. Jung, M. H. Lee, T. W. Kim, D. I. Son, A. N. Han and S. W. Kim, *Org. Electron.*, 2011, **12**, 1566.
79. C. Li, J. C. Hsu, K. Sugiyama, A. Hirao, W. C. Chen and R. Mezzenga, *Macromolecules*, 2009, **42**, 5793.
80. K. Sugiyama, A. Hirao, J. C. Hsu, Y. C. Tung and W. C. Chen, *Macromolecules*, 2009, **42**, 4053.
81. Ras, R. H. A. Laiho, S. Valkama, J. Ruokolainen, R. Österbacka and O. Ikkala, *Macromolecules*, 2006, **39**, 7648.
82. B. Ahn, D. M. Kim, J. C. Hsu, Y. G. Ko, T. J. Shin, J. Kim, W. C. Chen and M. Ree, *ACS Macro Lett.*, 2013, **2**, 555.
83. J. C. Hsu, C. Li, K. Sugiyama, R. Mezzenga, A. Hirao and W. C. Chen, *Soft Matter*, 2011, **7**, 8440.
84. D. I. Son, D. H. Park, J. B. Kim, J. W. Choi, T. W. Kim, B. Angadi, Y. Yi and W. K. Choi, *J. Phys. Chem. C*, 2011, **115**, 2341.
85. P. Y. Lai and J. S. Chen, *Org. Electron.*, 2009, **10**, 1590.
86. Q. Zhang, J. Pan, X. Yi, L. Li and S. Shang, *Organic Electronics*, 2012, **13**, 1289.
87. N.-G. Kang, B. Cho, B.-G. Kang, S. Song, T. Lee and J.-S. Lee, *Adv. Mater.*, 2012, **24**, 385.
88. K. Kim, Y. Y. Kim, S. Park, Y. G. Ko, Y. Rho, W. Kwon, T. J. Shin, J. Kim and M. Ree, *Macromolecules*, 2014, **47**, 4397.
89. B.-G. Kang, S. Song, B. Cho, N.-G. Kang, M.-J. Kim, T. Lee and J.-S. Lee, *J. Polym. Sci., Part A: Polym. Chem.*, 2014, **52**, 2625.
90. M. Tardi and P. Sigwalt, *Eur. Polym. J.*, 1972, **8**, 137.
91. N.-G. Kang, M. Changez and J.-S. Lee, *Macromolecules*, 2007, **40**, 8553.
92. N.-G. Kang, M. Changez, M.-J. Kim and J.-S. Lee, *Macromolecules*, 2014, **47**, 6706.
93. B.-G. Kang, J. Jang, Y. Song, M.-J. Kim, T. Lee and J.-S. Lee, *Polym. Chem.*, 2015, **6**, 4264.
94. B.-G. Kang, J. Jang, Y. Song, M.-J. Kim, T. Lee and J.-S. Lee, *Macromolecules*, 2014, **47**, 8625.
95. W. Elsawy, M. Son, J. Jang, M.-J. Kim, Y. Ji, T. W. Kim, H. C. Ko, A. Elbarbary, M. H. Ham and J.-S. Lee, *ACS Macro Lett.*, 2015, **4**, 322.

CHAPTER 9

Organic Transistor Memory Devices and Materials

CHIAO-WEI TSENG[a] AND YU-TAI TAO*[a]

[a]Institute of Chemistry, Academia Sinica, Taipei, Taiwan, Republic of China, 11529
*E-mail: ytt@chem.sinica.edu.tw

9.1 Basic Concepts of Organic Field-Effect Transistors

With their potential advantages of light weight and ease of device fabrication, organic semiconducting materials are promising candidates for the future trend of smart wearable and foldable electronic applications.[1,2] In contrast to their inorganic counterparts, the properties of organic materials can be readily tailored through rational design and chemical synthesis to meet what is required in each application. Among them, organic field-effect transistors (OFETs), which amplify and switch electric signals, are being developed for applications in displays,[3,4] electronic paper,[5] radio frequency identification tags (RFIDs),[6,7] complementary inverters and chemical/biological sensors.[8,9] Unlike complex traditional bipolar junction transistors,[10] OFETs are composed of an organic semiconductor channel, a dielectric layer, and electrodes (source, drain, and gate). In terms of the arrangement of these components, OFETs can be classified into four types of configuration: bottom gate/top contact (BGTC), bottom gate/bottom contact (BGBC), top gate/top contact (TGTC) and top gate/bottom contact (TGBC) (Figure 9.1). Charge carriers are

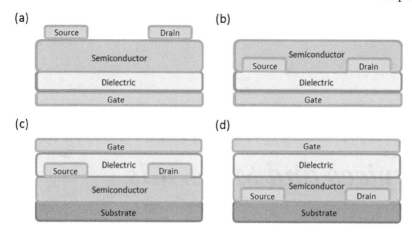

Figure 9.1 Organic field-effect transistor device configuration. (a) Bottom gate/top contact (BGTC), (b) bottom gate/bottom contact (BGBC), (c) top gate/top contact (TGTC) and (d) top gate/bottom contact (TGBC).

injected from the source electrode into the semiconductor upon the action of a gate bias, and migrate to the drain electrode upon application of a source–drain bias. The conductivity of the semiconductor depends on the density of charge carriers generated at the conducting channel/dielectric interface. The carrier density can be modulated by the application of a gate bias. The source–drain current (I_{DS}), the applied gate bias (V_G) and the source–drain bias (V_{DS}) in linear and saturation regions are related by eqn (9.1) and eqn (9.2), respectively.

$$(I_{DS})_{lin} = (W/L)\mu_{lin}C_i[(V_G - V_{th})V_{DS} - (1/2)V_{DS}^2] \quad (9.1)$$

$$(I_{DS})_{sat} = (W/2L)\mu_{sat}C_i(V_G - V_{th})^2 \quad (9.2)$$

where W is the channel width between source and drain electrodes, L is the channel length, C_i is the capacitance per unit area of the dielectric, V_{th} is the threshold voltage and μ is the field-effect charge mobility of the semiconductor. The linear current region is observed at initial low V_{DS} ($V_{DS} \ll V_G$), where the current increases with increasing V_G and V_{DS} (eqn (9.1)). The saturation region is observed under high V_{DS} ($V_{DS} \gg V_G$) (eqn (9.2)). In general, the key parameters for characterizing OFETs include the field-effect charge mobility (μ), which represents the average carrier drift velocity per unit electric field, the threshold voltage (V_{th}), which is the extrapolated voltage at which the transistor turns on, the ON/OFF ratio, which is defined as the ratio between the highest (ON) and the lowest (OFF) current value, and the subthreshold swing (S), defined by eqn (9.3), related to the sensitivity of the current to the gate bias change. It is estimated using eqn (9.4).[11]

$$S = dV_G/d(\log I_{DS}) \text{ at } V_G < V_{th} \quad (9.3)$$

$$S = kT\ln 10/e[1 + (e^2/C_i)N_{total}] \quad (9.4)$$

where N_{total} is the total trap density including bulk film traps and interfacial traps per unit area per unit energy. Optimization of these parameters is very important for the development of high performance OFETs. Besides the chemical structure of the organic semiconductor itself,[12] the metal electrodes[13] (which influence the contact resistance), the dielectric materials[14] (which affect the gate electric field, the gate current leakage, and the charge trapping) and the metal/organic semiconductor and organic semiconductor/dielectric interfaces[15] (which influence semiconductor film morphology and microstructure) all contribute to OFET performance. To date, OFETs with performance exceeding that of amorphous silicon ($\mu \sim 0.1$ to 1, ON/OFF ratio $\sim 10^6$ to 10^8) and approaching that of polycrystalline silicon ($\mu > 10$) have been reported. These developments and the eventual commercial production are bound to make a huge impact in future electronics applications.

9.2 Transistor Memory Devices

Memory devices are one of the key components for data storage in electronic circuity. There are volatile and non-volatile types. Volatile memory requires power to maintain stored data, *e.g.* dynamic/static random access memory (RAM). Non-volatile memory retains stored data without continued powering after a switching action, *e.g.* magnetic tapes, optical discs, hard disks and flash memory. Among these, flash memory devices composed of floating gate transistor arrays are widely used. For a memory device, two (or more) controlled states (for example, conductivity, magnetism, polarization, *etc.*) that can be switched between each other are required. In conventional Si-based transistor devices, memory effects can be achieved by inserting a floating gate (conductively-doped poly-Si) within the gate dielectric (Figure 9.2a). The floating gate is insulated all around by an oxide layer. Trapped charges in the floating gate create an additional electrical field which shields the applied gate field and shifts the threshold voltage, resulting in a response change, so that data can be stored by encoding "0" and "1" (the binary state detection) on the basis of different electric responses (for example, conductivity) at a specific voltage (Figure 9.2b).

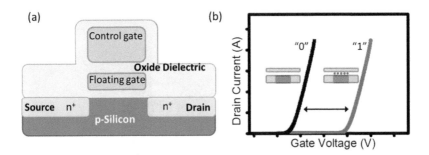

Figure 9.2 Schematic illustrations of (a) a cross-section of a Si-based floating gate transistor, and (b) the operational mechanism and the transfer characteristics (I_{DS} *vs.* V_G) for two binary states.

The processes in which charges are transferred to the floating gate and trapped there and the processes in which trapped charges are de-trapped back to the channel are referred to as programming and erasing (Figure 9.3). An insulating block (the tunneling dielectric layer) needed to retain the charges in the floating gate could also be a barrier for the charges to be trapped. Depending on the insulator thickness and applied gate bias, direct tunneling and Fowler–Nordheim (F–N) tunneling are possible mechanisms to program/erase devices (Figure 9.3d).[16,17] With a very thin insulating layer, the charges can readily reach the floating gate through direct tunneling under low gate bias (low operation voltage) to effect the conductivity change (thus a short response time). However, a thin insulator layer may also result in poor charge retention capability.

Several parameters are used to characterize a memory device. For example, the memory window, defined as the threshold voltage shift (ΔV_{th}) between the programming and erasing operations, indicates the ability for charge storage of the floating gate. The number of trapped charges (Δn) can be estimated according to eqn (9.5):[18,19]

$$\Delta n = -\Delta V_{th} \times C_i/e \qquad (9.5)$$

where ΔV_{th} is the threshold voltage shift, C_i is the dielectric capacitance and e is the elementary charge. The larger the memory window, the more accurate the data read-out will be. The charge retention characteristics, reflecting the data keeping, are another important parameter for memory devices. The retention capability of transistor memory is primarily influenced by the defect charge leakage from the gate dielectric, the lateral charge diffusion

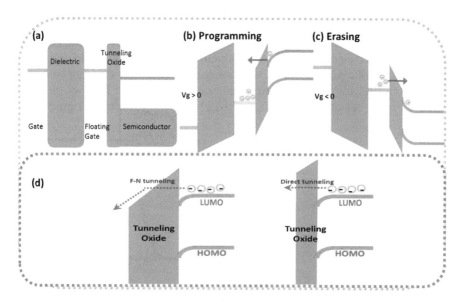

Figure 9.3 Schematic illustrations of energy band diagrams of a floating gate transistor memory under (a) zero gate voltage, (b) program mode, (c) erase mode, and (d) F–N tunneling and direct tunneling through a potential barrier.

between the floating gate and the charge loss during the read-out process. Furthermore, the programming/erasing speed, or the response time, refers to the minimum time required to program/erase a device. The reliability or durability is the capability of the memory device to endure repeated write (program)–read–erase–read (WRER) operation cycles.

For high performance, high density, and low cost applications, a great deal of effort has been devoted in recent years to scale down device size and build three-dimensional stacking of planar arrays of memory elements. One of the main limitations of conventional flash memory is that charge loss from the floating gate is facile due to the reduction in the thickness of the insulator and the spacing between adjacent floating gates.[20,21] Recently, flash memory relying on trapping charges in discrete charge trapping sites (charge-trap flash), such as non-conductive silicon nitride (Si_3N_4),[22,23] metal nanoparticles[24,25] or dispersed charge-accepting organic molecules,[26,27] instead of conventional floating gates (conductive doped poly-Si) have shown numerous advantages including improved device reliability and enhanced data retention for small feature size device applications.

9.3 Organic Transistor Memory Devices

Organic materials with unique properties such as molecular dipoles, discrete HOMO/LUMO energy levels, conjugated systems and appropriate electrochemical reduction/oxidation potentials have attracted much attention for building various electronic devices. Memory devices based on organic materials for data storage are yet another important component to consider in electronic circuitry. In order to achieve electrical bistability (the binary states of 1 and 0) in organic non-volatile memory (ONVM) based on OFETs, the charge carriers (after accumulating/dissipating) need to be maintained in the semiconductor channel after the gate bias is turned off. This has been achieved by intentionally incorporating ferroelectric polymers, polymer electrets or charge carrier traps in the conducting channel or in the dielectric layer to effect conductivity changes upon application of an external electric field to induce polarization change or charge trapping. The process can be reversed by inverting the external gate bias, as shown in Figure 9.4.[28] Depending on the way the carrier density is modulated, there are several types of organic transistor memory.

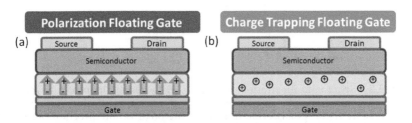

Figure 9.4 Schematic configuration of (a) polarization induced memory device, and (b) charge trapping induced memory device.

9.3.1 Organic Transistor Memory Devices Based on Ferroelectrics

Ferroelectrics are materials with intrinsic polarization that can be aligned or reversed by different external electric fields, as a result of re-orientation of the intrinsic dipoles. The electrical polarization response of ferroelectrics is analogous to the magnetic response in ferromagnets.[29] A Sawyer–Tower circuit can be used to characterize the ferroelectric polarization.[30] An idealized hysteresis loop based on the electric displacement as a function of field strength is shown in Figure 9.5. The electric displacement caused by free charges ($\varepsilon_r E$) and polarization (P) of the ferroelectric film under an externally applied electric field (E) can be described by eqn (9.6).[31]

$$D = \varepsilon_r E + P \tag{9.6}$$

where D the electric displacement (C m^{-2}), ε_r the relative dielectric permittivity, E the applied electric field and P is the polarization (C m^{-2}) due to intrinsic dipole moments of the material. At low electric field, the electric field is too small to trigger polarization, whereas with increasing electric field, the displacement increases linearly (eqn (9.6)). With continuous increase of the electric field, the film starts to be polarized by the electric field, resulting in rapidly decreasing/increasing dielectric displacement. A linear relation again is measured once the film is fully polarized. As the electric field shifts toward the negative direction, a reverse polarization takes place. The coercive field (E_c), which is the critical electric field for reversing the polarization of the ferroelectric film, is defined as the field needed to offset the polarization to zero. The saturation polarization is represented as P_s. The remnant polarization (P_r) is the holding polarization without the electric field. An idealized D–E hysteresis loop should be symmetrical ($+E_c = -E_c$ and $+P_r = -P_r$) and close to square shape (P_s equal to P_r). The coercive field (E_c), the remnant polarization (P_r)

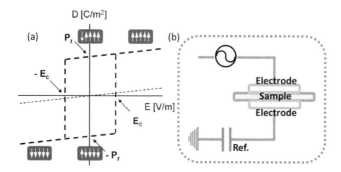

Figure 9.5 Schematic illustrations of (a) a D–E hysteresis loop of a ferroelectric thin film. The displacement (D) at zero applied electric field is the remnant polarization (P_r), switching polarization direction takes place under the coercive field (E_c). (b) The Sawyer–Tower circuit used for these measurements.

and the shape of the D–E loop, susceptible to the thickness, environmental temperature, mechanical stresses and charged defects in the film, are the main parameters used to characterize ferroelectrics.

In organic ferroelectric FET memory, a ferroelectric is used as a gate insulator or inserted at the interface between the semiconductor and gate insulator. The electric bistability states can be reached by polarization switching of the ferroelectric through the action of a gate bias to modulate the conductivity of the transistor. Hole or electron carriers are induced in the conductive channel primarily depending on the direction of the polarization; as a result a positive or negative threshold voltage shift occurs and the corresponding source–drain current changes. The film crystallinity of a ferroelectric is important, because intrinsic dipole moments cancel each other in amorphous films due to random orientations. To make the film ferroelectric, further optimization of the film crystallinity is necessary, such as by high temperature annealing or by introducing impurities or strain.[32,33] These additional treatments usually bring about a rough and inhomogeneous film surface, which is undesirable for high performance OFETs.[34] The ferroelectricity of poly(vinylidene fluoride) (PVDF) derives from the alignment of the C–F dipole moments, which can be altered by a gate bias applied through rotating the polymer backbones.[35] Compared to α or γ phase crystallites, the β phase crystallites of PVDF film have an all-*trans* chain conformation, giving superior polarization and ferroelectric response (Figure 9.6).[36] Thermal treatment can promote polymer chains to reorganize to form a highly crystalline β phase.[37] Enhancement of the β phase crystallites of the film by stretching the polymer to force a polymer conformation change was demonstrated as well.[38,39] In addition, the copolymerization of PVDF and poly(trifluoroethylene) (PTrFE) is another efficient way to induce the formation of β phase crystallites. By means of the additional fluorine atoms of PTrFE, steric hindrance assists in aligning the dipole moments and induces the all-*trans* conformation.[40,41]

In order to enhance the device performance for practical applications, it is necessary to address the surface roughness issue. An amorphous material, nylon, poly(*m*-xylylene adipamide) (MXD6), with ferroelectric-like properties was used as the gate insulator.[42] Without crystalline grain boundaries, amorphous nylon is potentially beneficial for inducing a smooth surface. Although a clear memory window with a distinct ON/OFF current ratio and a good retention capability was obtained, the program/erase speed (the switching time) is much slower in the MXD6-based device than in the PVDF/PTrFE device. Moreover, blending the ferroelectric PVDF with an amorphous polymer, such as poly(methyl methacrylate) (PMMA), to modulate the film morphology has also been demonstrated.[43] A device based on the PVDF/PMMA blend film with low surface roughness exhibited outstanding transistor performance (Figure 9.7a). The surface roughness (RMS) of the blend film apparently decreased with increasing content of PMMA. For example, the RMS of 5.1 nm in a pure PVDF film was reduced to 0.45 nm in a 60% PMMA blend film. However, the increasing PMMA content (which diluted the percentage of PVDF) resulted in lower ferroelectricity of the blend film

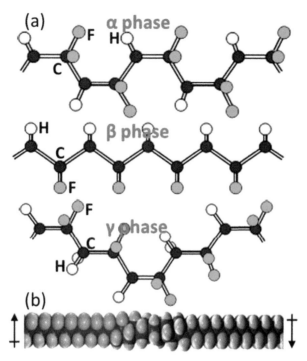

Figure 9.6 (a) Three PVDF crystalline conformation structures. (b) Schematic illustration of the dipole switching by rotating its axis in β phase PVDF. Reproduced with permission from ref. 41 © 2014 Nature Publishing Group and ref. 29 © 2009 WILEY-VCH Verlag GmbH & Co. KGaA, Weinheim.

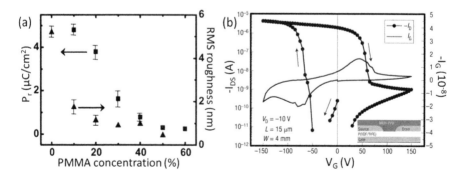

Figure 9.7 (a) Plots of remnant polarization (P_r) and RMS roughness as a function of the PMMA content in the blend films. (b) Hysteretic drain current as a function of gate voltage for MEH-PPV/P(VDF/TrFE) FeFET. Reproduced with permission from ref. 43 © 2009 WILEY-VCH Verlag GmbH & Co. KGaA, Weinheim and ref. 45 © 2005 Nature Publishing Group.

at the same time. In addition, adopting the top gate/bottom contact (TGBC) configuration offered an alternative to make a device with low surface roughness and remarkable ferroelectricity simultaneously. In this case, the surface roughness is determined by the semiconductor, rather than the ferroelectric. Thus a top gate ferroelectric-FET with PVDF/TrFE (dielectric) on top of P3HT (semiconductor) was fabricated.[44] A low surface roughness (RMS) of 0.7 nm can be obtained by optimizing the P3HT film growth process. High charge carrier mobility (0.1 cm^2 V^{-1} s^{-1}) and electric bistability can be obtained in this device. Alternatively, by using amorphous polymers, such as poly(2-methoxy-5-(2′-ethylhexyloxy)-*p*-phenylenevinylene) (MEH-PPV) (Figure 9.7b), or poly(9,9-dioctylfluorene-*co*-bithiophene) (F8T2) as semiconductors, the smooth surface in combination with the ferroelectric polymer led to improved device performance.[45,46] Recently, a graphene-based top gate ferroelectric-FET with PVDF/PTrFE as gate dielectric was demonstrated to give fast response and improved stability due to the ambipolar nature of graphene which compensates changes for both ferroelectric states.[47] Thus, fast switching response, long retention properties and high charge carrier mobility can be realized with ferroelectric-based memory devices.[48,49]

9.3.2 Organic Transistor Memory Devices Based on Charge Trapping

9.3.2.1 *Charge Trapping with Polymer Electrets as Trapping Sites*

An electret refers to a dielectric material with a quasi-permanent electrical field. In polymer electret transistor memory, charges are directly trapped in a polymer insulator instead of a floating gate. Therefore, a change of charging/discharging state in the polymer electret by applying a gate bias can bring about a bistability in the conductive channel. The device performance is strongly dependent on the nature of the polymer, such as its hydrophobicity/hydrophilicity, polarity/non-polarity, or even the functional groups on the polymer side-chains.[50,51] For instance, in 2002, a reversible threshold voltage shift was observed for both p- and n-channel transistors, in which the hydrophobic polymer insulators, poly(4-methylstyrene) (P4MS) and TOPAS (a cyclic olefin copolymer) were used. These systems exhibited better memory characteristics than devices with hydrophilic SiO$_2$ and glass resin (GR).[52] The threshold voltage shift was attributed to either the result of trapped static charges or induced dipole orientation of the polymer dielectric. Poly(α-methylstyrene) (PαMS) is another material shown to serve as a charge electret.[53] Incorporating PαMS in a pentacene-based transistor resulted in excellent memory performance, including a large memory window (90 V), a short switching time (~1 µs), and a long charge retention time (100 h). However, high operation voltages (60–100 V) were necessary. The memory effect was believed to originate from charge storage in PαMS by charge transfer from the pentacene to the PαMS, since the threshold voltage (V_{th}) shift was related to the gate bias applied during the programming process, and it was observed

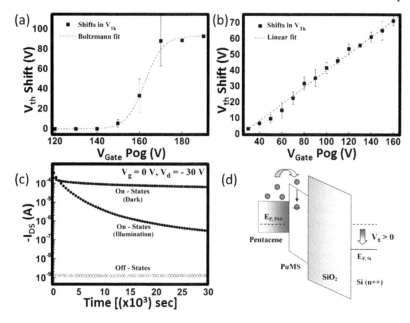

Figure 9.8 Threshold voltage shifts as a function of pulse gate bias (with the same pulse time of 1 μs) for the PαMS devices (a) without (b) with illumination with visible light during the programming process, where dashed lines are fitting results using Boltzmann and linear fits. (c) Retention characteristics of the PαMS devices with and without illumination. The ON and OFF states were reached by applying +200 V and −100 V for 1 μs, respectively. (d) Schematic illustration of charge transfer between pentacene and PαMS. Reproduced with permission from ref. 53 © 2006 WILEY-VCH Verlag GmbH & Co. KGaA, Weinheim.

that the charge retention capability decreased upon photo-illumination. This implies that there is a critical energy barrier for transfer and charge trapping (Figure 9.8). In addition, implanting a series of styrenic polymers between pentacene and a SiO_2 dielectric in a top-contact/bottom-gate OFET provided further insight into the real role that the polymer electret played.[54] The polymers used included non-polar and hydrophobic polymers (PS, PVN, PαMS, P4MS), moderately polar and hydrophilic polymers (PVP, PVPyr), and a highly polar and hydrophilic polymer (PVA) as shown in Figure 9.9a. The magnitude of the memory window and the charge retention time of memory devices strongly depend on the hydrophobicity (water contact angle) and polarity (dielectric constant) of the polymer electrets (Figure 9.9). It was observed that non-polar and hydrophobic polymers (PS, PVN and PαMS) are more efficient chargeable dielectrics than those hydrophilic and polar polymers (PVA, PVP and PVPyr). Additionally, the PVP thin film prepared using the solvents with different amounts of moisture, ions and impurities resulted in different transfer curve shifts. The results suggested that the hydrophilic and polar polymers had poor charge trapping capability due to rapid dissipation of

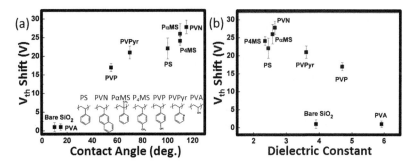

Figure 9.9 Threshold voltage shifts as a function of (a) contact angle (hydrophobicity), and (b) dielectric constant for the polymer electrets used. Reproduced with permission from ref. 54 © 2008 WILEY-VCH Verlag GmbH & Co. KGaA, Weinheim.

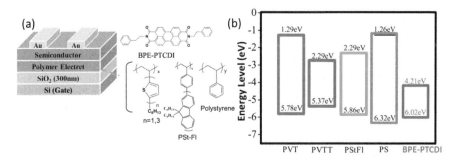

Figure 9.10 (a) Structure of the BPE-PTCDI thin film based polymer electret OFET memory device. (b) Energy-band diagram in zero-gate-voltage configuration of polymer electrets and the BPE-PTCDI semiconductor. Reproduced from ref. 58 © 2014 RSC.

stored charge through the conductive pathways derived from dipoles, moisture, ions and even impurities remaining in the polymer electrets. Recent studies indicated that in ambipolar transistor memory, both holes and electrons can be accumulated in the channel and subsequently trapped into the polymer dielectric.[55,56] The ambipolarity of the semiconductor can lead to bidirectional transfer curve shifts, a large memory window and enhanced device performance (such as shorter programming time). It is suggested that hydrophilic and polar polymer dielectrics may contain hydroxyl groups, which could trap electrons irreversibly to hamper electron carrier accumulation and suppress the ambipolarity, resulting in a poor memory effect.[57]

Polymer electrets with pendent conjugated side-chains could provide additional charge active sites which benefit the charge storage capability to enhance the memory effect. BPE-PTCDI-based transistors with polymer electrets prepared from vinylthiophene (PVT), vinylterthiophene (PVTT), styrene (PS), and styrene *para*-substituted with fluorene (PSt-Fl) were studied (Figure 9.10).[58,59] The longest thiophenyl side chain with its high HOMO level and smallest band gap facilitated hole injection from BPE-PTCDI due to a small barrier, so that a larger memory window can be observed in the PVTT device compared to the PVT one. In addition, the PVTT devices showed a larger memory window and better device performance than PSt-FI, which has a pendant of similar conjugation length, since the thiophene moiety is a stronger electron donor compared with the fluorene or phenyl moiety. These observations suggested that the charge storage capability of the device was primarily dominated by the conjugation length and the donor/acceptor properties of the side chains.

9.3.2.2 Charge Trapping with Molecule–Polymer Hybrids as Trapping Sites

A hybrid electret system that directly incorporates molecular units as the functional moieties into the polymer matrix has been proposed for building memory devices. Steep flanks of hysteresis with a large memory window can be obtained by embedding donor molecules, such as tetrathiafulvalene (TTF) and ferrocene (Fc), into a blend matrix, such as poly(ethylene oxide) (PEO), polycarbonate (PC), PS and PMMA, in the CuPc-based transistor.[60] The memory effect is attributed to the trapping of charge in the molecular donor moieties of the hybrid system, because the device without these donor molecules showed little hysteresis. The trapped charges remain in the polymer electret due to the tunneling barrier imposed by the insulating polymer matrix. In another study, BPE-PTCDI OFETs with hybrid polymer electrets composed of a star-shaped donor polymer matrix of N(PTPMA)$_3$ blended with the acceptor molecule PCBM, the donor molecule TIPS-pen, and the donor molecule Fc were examined (Figure 9.11).[61] In the PCBM/N(PTPMA)$_3$ blend electret device, the memory behavior changed from irreversible characteristics (write-once-read-many, WORM) to reversible characteristics after blending PCBM into N(PTPMA)$_3$. The irreversible behavior was thought to be caused by the strongly electron-donating nature of the triphenylamine moieties in the polymer matrix. The blending of PCBM enables negative charges to be trapped in the electret, so that the device could be erasable and switchable. The memory window increased with increasing PCBM percentage. Moreover, in comparison with the blend TIPS-pen/N(PTPMA)$_3$ and Fc/N(PTPMA)$_3$ electret devices, the memory features showed reversible and irreversible characteristics (WORM), respectively. Both N(PTPMA)$_3$ and Fc are strong electron-donors with high LUMO levels, giving rise to a larger energy barrier for electron injection between semiconductor BPE-PTCDI and Fc than that between BPE-PTCDI and TIPS-pen or PCBM (Figure 9.11). Reversible characteristics were obtained once the electron-trapping unit was included in the blend electret. The characteristics of the memory device are strongly affected by the energy levels and charge transfer in the molecule–polymer hybrid electret.

Organic Transistor Memory Devices and Materials 307

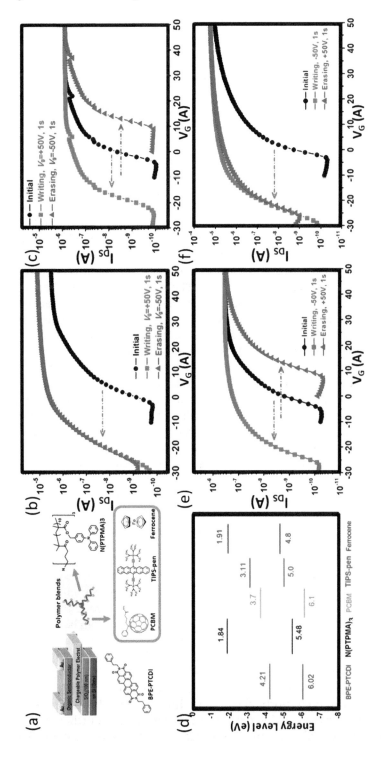

Figure 9.11 (a) Device structure and chemical structures of compounds used in the memory device. Transfer curve shifts of BPE-PTCDI-based transistor devices with (b) N(PTPMA)$_3$, (c) 5 wt% PCBM:N(PTPMA)$_3$, (e) 5 wt% TIPS-pen:N(PTPMA)$_3$ and (f) 5 wt% ferrocene:N(PTPMA)$_3$ as electrets. (d) Energy-band diagram of the materials studied. Reproduced with permission from ref. 61 © 2014 American Chemical Society.

9.3.2.3 Charge Trapping with Redox-Active Molecules as Trapping Sites

Redox-active molecules are capable of gaining or losing charges (electrons) upon application of a driving potential. For instance, ferrocene can lose an electron to undergo one electron oxidation reaction,[62] while Zn(II) porphyrin (ZnPor) can lose up to two electrons so that it can exist in three different forms (ZnPor^0, ZnPor^{+1} and ZnPor^{+2}).[63,64] If a molecule can reversibly exist in different oxidization/reduction states, it has the potential to be used as a charge trap in a memory device.[65] While redox chemistries commonly occur in solution, similar behavior can be observed when molecules are confined on a surface in a solid state environment. A self-assembled monolayer (SAM) of a redox-active molecule on a variety of substrates provides a simple and useful way to provide a platform to study its electronic properties in a solid state environment. Prior work based on the SAMs of ferrocene and ZnPor derivatives grafted on silicon (Si) substrates showed one and two reversible redox states, respectively,[66,67] and the redox reaction is suggested to be due to charge (electron) transfer between the molecules and the Si substrate. The transfer can be blocked by a tunnel barrier such as a long alkyl-chain linker between the redox moiety and the Si substrate[68,69] or a thin silicon oxide layer.[70,71] Tetrathiafulvalene (TTF) is another example which can be easily and reversibly oxidized to the radical cation TTF$^{·+}$ and the dication TTF^{2+} ($E^1_{1/2}$ (0.34 V) and $E^2_{1/2}$ (0.78 V)). The redox switching capability of a TTF-based SAM grafted on indium tin oxide (ITO) showed similar redox properties to those in solution.[72] A series of Write 1–Read–Write 2–Read–Erase–Read cycles employing an electrochemical input for writing and optical (absorption spectrum at 420 nm) and magnetic (EPR signal) outputs as read-out signals for TTF SAMs have been demonstrated. The three redox states of the TTF SAM presented excellent reversible switching characteristics and a high stability under repeated external electrical stimuli.

Incorporating molecular thin films into metal-oxide–semiconductor (MOS) capacitors to realize charge storage behavior of organic molecules in practical device applications has been carried out.[73] A molecular thin film-based capacitor with SiO_2 and Al_2O_3 as the tunneling and control oxide, respectively, was fabricated by thermal evaporation of a series of molecules with different energy levels (HOMO and LUMO) and charge carrier mobilities, as shown in Figure 9.12. Capacitance–voltage measurements were used to study charge storage, retention and programming/erasing (P/E) endurance characteristics. High charge storage density (around 5×10^{12} cm^{-2}) and good P/E cycling endurance (only 10% variation after more than 10^5 P/E cycles) can be observed in PTCDA-, PTCBI- and C_{60}-based molecular thin film devices. A charge retention capability determined by the lateral charge mobility of the thin films was suggested. The high charge retention time obtained in the PTCBI-containing capacitor is attributed to the low charge mobility (10^{-4} to 10^{-5} cm^2 V^{-1} s^{-1}) of the PTCBI films with poor π-orbital overlap, which prevents lateral transport of trapped charges to the existing defects in the oxide layers. Further applications of redox-active molecules in nanodevices such as nanowire field-effect

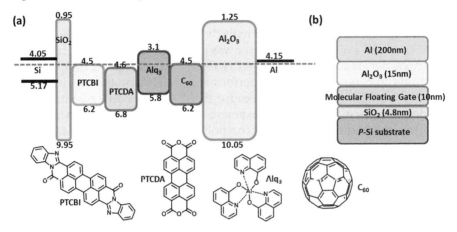

Figure 9.12 (a) Energy-band diagram in zero-gate-voltage configuration of four different metal-oxide–semiconductor (MOS) capacitors with the chemical structure of each material. (b) Schematic cross section of the device structure.

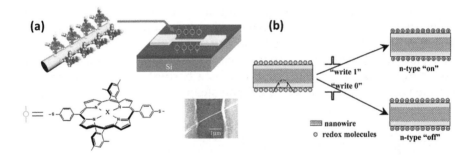

Figure 9.13 (a) NW-based FET functionalized with porphyrin molecules (the inset shows the SEM image of the device). (b) Schematic illustration of positive charges creating an ON or logic "1" state, and negative charges producing an OFF or logic "0" state. Reproduced with permission from ref. 74 and 76 © 2004 and 2002 American Chemical Society.

transistors (NW-FETs) have been demonstrated.[74,75] n-Channel InP NW-FETs with an oxide shell functionalized with cobalt-phthalocyanine (CoPc) molecules by spin-coating or chemisorption onto the conduction channel were fabricated, as shown in Figure 9.13.[76] Sweeping/pulsing the gate bias enables conversion of the CoPc molecules into different redox states ($Co^{2+/3+}$), which subsequently alters the conductance of the FET and gives the device electric hysteresis/bistability. For example, a negative gate bias pulse caused accumulation of positive charge in the NWs, leading to oxidation of the CoPc molecules. These oxidized molecules working as positive floating gates led to a high conductance of the n-type InP NW-FET. The NW device has an ON/OFF ratio around 10^4 and a charge retention time in excess of 20 min.

A copper phthalocyanine (CuPc)-based OFET with electron-donating TTF derivatives as the buffer layer thermally deposited at the interface between a CuPc layer and a SiO$_2$ dielectric layer was studied.[60] A high concentration of charge carriers is suppressed by the TTF layer due to charge transfer between TTF and CuPc, resulting in poor performance of the transistors with a very low ON/OFF ratio, presumably because the CuPc is in direct contact with the donor molecules (TTF). Improvement of the device performance can be achieved by blending the donors into a polymer matrix such as PMMA, which can weaken the charge transfer by the inserted polymer chains. While electron-donors or electron-acceptors can be expected to trap mainly holes or electrons, respectively, an extended conjugated system which can delocalize both positive charges and negative charges has been used for the purpose of charge trapping. Thus poly(10,12-pentacosadiynoic acid, PDA) with alternating ene–yne conjugated backbones was used as the charge storage vehicle in a pentacene-based thin film transistor.[77] The polymer was prepared by thermal deposition/UV-irradiation of these diacetylenic acid molecules on a silica substrate. Depending on the surface modification (unmodified or OTS-modified), the molecules self-assembled into unique rings of H-bonded dimers or discrete multilayer clusters. The conjugated backbones obtained upon topochemical polymerization[78] of the monomers serve to trap both holes and electrons. It was observed that, on a bare silica surface, the polyPDA formed rather extended ring patterns of the dimer layer, with few molecular contacts between the pentacene layers and the conjugated backbones (with the alkyl chain interposed as a barrier) (Figure 9.14a and b). On the OTS-modified surface, small and steep clusters of multilayers allow

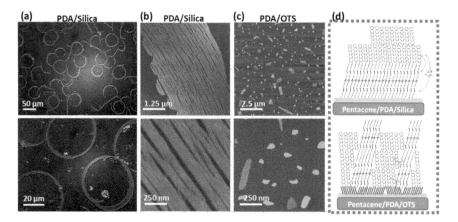

Figure 9.14 (a and b) Optical micrographs and AFM topographic images of polyPDA films formed on a bare silica surface. (c) AFM topographic images of polyPDA films deposited on OTS-modified silica surfaces. (d) Schematic illustration of the charge transfer between the pentacene film and the polyPDA layer on the silica surface and OTS-modified silica surface, respectively. Reproduced with permission from ref. 78 © 2010 American Chemical Society.

more molecular contacts between the pentacene layers and the conducting backbones (Figure 9.14c). The system with greater molecular contact shows faster responses in the trapping/de-trapping processes so that a larger memory window and lower charge retention characteristics were obtained, since charge transport is more facile in such a system (Figure 9.14d).

A PPOM (push–pull organic molecule) containing triarylamine as an electron-donating group, thiophene as a spacer and malononitrile as an electron-withdrawing group, with different dihedral angles induced by steric hindrance of side chains within the molecule is shown in Figure 9.15a. An OFET with PPOMs as charge traps between the PMMA dielectric and the pentacene semiconductor was studied.[79] The charge trapping capability of different PPOMs highly depends upon the tilted dihedral angles. PPOMs with larger dihedral angles trapped more charges (larger memory window), and gave longer charge retention times. This was attributed to the high potential energy barrier needed for the conformational change between the anion state (ON state) and the neutral state (OFF state). Furthermore, light-charge organic memory (LCOM) with the functions of a photo-sensor and non-volatile memory has been demonstrated by incorporating an M-C10 molecule, which is an electron-acceptor with dicyanomethylene groups and n-decyl chains attached to its conjugated core (as shown in Figure 9.15b), into a pentacene-based transistor.[80] The M-C10 modification layer was formed by spin-coating and then thermally annealed. The annealed M-C10 layer with well-organized alkyl chains not only enhances the performance of the pentacene transistor (high carrier mobility of 2.74 cm^2 V^{-1} s^{-1}), but also assists in maintaining the stored charges (92% of the stored charges after 20 000 s). The device was programmed by applying a positive gate bias under photo-illumination. The shift in transfer curve increased with increasing incident light power. It was suggested that the photo-excited electrons transfer from pentacene into the M-C10 layer and are trapped there. The stored electrons could be neutralized by hole carriers accumulated at the interface when a negative gate bias is applied (erasing). A system involving both single crystal fullerene C_{60} needles (N–C_{60})[81] and CuPc nanoparticles (CuPc NPs) is an example of a double floating gate memory device (Figure 9.15c).[82] The device was based on a pentacene thin film transistor using an ITO glass substrate with high dielectric constant (high k) HfO_2 as the insulator, and cross-linked poly(4-vinylphenol) (c-PVP) (covering the N–C_{60}/CuPc double floating gate) as the tunneling layer. The results showed that the double floating gate transistor memory with discrete p-type CuPc NPs and n-type N–C_{60} as hole and electron charge traps could enhance the memory window and data storage capacity.

9.3.2.4 Charge Trapping with Self-Assembled Monolayers as Trapping Sites

Organic transistor memory with a single layer of molecules serving as floating gate has been demonstrated.[83] This approach allows devices to be operated at low power, permits low temperature processing and only requires a very

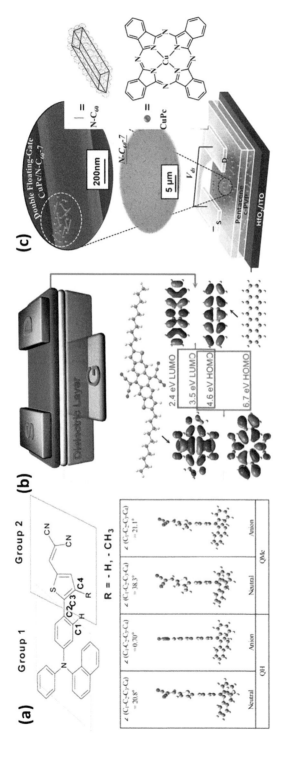

Figure 9.15 (a) Molecular structure of the PPOMs, DFT optimized structures and dihedral angles (C1–C2–C3–C4) of neutral and electronically charged anionic QH (R = H) and QMe (R = CH$_3$). (b) Device configuration of LCOM, and the molecular structures, HOMO–LUMO energy level and corresponding calculated electronic structures of M-C10, as well as pentacene molecules. (c) Schematic illustration of the pentacene based memory device with the CuPc/N-C$_{60}$ double floating gate, along with optical microscopy and TEM images. Reproduced with permission from ref. 79 © 2010 RSC, ref. 80 © 2013 Nature Publishing Group and ref. 82 © 2014 WILEY-VCH Verlag GmbH & Co. KGaA, Weinheim.

simple process to install the floating gate (usually adsorption from solution). One example used a monolayer floating gate consisting of C_{60}-functionalized octadecylphosphonic acids ($C_{60}C_{18}$-PA) mixed with insulating alkyl phosphonic acids (C_{18}-PA) in different ratios for dispersing the charges, as shown in Figure 9.16a. The device with 100% C_{60}-carrying SAM has the largest memory window yet poor retention characteristics. The device with the lowest 1% C_{60}-carrying SAM has the smallest memory window nevertheless it has better retention characteristics than others. The drain current remained 50% of its initial value after 6 h. The longer charge retention time is suggested to be due to the formation of small islands of charge trap moieties (C_{60}-PA) or even single trap sites isolated from each other by the insulating alkyl chains (C_{10}-PA). This would facilitate pinning of the trapped charges. To further improve the memory performance C_{60} tethered to a short alkyl chain ($C_{60}C_6$-PA) was mixed with C_{18}-PA and an even more extended insulating molecule (12-dodeylphosphonic acid) N,N-dimethyl-N-octadecyl ammonium chloride ($C_{18}N^+(Me)_2C_{12}$-PA) to form 5% C_{60} mixed SAMs (Figure 9.16b and c).[84] The C_{60} moieties tend to be deeply buried in the insulator or to be partially shielded by the alkyl chains of the surrounding molecules (C_{18}-PA and ($C_{18}N^+(Me)_2C_{12}$-PA)). Moreover, the alkyl side chain length of the semiconductor (sexithiophene, 6T) was varied from C_2 to C_6 to C_{10}, in the hope that the distance between the semiconductor and the C_{60} trap site could be tuned (Figure 9.16d). The device with mixed SAMs of $C_{60}C_6$-PA and $C_{18}N^+(Me)_2C_{12}$-PA showed a large improvement in the memory window (1.0 V) and charge retention time (10^5 s), compared with the device with mixed SAMs of $C_{60}C_6$-PA and C_{18}-PA (retention, 10^3 s). More effective charge retention is believed to be due to the covering of C_{60} moieties by the flexible octadecyl alkyl chains, so that the charge trapping sites are isolated from the semiconductor channel. A variation of the side chain length of 6T did not impact the memory behavior significantly.

Another example used a monolayer of azobenzene (Azo) dyes as the floating gate in a pentacene-based OFET (Figure 9.17).[85] Electric bistability was triggered by not only electric bias but also photo-irradiation. *Cis–trans* isomerization of the Azo dyes was excluded in this closely packed monolayer. Rather, photo-irradiation results in the excitation of the Azo moieties,[86] followed by charge transfer between the pentacene and the Azo dyes. Depending on the substituent on the Azo moieties, either electron or hole charges can be accommodated, leading to transfer curve shifts in different directions. Electric bias can induce charge trapping on the Azo sites as well, with the efficiency of charge trapping depending on the p-substituent's ability to stabilize a specific charge (hole/electron). For example, with CF_3-substituted Azo, the electrons are more readily trapped and stabilized in this SAM than in others (Figure 9.17b). Hole carriers are more readily transferred to a SAM with CH_3 substituents, which can stabilize the positive charges (Figure 9.17c). In addition, the morphology effect of the molecular floating gate has also been studied. Compared with devices with discrete multilayer clusters of Azo moieties prepared by thermal deposition on silica substrates, the charging of a monolayer of Azo dyes on the dielectric layer is slower, yet the trapped

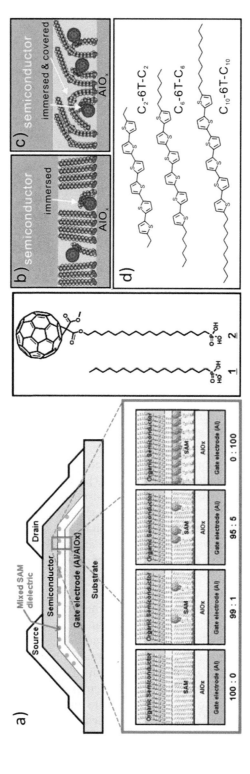

Figure 9.16 Schematic illustration of an OFET with a mixed SAM of (a) C_{18}-PA/$C_{60}C_{18}$-PA, inset shows devices with different stoichiometry of C_{18}-PA and $C_{60}C_{18}$-PA and their chemical structures, (b) C_{18}-PA/$C_{60}C_6$-PA and (c) C_{18}-PA/$C_{18}N^+(Me)_2C_{12}$-PA. (d) The derivatives of sexithiophene semiconductor with alkyl side chains of C_2, C_6 and C_{10}. Reproduced with permission from ref. 83 and 84 © 2010 and 2014 WILEY-VCH Verlag GmbH & Co. KGaA, Weinheim.

Organic Transistor Memory Devices and Materials 315

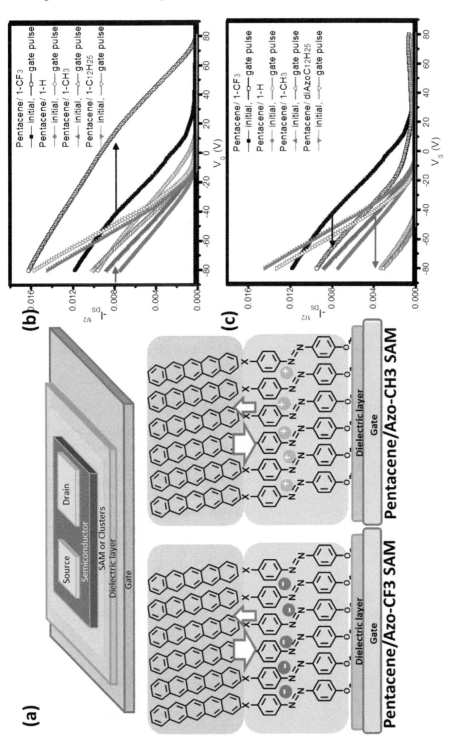

Figure 9.17 Schematic illustration of (a) the transistor memory device configuration and the charge trapping process in the azobenzene (Azo)-based SAMs with CH$_3$ and CF$_3$ functional groups respectively, (b and c) transfer characteristics of the devices after gate bias pulse of 100 V and −100 V for 60 s. Reproduced with permission from ref. 85 © 2012 American Chemical Society.

charges can be retained better. The charging of clusters is faster (as evidenced by a larger memory window under the same gate pulse time with the SAM device) but de-trapping of charges is also faster. This is attributed to the proximity of the trapping sites and conducting channel, which makes charge transfer between each other easier. In contrast, the trapping sites in a monolayer are separated from the semiconductor by alkyl chain spacers in a closely packed monolayer.

9.4 Organic Transistor Memory Devices Incorporating Nanoparticles

Metallic or semiconductor nanoparticles (NPs) are perceived to be good charge carrier traps. Compared to traditional planar floating gates, discrete traps of a NP-containing device could prevent charge leakage from the floating gate and enable scaling down the device for nanometer-sized device applications. Generally, NPs can be created by chemical methods such as the Turkevich synthesis,[87,88] or physical methods such as thermal deposition.[89] Because their size is usually larger than a single molecule, various impacts on the film morphology of a transistor can result, depending on the location of the NPs. The memory effect can be induced by inserting NPs into the gate dielectric, in the semiconductor layer, or at the dielectric/semiconductor interface (Figure 9.18) as described below.

9.4.1 Inserting NPs into the Gate Dielectric

A P3HT-based OFET with Au-NPs located on a SiO_2/poly(4-vinyl phenol) (PVP) dielectric insulator has been reported.[90] Citrate-reduced Au-NPs were deposited on a silica surface by electrostatic adsorption using polyions (PAH/PSS) as the bonding agents.[91,92] After capping with the same polyions and then an insulating PVP layer, P3HT was spin-coated on top (process shown in Figure 9.19a). The memory device has an ON/OFF ratio over 1500 and a data retention time of about 200 s. Further enhancement of the charge retention characteristics was achieved by increasing the thickness of the

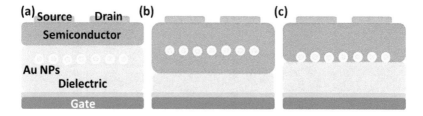

Figure 9.18 Schematic diagram of NP-based transistor memory devices with NPs (a) inserted into the gate dielectrics, (b) incorporated in the semiconductor layer, and (c) at the dielectric/semiconductor interface.

PVP layer or by using other insulating polymers such as PMMA.[93,94] Block copolymer-wrapped Au-NPs were also used as charge carrier traps in OFET memory.[95,96] The Au-NPs were synthesized *in-situ* in the block copolymer, polystyrene-*block*-poly(4-vinylpyridine) (PS-*b*-P4VP), which acted as a template for the growth of ordered NP arrays (Figure 9.19b).[97,98] This method can provide optimum control over NP size and isolation by modulating the molecular weight of each block polymer. By using this method, the devices were shown to have stable and high charge capacity, programmable–erasable properties and a long charge retention ability (>1000 s).

An attempt to enhance the memory effect by increasing the number of trapping sites has been reported. A closely packed Au-NP monolayer embedded in Al_2O_3 dielectrics as charge carrier traps in a pentacene OFET was demonstrated.[99] A Langmuir film of self-assembled Au-NPs (dodecanethiol (DT)-passivated Au-NPs) was transferred to the device substrate using a micro-contact printing technique (μCP) using a poly(dimethylsiloxane) (PDMS) pad, as shown in Figure 9.20.[100] In contrast to the reference devices with Au-NPs created by thermal deposition or electrostatic adsorption, the device containing a micro-contact printed Au-NP array showed a larger memory window

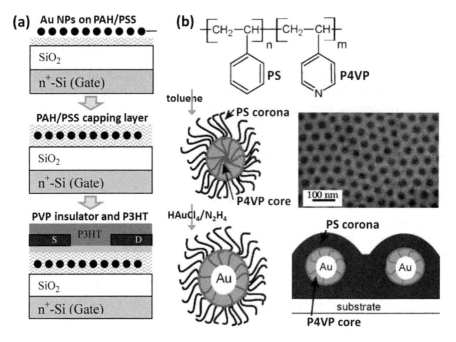

Figure 9.19 Schematic illustration of (a) the fabrication processes of the device with PAH/PSS as the bonding agents, and (b) the chemical structure of the PS-*b*-P4VP copolymer and *in situ* synthesis of Au-NPs in the block copolymer micelle. The TEM image of the copolymer film includes dark circles for the P4VP cores and the bright PS matrix. Reproduced with permission from ref. 95 © 2008 WILEY-VCH Verlag GmbH & Co. KGaA, Weinheim.

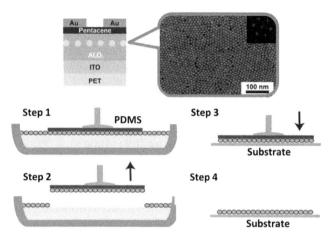

Figure 9.20 The structure of the memory device and SEM image of an Au-NP monolayer formed by μCP. Schematic illustration of the process of micro contact printing of a NP array on the device substrate. Reproduced with permission from ref. 99 © 2012 WILEY-VCH Verlag GmbH & Co. KGaA, Weinheim.

(about 16.5 V), longer retention time (>10^5 s) and good endurance properties (>1000 repeating cycles). Direct blending of Au-NPs into the polymer insulator PMMA was also studied.[101] The device performance was strongly dependent upon the composition of the blend film. Increasing the content of Au-NPs led to an improvement in the memory characteristics but a decrease in the transistor performance at the same time. This can be attributed to the roughness of the insulator surface created by the aggregation of NPs, which results in poor crystallinity of the pentacene film in the device.

9.4.2 Incorporating NPs in the Semiconductor

NPs buried in the dielectric layer will have a long distance from the channel materials where the charge carriers reside. For closer proximity to the conductive channel or less impact on the crystallization of the semiconductor, introducing charge carrier traps into the semiconductor has been tried.[102,103] A CuPc-based thin film transistor with a thin MoO_3 layer inserted into the semiconductor layer is able to shift the transfer curve.[104] The device exhibited a transfer curve shift of about 40 V, an ON/OFF ratio exceeding 10^3 and a long charge retention time (>10^5 s). The inserted MoO_3 layer has only a small effect on the field-effect charge mobility. It was supposed that electrons could be trapped at the MoO_3 surface under a positive gate bias and then retained there to serve as the floating gate. The charges could be de-trapped by applying an inverse gate bias. An OFET-based memory device composed of Ag-NPs embedded in a pentacene layer was also studied (Figure 9.21).[105,106] The memory effect was dependent upon the location of

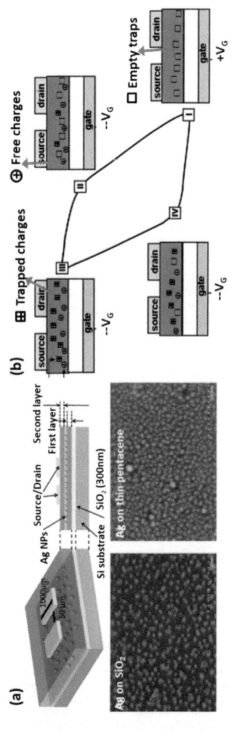

Figure 9.21 (a) The structure of the device incorporating Ag-NPs into a pentacene film. SEM images of Ag-NPs on silica and a silica substrate with 15 nm pentacene deposited. (b) Schematic illustration of a current hysteresis loop and its corresponding trap states. Reproduced with permission from ref. 105 © 2010 Elsevier B.V.

the Ag-NPs within the pentacene film. The maximum memory window of 90 V is clearly larger compared with the results from devices composed of Ag-NPs deposited on the dielectrics. It was suggested that Ag-NPs introduced at different locations in the semiconductor layer would collect different amounts of charge carriers.

Blending Au-NPs within a poly(9,9-dioctylfluorene-*alt*-bithiophene) (F8T2) semiconductor-based polymer FET was studied (Figure 9.22a).[107] The memory characteristics were strongly dependent upon the loading concentration (0–9%) of Au-NPs and the chain-length (C_6, C_{10} and C_{18}) of the thiolate molecule which covers the Au-NPs' surface. The device comprising the highest content (9%) of Au-NPs and shorter carbon-chain (C_6) molecules exhibited the largest memory effect with a memory window of around 41 V and an ON/OFF ratio of 10^3, which can be maintained for at least 10^4 s. Utilizing light-controlled electrical behavior was demonstrated by incorporating CdSe quantum dots (QDs) into the P3HT matrix as the active layer.[108,109] The electron trapping in QDs is induced by illumination, which serves as the writing process. The trapped charge was maintained for 1 h. The device performance can be further improved by using core–shell CdSe@ZnSe QDs instead of CdSe QDs (Figure 9.22b).[110] The quantum well-structured core–shell QDs led to the enhancement of the ON/OFF ratio (2700), which was maintained for 8000 s without noticeable decay.

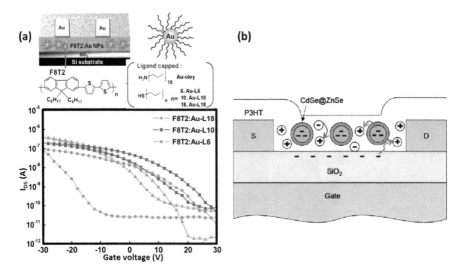

Figure 9.22 (a) Chemical structure of F8T2 semiconductor and the device structure of a hybrid SAM-capped Au-NPs/F8T2 based memory device. Transfer characteristics with bidirectional scan of gate bias for hybrid devices with different hydrocarbon chain lengths of SAM including C_{18}, C_{10} and C_6. (b) The device structure for hybrid quantum dot/P3HT memory. Reproduced with permission from ref. 107 © 2013 American Chemical Society and ref. 110 © 2009 Elsevier B.V.

9.4.3 Embedding NPs at the Semiconductor/Dielectric Interface

In contrast to other planar floating gates, surrounding NPs with channel materials has the advantage of maximized contact between the charge carrier traps and the semiconductor, as well as minimized damage to the gate dielectrics caused by the inserted NPs. The incorporation of NPs at the semiconductor/dielectric interface has been demonstrated as well.[111,112] However, the presence of NPs may influence the crystallinity and morphology of the semiconductor films, which can significantly affect the OFET performance.[113] This issue was addressed in a study employing pentacene-based transistors with Au-NPs as the charge carrier traps at the dielectric surface.[114] The crystallinity and packing orientation of the pentacene film depend critically upon the surface properties of the NPs, which can be modulated by a SAM on the NPs (Figure 9.23a and b). The crystallinity of the pentacene film depends on the chain length of the SAM used to modify the Au-NPs. Thus the more "hydrophobic" and "oleophobic" the better the crystallinity. The SAM also modulates the charge trapping capability of the NPs through gold work function modulation and the tunneling distance imposed by the

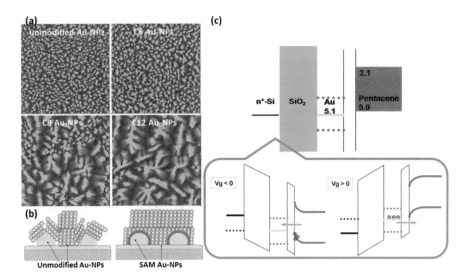

Figure 9.23 (a) AFM images of pentacene film deposited on different substrates including bare Au-NPs/silica and n-alkanethiol (C_6, C_8 and C_{12})-modified Au-NPs/silica substrates, respectively. (b) Schematic illustration of pentacene molecules deposited on bare Au-NPs and SAM-modified Au-NP substrates respectively. (c) Schematic energy-band diagrams of pentacene/SAM/Au-NPs/silica/n$^+$-Si under zero gate bias, negative gate bias and positive gate bias. The red and blue dotted lines represent the work function for alkanethiol-modified Au and fluorinated alkanethiol-modified Au, respectively. Reproduced with permission from ref. 114 © 2009 American Chemical Society.

insulating hydrocarbon chains. The maximum memory window is obtained with a lower work function Au-NP capped with a small tunneling barrier (Figure 9.23c). Nevertheless, the smaller tunneling barrier associated with the shorter chain SAM leads to a shorter retention time. A longer chain SAM on the Au-NPs improves the retention time, but nevertheless slows down the trapping process (longer response time). A memory effect was also obtained by using Al NPs.[115] A thicker alumina shell created by O_2 plasma treatment increased the tunneling distance and the work function of Al. Capping the NPs with a short hydrocarbon chain SAM (C_4PA) reduced the work function, which improved the charge trapping ability and ON/OFF ratio, as well as the charge retention characteristics.

In general, the insulating coating needed to maintain the charge in NPs is also a barrier for the charges to be trapped. Thus, increasing the thickness of the insulating layer may increase the charge retention time but slow down the response speed, thus reducing the memory window at the same time. Recently, a study demonstrated that a memory device with fast response, large memory window and long charge retention can be achieved using Au-NPs capped with a SAM of azobenzene derivatives. The functionalized Au-NPs serve as a hybrid organic–metal double floating gate in a pentacene-based transistor.[116] The NPs provide a reservoir for charge carriers, whereas the azobenzene monolayer on the particle surface acts as a barrier layer and a work function modulator, as well as providing additional charge trapping sites at the Au-NP/semiconductor interface (Figure 9.24a). In comparison with simple *n*-alkanethiol-modified Au-NPs, CH_3 substituted azobenzene functionalized Au-NPs trapped around 70% more charges, had much faster response time and higher retention time. These characteristics were suggested to result from the promotion of charge trapping by the monolayer on the particle surface, as well as the shielding effect of the charged monolayer, so that the trapped charges are retained longer. A double floating gate utilizing a layer of rGO sheets covering Au-NPs in a pentacene transistor was also demonstrated (Figure 9.24b).[117] In contrast to single floating gate Au-NPs devices, an improvement of the field-effect charge mobility, the memory window and the charge retention time can be obtained in a system involving graphene oxide (rGO) sheets/Au-NPs. This is attributed to the introduction of the rGO sheet film, which can decrease the surface roughness caused by the Au-NPs and provide extra trapping sites. The energy barrier between the rGO and Au-NPs prevents charge leakage from the NP trap sites.

9.5 Conclusion

In a thin film transistor, charge carriers and thereafter the current are generated upon the application of a gate bias. To achieve electric bistability, an additional handle is needed to modulate the carrier density "reversibly" at the semiconductor/dielectric interface. Besides the compatibility with other organic electronics, organic memory devices offer great opportunities for tailorability in structure through synthesis to fine tune the properties

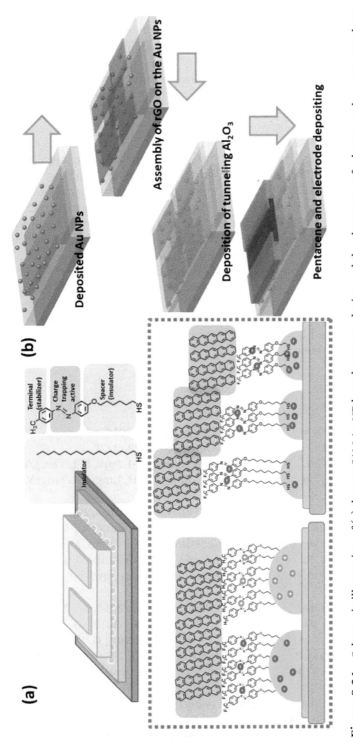

Figure 9.24 Schematic illustration of (a) the azo-SAM Au-NP based memory device and the charge transfer between the pentacene layer and the various azo-functionalized Au-NPs; (b) the fabrication process for rGO–Au-NPs memory devices. Reproduced with permission from ref. 116 © 2013 American Chemical Society and ref. 117 © 2013 WILEY-VCH Verlag GmbH & Co. KGaA, Weinheim.

required. Light and electric field have been considered to create such a handle, with the latter being more popular since the gate electrode is already present. Ferroelectrics and carrier traps have been introduced to create environmental changes near the conducting channel to modulate the carrier density at the interface. For a memory device to be practical, the switching between the two states (ON and OFF) has to be fast and the state has to be retained for long enough, besides the changes being large enough. The requirements of fast switching (response time) and staying in the state for a long time (retention time) appear to be contradictory. Systems with ferroelectrics involve reorganization of the conformation of organic polymer systems, which is difficult to be fast, and the system tends to relax when the electric field is turned off. Efforts to make the reorganization easier will also make the relaxation easier. Systems with carrier traps involve charge transfer between the conducting channel and the trapping sites. The trapped charges will dissipate or leak from the trapping sites. Although protecting the sites with insulating layers helps to retain the charges, the same layer will also hinder the trapping process. While there has been great progress in the past years in improving the different characteristics, a mechanism is necessary to trigger a change in the surroundings when the charges are trapped or the dipoles are aligned. This change can be reverted upon application of a reverse bias. This will be the challenge for the further development of organic memory devices.

References

1. J. A. Rogers, T. Someya and Y. Huang, *Science*, 2010, **327**, 1603.
2. T. Sekitani and T. Someya, *Adv. Mater.*, 2010, **22**, 2228.
3. J. Lee, D. H. Kim, J.-Y. Kim, B. Yoo, J. W. Chung, J.-I. Park, B.-L. Lee, J. Y. Jung, J. S. Park, B. Koo, S. Im, J. W. Kim, B. Song, M.-H. Jung, J. E. Jang, Y. W. Jin and S.-Y. Lee, *Adv. Mater.*, 2013, **25**, 5886.
4. J. Liang, L. Li, X. Niu, Z. Yu and Q. Pei, *Nat. Photonics*, 2013, **7**, 817.
5. G. H. Gelinck, H. E. A. Huitema, E. van Veenendaal, E. Cantatore, L. Schrijnemakers, J. B. P. H. van der Putten, T. C. T. Geuns, M. Beenhakkers, J. B. Giesbers, B.-H. Huisman, E. J. Meijer, E. M. Benito, F. J. Touwslager, A. W. Marsman, B. J. E. van Rens and D. M. de Leeuw, *Nat. Mater.*, 2004, **3**, 106.
6. Y. Zhan, Y. Mei and L. Zheng, *J. Mater. Chem. C*, 2014, **2**, 1220.
7. K. J. Baeg, M. Caironi and Y. Y. Noh, *Adv. Mater.*, 2013, **25**, 4210.
8. O. Knopfmacher, M. L. Hammock, A. L. Appleton, G. Schwartz, J. Mei, T. Lei, J. Pei and Z. Bao, *Nat. Commun.*, 2014, **5**, 2954.
9. M. Magliulo, K. Manoli, E. Macchia, G. Palazzo and L. Torsi, *Adv. Mater.*, 2014, **26**, 1.
10. J. S. Yuan and J. J. Liou, *Semiconductor Device Physics and Simulation*, Springer Science+Business Media, New York, 1st edn, 1998, ch. 3, p. 53.
11. J. Zaumseil and H. Sirringhaus, *Chem. Rev.*, 2007, **107**, 1296.

12. J. Mei, Y. Diao, A. L. Appleton, L. Fang and Z. Bao, *J. Am. Chem. Soc.*, 2013, **135**, 6724.
13. M. Gruber, E. Zojer, F. Schürrer and K. Zojer, *Adv. Funct. Mater.*, 2013, **23**, 2941.
14. R. P. Ortiz, A. Facchetti and T. J. Marks, *Chem. Rev.*, 2010, **110**, 205.
15. C.-A. Di, Y. Liu, G. Yu and D. Zhu, *Acc. Chem. Res.*, 2009, **42**, 1573.
16. J. C. Ranuárez, M. J. Deen and C.-H. Chen, *Microelectron. Reliab.*, 2006, **46**, 1939.
17. S.-T. Han, Y. Zhou and V. A. L. Roy, *Adv. Mater.*, 2013, **25**, 5425.
18. S. M. Sze and K. K. Ng, *Physics of Semiconductor Devices*, John & Sons, New York, 3rd edn, 2007.
19. P. Heremans, G. H. Gelinck, R. Müller, K.-J. Baeg, D.-Y. Kim and Y.-Y. Noh, *Chem. Mater.*, 2011, **23**, 341.
20. G. W. Burr, B. N. Kurdi, J. C. Scott, C. H. Lam, K. Gopalakrishnan and R. S. Shenoy, *IBM J. Res. Dev.*, 2008, **52**, 449.
21. D. Ielmini, *Microelectron. Eng.*, 2009, **86**, 1870.
22. P. C. Y. Chen, *IEEE Trans. Electron Devices*, 1977, **24**, 584.
23. M. H. White, Y. Yang, A. Purwar and M. L. French, *IEEE Trans. Compon., Packag., Manuf. Technol., Part A*, 1997, **20**, 190.
24. E. Leobandung, L. Guo and S. Y. Chou, *Appl. Phys. Lett.*, 1995, **67**, 2338.
25. Z. Liu, C. Lee, V. Narayanan, G. Pei and E. C. Kan, *IEEE Trans. Electron Devices*, 2002, **49**, 1606.
26. S. Paydavosi, H. Abdu, G. J. Supran and V. Bulović, *IEEE Trans. Nanotechnol.*, 2011, **10**, 1536.
27. J. S. Lindsey and D. F. Bocian, *Acc. Chem. Res.*, 2011, **44**, 638.
28. B. M. Dhar, R. Özgün, T. Dawidczyk, A. Andreou and H. E. Katz, *Mater. Sci. Eng., R*, 2011, **72**, 49.
29. R. C. G. Naber, K. Asadi, P. W. M. Blom, D. M. de Leeuw and B. de Boer, *Adv. Mater.*, 2010, **22**, 933.
30. C. B. Sawyer and C. W. Tower, *Phys. Rev.*, 1930, **35**, 269.
31. D. Damjanovic, *Rep. Prog. Phys.*, 1998, **61**, 1267.
32. H. Guo, Y. Zhang, F. Xue, Z. Cai, Y. Shang, J. Li, Y. Chen, Z. Wu and S. Jiang, *CrystEngComm*, 2013, **15**, 1597.
33. C. Xing, M. Zhao, L. Zhao, J. You, X. Cao and Y. Li, *Polym. Chem.*, 2013, **4**, 5726.
34. H. Dong, X. Fu, J. Liu, Z. Wang and W. Hu, *Adv. Mater.*, 2013, **25**, 6158.
35. H. S. Nalwa, *Ferroelectric Polymers: Chemistry, Physics and Applications*, Marcel Dekker, New York, 1995.
36. A. J. Lovinger, *Science*, 1983, **220**, 1115.
37. R. Gregorio Jr and E. M. Ueno, *J. Mater. Sci.*, 1999, **34**, 4489.
38. V. Sencadas, R. Gregório Jr and S. Lanceros-Mendéz, *J. Macromol. Sci., Part B: Phys.*, 2009, **48**, 514.
39. L. Li, M. Zhang, M. Rong and W. Ruan, *RSC Adv.*, 2014, **4**, 3938.
40. R. G. Kepler and R. A. Anderson, *Adv. Phys.*, 1991, **41**, 1.
41. A. Itoh, Y. Takahashi, T. Furukawa and H. Yajima, *Polym. J.*, 2014, **46**, 207.
42. R. Schroeder, L. A. Majewski and M. Grell, *Adv. Mater.*, 2004, **16**, 633.

43. S. J. Kang, Y. J. Park, I. Bae, K. J. Kim, H.-C. Kim, S. Bauer, E. L. Thomas and C. Park, *Adv. Funct. Mater.*, 2009, **19**, 2812.
44. R. C. G. Naber, M. Mulder, B. de Boer, P. W. M. Blom and D. M. de Leeuw, *Org. Electron.*, 2006, **7**, 132.
45. R. C. G. Naber, C. Tanase, P. W. M. Blom, G. H. Gelinck, A. W. Marsman, F. J. Touwslager, S. Setayesh and D. M. deLeeuw, *Nat. Mater.*, 2005, **4**, 243.
46. S.-W. Jung, J.-K. Lee, Y. S. Kim, S.-M. Yoon, I.-K. You, B.-G. Yu and Y.-Y. Noh, *Curr. Appl. Phys.*, 2010, **10**, e58.
47. Y. Zheng, G.-X. Ni, C.-T. Toh, M.-G. Zeng, S.-T. Chen, K. Yao and B. Özyilmaz, *Appl. Phys. Lett.*, 2009, **94**, 163505.
48. G. H. Gelinck, A. W. Marsman, F. J. Touwslager, S. Setayesh, D. M. de Leeuw, R. C. G. Naber and P. W. M. Blom, *Appl. Phys. Lett.*, 2005, **87**, 092903.
49. R. C. G. Naber, P. W. M. Blom, G. H. Gelinck, A. W. Marsman and D. M. de Leeuw, *Adv. Mater.*, 2005, **17**, 2692.
50. Q.-D. Linga, D.-J. Liaw, C. Zhu, D. S.-H. Chan, E.-T. Kang and K.-G. Neoh, *Prog. Polym. Sci.*, 2008, **33**, 917.
51. Y.-H. Chou, H.-C. Chang, C.-L. Liu and W.-C. Chen, *Polym. Chem.*, 2015, **6**, 341.
52. H. E. Katz, X. M. Hong, A. Dodabalapur and R. Sarpeshkar, *J. Appl. Phys.*, 2002, **91**, 1572.
53. K.-J. Baeg, Y.-Y. Noh, J. Ghim, S.-J. Kang, H. Lee and D.-Y. Kim, *Adv. Mater.*, 2006, **18**, 3179.
54. K.-J. Baeg, Y.-Y. Noh, J. Ghim, B. Lim and D.-Y. Kim, *Adv. Funct. Mater.*, 2008, **18**, 3678.
55. M. Debucquoy, M. Rockelé, J. Genoe, G. H. Gelinck and P. Heremans, *Org. Electron.*, 2009, **10**, 1252.
56. M. Debucquoy, D. Bode, J. Genoe, G. H. Gelinck and P. Heremans, *Appl. Phys. Lett.*, 2009, **95**, 103311.
57. L.-L. Chua, J. Zaumseil, J.-F. Chang, E. C.-W. Ou, P. K.-H. Ho, H. Sirringhaus and R. H. Friend, *Nature*, 2005, **434**, 194.
58. Y. H. Chou, S. Takasugi, R. Goseki, T. Ishizone and W. C. Chen, *Polym. Chem.*, 2014, **5**, 1063.
59. J.-C. Hsu, W.-Y. Lee, H.-C. Wu, K. Sugiyama, A. Hirao and W.-C. Chen, *J. Mater. Chem.*, 2012, **22**, 5820.
60. W. Wu, H. Zhang, Y. Wang, S. Ye, Y. Guo, C. Di, G. Yu, D. Zhu and Y. Liu, *Adv. Funct. Mater.*, 2008, **18**, 2593.
61. Y. C. Chiu, T. Y. Chen, Y. Chen, T. Satoh, T. Kakuchi and W. C. Chen, *ACS Appl. Mater. Interfaces*, 2014, **6**, 12780.
62. R. R. Gagne, C. A. Koval and G. C. Lisensky, *Inorg. Chem.*, 1980, **19**, 2854.
63. D. G. Davis and D. J. Orleron, *Anal. Chem.*, 1966, **38**, 179.
64. G. Balducci, G. Chottard, C. Gueutin, D. Lexa and J.-M. Savéant, *Inorg. Chem.*, 1994, **33**, 1972.
65. N. Fuentes, A. Martín-Lasanta, L. Á. de Cienfuegos, M. Ribagorda, A. Parra and J. M. Cuerva, *Nanoscale*, 2011, **3**, 4003.

66. K. M. Roth, A. A. Yasseri, Z. Liu, R. B. Dabke, V. Malinovskii, K.-H. Schweikart, L. Yu, H. Tiznado, F. Zaera, J. S. Lindsey, W. G. Kuhr and D. F. Bocian, *J. Am. Chem. Soc.*, 2003, **125**, 505.
67. J. S. Lindsey and D. F. Bocian, *Acc. Chem. Res.*, 2011, **44**, 638.
68. T. Pro, J. Buckley, K. Huang, A. Calborean, M. Gély, G. Delapierre and G. Ghibaudo, *IEEE Trans. Nanotechnol.*, 2009, **8**, 204.
69. T. Pro, J. Buckley, R. Barattin, A. Calborean, V. Aiello, G. Nicotra, K. Huang, M. Gély, G. Delapierre, E. Jalaguier, F. Duclairoir, N. Chevalier, S. Lombardo, P. Maldivi, G. Ghibaudo and S. Deleonibus, *IEEE Trans. Nanotechnol.*, 2011, **10**, 275.
70. Q. Li, S. Surthi, G. Mathur, S. Gowda, V. Misra, T. A. Sorenson, R. C. Tenent, W. G. Kuhr, S.-I. Tamaru, J. S. Lindsey, Z. Liu and D. F. Bocian, *Appl. Phys. Lett.*, 2003, **83**, 198.
71. Q. Li, S. Surthi, G. Mathur, S. Gowda, Q. Zhao, T. A. Sorenson, R. C. Tenent, K. Muthukumaran, J. S. Lindsey and V. Misra, *Appl. Phys. Lett.*, 2004, **85**, 1829.
72. C. Simão, M. Mas-Torrent, J. Casado-Montenegro, F. Otón, J. Veciana and C. Rovira, *J. Am. Chem. Soc.*, 2011, **133**, 13256.
73. S. Paydavosi, H. Abdu, G. J. Supran and V. Bulović, *IEEE Trans. Nanotechnol.*, 2011, **10**, 594.
74. C. Li, J. Ly, B. Lei, W. Fan, D. Zhang, J. Han, M. Meyyappan, M. Thompson and C. Zhou, *J. Phys. Chem. B*, 2004, **108**, 9646.
75. C. Li, W. Fan, D. A. Straus, B. Lei, S. Asano, D. Zhang, J. Han, M. Meyyappan and C. Zhou, *J. Am. Chem. Soc.*, 2004, **126**, 7750.
76. X. Duan, Y. Huang and C. M. Lieber, *Nano Lett.*, 2002, **2**, 487.
77. C.-W. Tseng and Y.-T. Tao, *ACS Appl. Mater. Interfaces*, 2010, **2**, 3231.
78. G. M. J. Schmidt, *Solid State Photochemistry*, Verlag Chemie, Weinheim, Germany, 1976.
79. M.-H. Jung, K. H. Song, K. C. Ko, J. Y. Lee and H. Lee, *J. Mater. Chem.*, 2010, **20**, 8016.
80. L. Zhang, T. Wu, Y. Guo, Y. Zhao, X. Sun, Y. Wen, G. Yu and Y. Liu, *Sci. Rep.*, 2013, **3**, 1080.
81. H. Li, B. C.-K. Tee, J. J. Cha, Y. Cui, J. W. Chung, S. Y. Lee and Z. Bao, *J. Am. Chem. Soc.*, 2012, **134**, 2760.
82. H.-C. Chang, C. Lu, C.-L. Liu and W.-C. Chen, *Adv. Mater.*, 2015, **27**, 27.
83. M. Burkhardt, A. Jedaa, M. Novak, A. Ebel, K. Voïtchovsky, F. Stellacci, A. Hirsch and M. Halik, *Adv. Mater.*, 2010, **22**, 2525.
84. A. Khassanov, T. Schmaltz, H.-G. Steinrück, A. Magerl, A. Hirsch and M. Halik, *Adv. Mater. Interfaces*, 2014, **1**, 1400238.
85. C. W. Tseng, D. C. Huang and Y. T. Tao, *ACS Appl. Mater. Interfaces*, 2012, **4**, 5483.
86. M. M. Russew and S. Hecht, *Adv. Mater.*, 2010, **22**, 3348.
87. J. Turkevich, P. C. Stevenson and J. Hillier, *Discuss. Faraday Soc.*, 1951, **11**, 55.
88. J. Kimling, M. Maier, B. Okenve, V. Kotaidis, H. Ballot and A. Plech, *J. Phys. Chem. B*, 2006, **110**, 15700.

89. D. Gaspar, A. C. Pimentel, T. Mateus, J. P. Leitão, J. Soares, B. P. Falcão, A. Araújo, A. Vicente, S. A. Filonovich, H. Águas, R. Martins and I. Ferreira, *Sci. Rep.*, 2013, **3**, 1469.
90. Z. Liu, F. Xue, Y. Su, Y. M. Lvov and K. Varahramyan, *IEEE Trans. Nanotechnol.*, 2006, **5**, 379.
91. J. Schmitt, G. Decher, W. J. Dressick, S. L. Brandow, R. E. Geer, R. Shashidhar and J. M. Calvert, *Adv. Mater.*, 1991, **9**, 61.
92. J.-S. Lee, J. Cho, C. Lee, I. Kim, J. Park, Y.-M. Kim, H. Shin, J. Lee and F. Caruso, *Nat. Nanotechnol.*, 2007, **2**, 790.
93. S.-J. Kim, Y.-S. Park, S.-H. Lyu and J.-S. Lee, *Appl. Phys. Lett.*, 2010, **96**, 033302.
94. Y. M. Kim, S. J. Kim and J. S. Lee, *IEEE Electron Device Lett.*, 2010, **31**, 503.
95. W. L. Leong, P. S. Lee, A. Lohani, Y. M. Lam, T. Chen, S. Zhang, A. Dodabalapur and S. G. Mhaisalkar, *Adv. Mater.*, 2008, **20**, 2325.
96. Q. Wei, Y. Lin, E. R. Anderson, A. L. Briseno, S. P. Gido and J. J. Watkins, *ACS Nano*, 2012, **6**, 1188.
97. C. Lee, J.-H. Kwon, J.-S. Lee, Y.-M. Kim, Y. Choi, H. Shin, J. Lee and B.-H. Sohn, *Appl. Phys. Lett.*, 2007, **91**, 153506.
98. C.-M. Chen, C.-M. Liu, K.-H. Wei, U.-S. Jeng and C.-H. Su, *J. Mater. Chem.*, 2012, **22**, 454.
99. S.-T. Han, Y. Zhou, Z.-X. Xu, L.-B. Huang, X.-B. Yang and V. A. L. Roy, *Adv. Mater.*, 2012, **24**, 3556.
100. T. Wen and S. A. Majetich, *ACS Nano*, 2011, **5**, 8868.
101. Y. Zhou, S.-T. Han, Z.-X. Xu and V. A. L. Roy, *Nanotechnology*, 2012, **23**, 344014.
102. S.-T. Han, Y. Zhou, Z.-X. Xu and V. A. L. Roy, *Appl. Phys. Lett.*, 2012, **101**, 033306.
103. C. Raimondo, N. Crivillers, F. Reinders, F. Sander, M. Mayor and P. Samorì, *Proc. Natl. Acad. Sci. U. S. A.*, 2012, **109**, 12375.
104. Y. Guo, Y. Liu, C.-A. Di, G. Yu, W. Wu, S. Ye, Y. Wang, X. Xu and Y. Sun, *Appl. Phys. Lett.*, 2007, **91**, 263502.
105. S. Wang, C.-W. Leung and P. K. L. Chan, *Org. Electron.*, 2010, **11**, 990.
106. X. C. Ren, S. M. Wang, C. W. Leung, F. Yan and P. K. L. Chan, *Appl. Phys. Lett.*, 2011, **99**, 043303.
107. H.-C. Chang, C.-L. Liu and W.-C. Chen, *ACS Appl. Mater. Interfaces*, 2013, **5**, 13180.
108. C.-C. Chen, M.-Y. Chiu, J.-T. Sheu and K.-H. Wei, *Appl. Phys. Lett.*, 2008, **92**, 143105.
109. W. U. Huynh, J. J. Dittmer and A. P. Alivisatos, *Science*, 2002, **295**, 2425.
110. M.-Y. Chiu, C.-C. Chen, J.-T. Sheu and K.-H. Wei, *Org. Electron.*, 2009, **10**, 769.
111. C. Novembre, D. Guérin, K. Lmimouni, C. Gamrat and D. Vuillaume, *Appl. Phys. Lett.*, 2008, **92**, 103314.
112. Y.-S. Park and J.-S. Lee, *Adv. Mater.*, 2015, **27**, 706.
113. F. Alibart, S. Pleutin, D. Guérin, C. Novembre, S. Lenfant, K. Lmimouni, C. Gamrat and D. Vuillaume, *Adv. Funct. Mater.*, 2010, **20**, 330.

114. C.-W. Tseng and Y.-T. Tao, *J. Am. Chem. Soc.*, 2009, **131**, 12441.
115. C.-W. Tseng, Y.-L. Chen and Y.-T. Tao, *Org. Electron.*, 2012, **13**, 1436.
116. C.-W. Tseng, D.-C. Huang and Y.-T. Tao, *ACS Appl. Mater. Interfaces*, 2013, **5**, 9528.
117. S.-T. Han, Y. Zhou, C. Wang, L. He, W. Zhang and V. A. L. Roy, *Adv. Mater.*, 2013, **25**, 872.

CHAPTER 10

Organic Floating Gate Transistor Memory Devices

HUNG CHIN WU[a], YING-HSUAN CHOU[a], HSUAN-CHUN CHANG[a], AND WEN-CHANG CHEN*[a]

[a]Department of Chemical Engineering, National Taiwan University, Taipei 10617, Taiwan
*E-mail: chenwc@ntu.edu.tw

10.1 Nanoparticle Embedded Materials as Charge Storage Layers

Digital information storage devices based on organic/polymeric materials are attracting a large amount of research interest owing to their advantages of low-cost, flexibility in molecular design, and good solution processability with respect to their inorganic counterparts.[1–3] Memory devices based on three terminal organic field-effect transistors (OFETs), typically, are assembled to switch the devices between the high/low conductance states, and such devices can be classified as floating gate-, polymer electret-, or ferroelectric-type memory with different kinds of charge storage layers.[4–6] Among these, floating gate memory can precisely control the amount of charge stored in specific trapping sites, which can overcome the limitations of reducing device size and meet the requirements for high density memory cells.

Floating gate transistor memory devices were first explored in 1967 by Kahng and Sze.[7] To memorize digital information, the charges in floating

Organic Floating Gate Transistor Memory Devices 331

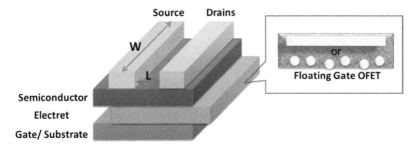

Figure 10.1 OFET memory configurations with floating gate dielectric layers.

gate devices are stored in a conducting/semiconducting layer, which is fully covered by a dielectric matrix (Figure 10.1).[8–10] The dielectric layer between the floating gate and the semiconductor should be very thin in order to obtain a sufficiently high electric field for charge injection; while the dielectric layer, on the other hand, must be thick enough to prevent current leakage and automatically discharge as the gate voltage is removed. Moreover, the stored charge in floating gates can be removed using a reverse bias, showing typical non-volatile memory characteristics in OFET memory.[11] Floating gates in transistor memory devices can be concrete (*i.e.* polysilicon) or discrete (*i.e.* Si, Au, and other metal nanoparticles (NPs)), and Au NPs are the most appropriate materials due to their chemical stability and high work function.[12] NP floating gates can be fabricated using several methods, including thermal evaporation, chemical self-assembly, and polymer–nanoparticle hybrids.[13–19] The following sections will categorize the examples based on NPs or functional moiety embedded materials for floating gate OFET memory applications. Table 10.1 provide a summary of OFETs and memory properties for representative floating gate materials.

10.1.1 Thermal Evaporation

Thermally deposited metallic nanoparticles have been largely used in organic floating gate FET memory devices on gate substrates, and are normally embedded in polymer dielectrics. Kim and his coworkers revealed a solution-processed F8T2 floating gate memory in 2010 (Figure 10.2).[13] Au NPs were thermally evaporated at the interface between polystyrene and cross-linked poly(4-vinylphenol), which acted as electron trapping sites. Reliable memory characteristics were then obtained with a field-effect mobility of 0.02 cm^2 V^{-1} s^{-1} and a memory window of 30 V. This F8T2 floating gate based on Au NPs, moreover, possessed a high on/off ratio (more than 10^4) during programming and erasing (under a gate bias of 80 V) in a relatively short timescale (less than 1 s) along with a long retention time of a few hours.

Not only Au NPs, but also Ag, Cu, and Al NPs have also been demonstrated as charge trapping materials in transistor memory. Wang *et al.*

Table 10.1 OFET memory device performance based on floating gate electrets.

Electret	Process	Device structure	Device characteristics	Ref.
Au NPs	Thermal evaporation	cPVP/Au NPs/PS	Large ΔV_{th} (~30 V) $I_{on}/I_{off} \sim 10^4$ Flexible device	13
Metal NPs (Au, Ag, Cu, Al)	Thermal evaporation	PS/PVP/metal NPs PS/PMMA/metal NPs	Large ΔV_{th} (~52 V) $I_{on}/I_{off} \sim 10^4$ Long retention time (>10^7 s) Flexible device	20
Au NPs	Chemical assembly	Al_2O_3/Au NP/Al_2O_3	Large ΔV_{th} (~16.5 V) Long retention time (>10^5 s) Good endurance properties Flexible device	32
Small molecule		Small molecule: polymer blend film	Steep flanks of hysteresis with $I_{on}/I_{off} \sim 10^4$ Long retention time (>24 h)	59
Small molecule (PCBM)		PCBM:PVP composite layer	$I_{on}/I_{off} \sim 10^4$ ΔV_{th} (~20 V) Long retention time (>40 h)	60
Au NPs	Chemical assembly	Au NPs/PS-P2VP	ΔV_{th} increased with the density of Au Long retention time (>40 h)	47
Au NPs		PMMA/(Au NPs/PEs)	ΔV_{th} (~34 V) Long retention time (>1 year)	16
Au NPs	Hybrid	Au NPs embedded in active layer	ΔV_{th} increased with the size of Au NPs ΔV_{th} (C_n–Au NPs > FC_n–Au NPs)	56
Au NPs		cPVP/(APTES/Au NPs)/cPVP	ΔV_{th} (>10 V) Long retention time (>1 year) Flexible device	82
QDs (CdSe–ZnS)	Hybrid	QD:PMMA	Long retention time WRER cycle	57
Double floating (rGO, Au NPs)		Hybrid double hybrid	$\Delta V_{th} > 1.95$ V $I_{on}/I_{off} \sim 3 \times 10^4$ Long retention time (>10^5 s) Flexible device	63

Ag NPs	Thermal evaporation	Pentacene/Ag NPs/pentacene	$\Delta V_{th} \sim 60$ V; $I_{on}/I_{off} \sim 10^5$	14
Au NPs	Chemical assembly	P3HT/PVP/Au NPs/[PHA/PSS]	$I_{on}/I_{off} \sim 1500$; Retention time ~ 200 s	31
Au NPs	Chemical assembly	PVP/Au NPs/APTES	$I_{on}/I_{off} \sim 10^3$; Good endurance properties; Retention time $> 10^5$ s	33
Au NPs	Chemical assembly	Au NPs/PS-P4VP	$\Delta V_{th} \sim 2.1$ V; Retention time $> 10^5$ s	18
Au NPs	Chemical assembly	PEs/Au NPs/PMMA	ΔV_{th} (~34 V); Long retention time (>1 year)	16
Au NPs	Chemical assembly	PEs/Au NPs	ΔV_{th} (~7.7 V); Long retention time (>1 year)	48
Au NPs	Chemical assembly	cPVP/Au NPs/F8T2	Long retention time (>10^3 s); $I_{on}/I_{off} \sim 10^6$	54
C_{60}	Chemical assembly	SAM/C_{60}/aliphatic	Small operating voltage (~2 V)	61
rGO		AuCl$_3$ doped rGO	Large $\Delta V_{th} \sim 2.7$ V; Flexible device	71
Au NPs	Chemical assembly	PVP/APTES/Au NPs	ΔV_{th} (~15 V); Long retention time (>10^5 s); Bending test (>2000 cycles); Flexible device	85
Au NPs	Chemical assembly	SiO$_2$-SAM/Au NPs	ΔV_{th} (~22 V); $I_{on}/I_{off} \sim 10^4$	17
Au NPs	Chemical assembly	Au NPs/PMMA	Long retention time	34
Au NPs	Chemical assembly	Multi-stack charge trapping layer	ΔV_{th} (~14.6 V); Long retention time; Good endurance properties	49

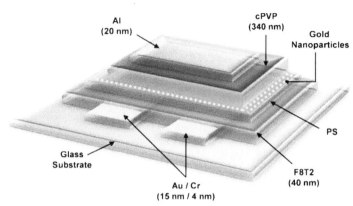

Figure 10.2 Schematic top-gate/bottom-contact polymer field-effect transistor memory device using a Au NP floating gate. Reproduced with permission from ref. 13. Copyright 2010 WILEY-VCH Verlag GmbH & Co. KGaA, Weinheim.

reported the modification of the memory properties by adjusting the thickness of an Ag NP layer embedded in an organic semiconductor.[14] The memory window could be enlarged to 90 V with 5 nm Ag NPs deposited. In addition, Kim *et al.* explored the effects of using various metallic NPs, including Au, Ag, Cu, and Al, on the characteristics of organic floating gate memory devices, and such NPs were embedded within polymer dielectrics (Figure 10.3).[20] Among different kinds of NPs, Au NPs have been primarily used in device fabrication because of their good chemical stability, small energy perturbation, and the fact that they can be formed in a quite uniform manner using thermal evaporation.[21,22] Subsequently, a variety of metallic NPs, such as Ag, Cu, and Al, were studied to optimize the performance of floating gate memory. The experimental results showed that the memory windows were increased in proportion to the surface number density of NPs. Hence, the Au NP-containing memory devices showed the largest memory window of 52 V due to their highest density of NPs (around 3.0×10^{12} cm^{-2}). The devices with Al NPs, however, exhibited very narrow memory windows because most of the thermally deposited Al formed large aggregates of approximately 200 nm, leading to a low number density of NPs on the surface. The size and distribution (both horizontally and vertically) of the thermally deposited metallic NPs is mainly determined by the atomic surface mobility, chemical reactivity, and density of the metals during the evaporation process.[23,24] Therefore, the size of the NPs could be precisely controlled by the deposition rate, and a larger memory window of the floating gate devices could be obtained by depositing smaller sized NPs (*i.e.* higher surface density).[25] High surface density metallic NP trapping sites, however, may dissipate easily through neighboring NPs, resulting in current leakage in the device. Thus, for such NP-based floating gate memory, improvement of the long term stability of the device is still the main challenge at present.

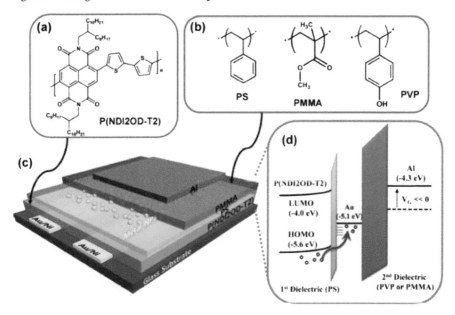

Figure 10.3 Chemical structures of (a) the polymer semiconductor P(NDI2OD-T2) and (b) the polymer dielectrics PS, PMMA and PVP. Schematic illustration of (c) the organic nano-floating gate memory device and (d) its operating mechanism. Reproduced with permission from ref. 20. Copyright 2013 WILEY-VCH Verlag GmbH & Co. KGaA, Weinheim.

10.1.2 Chemical Assembly

Floating gate memory devices containing charge-trapping NPs with a higher surface density usually exhibit better memory performances. The thermal evaporation method, however, is not suitable for obtaining high density NPs because the size of the NPs would simply increase as the film thickness increases.[13] Hence, solution processable methods have been explored to generate NPs with uniform size distribution and ultrahigh surface density.[26,27] Ordered and close-packed Au NPs could be generated using chemical self-assembly techniques, such as Langmuir–Schaefer deposition and electrostatic layer-by-layer self-assembly.[28–31] Moreover, the uniform Au NPs then acted as charge storage sites in floating gate devices with a polymer dielectric layer. A printable Au NP charge trapping layer with ultrahigh density on a flexible polyethylene terephthalate (PET) substrate was reported by Han and his coworkers in 2012 (Figure 10.4).[32] By embedding a close-packed Au NP monolayer as trapping sites, multilevel data storage could be exhibited under an external gate bias. A large memory window (16.5 V) and a long retention time (10^5 s) was achieved owing to the high density of NPs and low lateral charge leakage, and good device endurance (>1000 cycles) and mechanical stability (>500 bending cycles) were also observed. More importantly, the chemical assembly method is compatible with a low temperature process, which is extendable to the fabrication of large area memory devices.

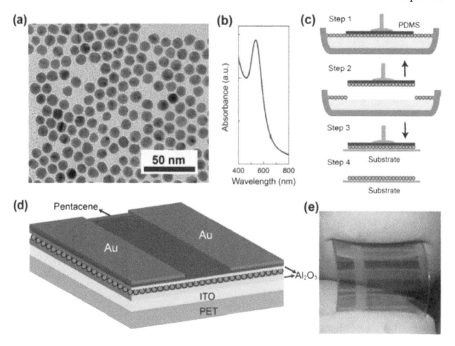

Figure 10.4 (a) TEM image of Au NPs coated with DDT with average diameter of 10 nm. (b) UV-vis absorption spectrum of Au NPs in toluene solution. (c) Schematic illustration of the process for forming a NP array. (d) 3D schematic diagram of the memory device architecture. (e) Optical photograph of fabricated flexible memory device. Reproduced with permission from ref. 32. Copyright 2012 WILEY-VCH Verlag GmbH & Co. KGaA, Weinheim.

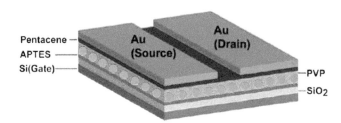

Figure 10.5 Schematic diagram of an Au NP floating gate memory device with a 3-aminopropyltriethoxysilane (APTES) functionalized substrate.

Furthermore, the size of the Au NPs could also influence the memory behavior. The performance of OFET floating gate memory devices with various sized Au NPs was investigated by Han *et al.* They synthesized an Au NP monolayer with particle diameters of 15 nm, 20 nm, and 25 nm through a citrate-reduction method on a 3-aminopropyltriethoxysilane (APTES) functionalized substrate (Figure 10.5).[33] In the programming/erasing operation, reversible threshold voltage (V_{th}) shifts and reliable memory performance were observed. Moreover, a strong size-dependent effect on V_{th} shifts as well as memory switching properties, such as field-effect mobility, on/off current ratio, subthreshold swing, data retention characteristics (over 10^5 s), and operating endurance (800 cycles) was revealed.

Self-assembled nano-floating gates can be generated not only through chemical reaction,[17,33,34] but also using a block copolymer as a template. Block copolymers have unique associative properties, which can facilitate self-assembly of nanostructures under thermal or solvent treatment. The volume fraction of each block as well as the total molecular weight of the block copolymer can lead to different types of self-assembled molecular packings.[35–38] Based on the controllable self-assembly morphology of block copolymers, nanostructured hybrid materials (e.g. nanoparticle arrays) can be produced efficiently,[39–44] and such ordered NPs could be introduced into non-volatile organic floating gate memory to improve the data storage functionality.

Lee's group is one of the pioneers in synthesizing Au NPs in self-assembled block copolymers of poly(styrene-b-4-vinylpyridine) (PS-b-P4VP).[19] The work function of the NPs, the particle size, and NP loading density were investigated to design effective organic memory devices. Subsequently, Leong et al. demonstrated a polymeric memory device formed by in situ synthesis of Au NPs in PS-b-P4VP. The PS-b-P4VP micelles turned out to be an excellent model system, which formed a self-assembled ordered nanostructure simply, and the size of the NPs could be controlled precisely.[45] Both p-type pentacene and n-type F_{16}CuPc OFET-based memory devices have been demonstrated using Au NPs based on block copolymer templates as the trapping elements, and high charge capacity, and stable and programmable/erasable properties were explored, as shown in Figure 10.6.

The synthesis of NPs in a block copolymer directly, however, is generally done at a very low nanoparticle concentration to minimize the particle aggregation. Such a low concentration of NPs in the block copolymer leads to a small memory window for the memory devices. Watkins and his coworkers explored a simple approach for the preparation of well-ordered polymer–NP composites through an additive-driven assembly method,[46] and introduced such composites for organic floating gate memory devices.[47] Well-ordered Au NPs–PS-b-P2VP hybrid films were used as the charge trapping layers in the floating gate memory, and poly(3-hexylthiophene) served as the semiconductor layer (Figure 10.7). Since PS-b-P2VP could provide hydrogen bonding from the pyridine segment to the Au NPs, well-ordered hybrid materials with an Au NP loading up to 40 wt% were obtained. The memory characteristics were tuned effectively by controlling the concentration of Au NPs. In addition, this approach enables the fabrication of well-ordered charge storage layers by solution processing using block copolymers, which is extendable to innovations in large area and high density devices.

Layer-by-layer (LbL) self-assembly is another strategy to improve data storage properties by creating a multilayer charge trapping environment.

(a) In-situ synthesis of Gold Nanoparticles in PS-b-P4VP block copolymer

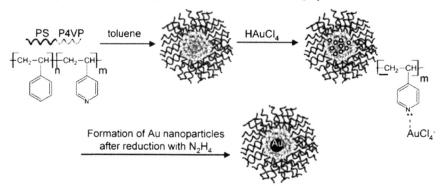

(b) Self-assembling of PS-b-(P4VP/AuNPs) on substrate

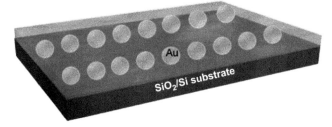

(c) Organic Transistor Floating Gate Memory

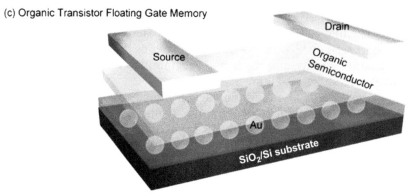

Figure 10.6 Schematic illustration of (a) micellization process with *in situ* synthesis of Au nanoparticles in PS-*b*-P4VP block copolymer; PS-*b*-(P4VP/Au NPs), (b) the self-assembly of PS-*b*-(P4VP/Au NPs) micellar film on a substrate and (c) cross-section of an OFET using PS-*b*-(P4VP/Au NPs) as floating gate memory elements. Reproduced from ref. 45 with permission from The Royal Society of Chemistry.

Self-assembled Au NP/polyelectrolyte (PE) multilayer films composed of Au NPs and PEs as charge-trap and insulating layers, respectively, for floating gate memory devices were also revealed by Lee *et al.*[15] By controlling the thickness of the tunneling oxide layer and the number of PE/Au NP multilayers, the charge trapping of Au NPs as well as the memory behavior can be significantly

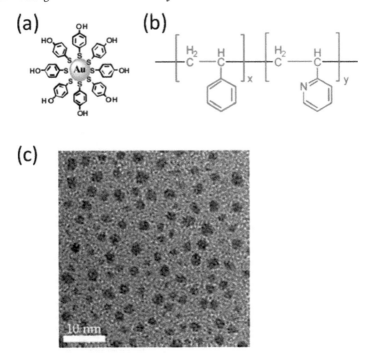

Figure 10.7 (a) Schematic representation of Au NPs with multiple H-bond-donating groups, (b) chemical structure of PS-P2VP, and (c) TEM image of synthesized Au NPs. Reproduced with permission from ref. 47. Copyright 2012 American Chemical Society.

changed with an optimal memory window of 14.6 V.[48,49] Subsequently, a variety of materials that serve as charge-trap elements were embedded in multilayer films for memory applications. For example, the use of 5.8 nm Pt NPs as charge-trap sites can yield a number density of approximately 10^{13} cm^{-2} in four bilayers; Ryu et al. investigated a double-stacked metal nanocrystal (e.g. Ni/Ni, Au/Au, Ni/Au, and Au/Ni) layer for charge trapping, and the work function of the metal nanocrystal could affect the non-volatile memory characteristics.[50]

Recently, self-assembled monolayers (SAM) have been attracting intense interest due to the possibility of producing molecule-based electronics and their unique interfacial properties.[51-53] Such SAM layers can be used to improve the adhesion, compatibility, and the charge transfer properties at the interface, leading to the improvement of OFET memory properties. Chen et al. demonstrated a novel approach to improve the characteristics of Au NP-based organic transistor memory devices by using SAMs with different functional groups as interfacial modifiers (Figure 10.8).[54] SAM-based interfacial engineering significantly improved the hysteresis, memory window (190 V), and on/off ratio (10^5) of the floating gate memory with an operation voltage of 100 V, and stable data retention was also observed in this memory system.

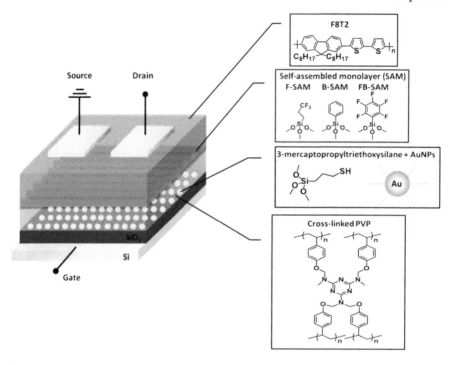

Figure 10.8 Schematic device configuration of a polymer FET memory device with a functionalized self-assembled monolayer.

10.1.3 Polymer–Nanoparticle Hybrids

Hybrids of metal nanoparticles (NPs)–polymer materials can also be employed as charge trapping elements of transistor memory devices. Typically, such hybrid layers are inserted between a thin tunneling dielectric and a thick blocking dielectric, or directly as a polymer semiconductor channel. Floating gate transistor memory devices with well-distributed metal NPs in charge trap hybrid layers allow modulation of high density memory devices with discrete charge storage sites in a low voltage range.[47] Han and his coworkers demonstrated controllable threshold voltage (V_{th}) shifts of Au NP–poly(3-hexylthiophene) (P3HT)-based composite transistors, as depicted in Figure 10.9.[55] By tuning the doping concentration of Au NPs in the P3HT matrix, V_{th} was changed from 12 to 27 V without device degradation. Using this technique, the switching voltages of unipolar inverters have also been adjusted systematically. In addition, efficient hole conduction and P3HT crystallinity changes were observed due to different concentrations of Au NPs, which eventually shifted the threshold voltage of the transistor memory in a controlled manner.

Moreover, Chen *et al.* presented a novel flexible non-volatile flash transistor memory device on a polyethylene naphthalate (PEN) substrate using 1D electrospun nanofibers of P3HT–Au NP hybrids (Figure 10.10).[56] The Au NPs

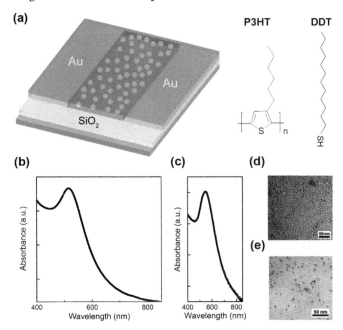

Figure 10.9 (a) 3D schematic diagram of the OFET device architecture and molecular structures of P3HT and DDT. UV-vis absorption spectra of (b) Au NPs in hexane solution and (c) solid state Au NPs obtained by drop casting. (d) TEM image of Au NPs coated with DDT. (e) TEM micrograph of P3HT–Au NP composite film. Reproduced with permission from ref. 55. Copyright 2012 American Institute of Physics.

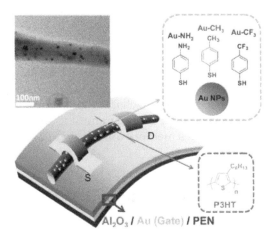

Figure 10.10 Schematic configuration of the hybrid nanofiber based transistor memory devices, chemical structures of P3HT and surface-modified Au NPs, and representative plan-view TEM image of the P3HT:Au hybrid nanofiber. Reproduced with permission from ref. 56. Copyright 2013 WILEY-VCH Verlag GmbH & Co. KGaA, Weinheim.

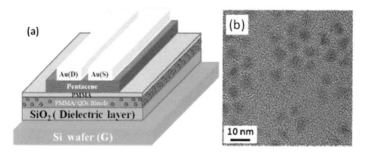

Figure 10.11 (a) Configuration of non-volatile FET memory device with QD floating gate and (b) TEM image of the QDs. The average diameter of the QDs was about 5.4 nm. Reproduced with permission from ref. 57. Copyright 2012 American Institute of Physics.

were functionalized with self-assembled *para*-substituted amino (Au–NH$_2$), methyl (Au–CH$_3$) or trifluoromethyl (Au–CF$_3$) tail groups on the benzenethiol moiety. Localized charge traps across the nanofiber channel were induced, and exhibited memory behavior under the applied electrical field (*i.e.* programming/erasing). With a low operation voltage of ±5 V, the hybrid nanofiber transistor memory devices exhibited threshold voltage shifts of 3.5–10.6 V with data retention for at least 10^4 s. The dipoles of the functionalized tail groups probably modified the work function of the Au NPs associated with the P3HT nanofiber channel, and Au–NH$_2$ possessed the largest negative threshold voltage shift. Also, the flexible devices remained reliable and stable even under a 5 mm bending radius or 1000 repetitive bending cycles.

Quantum dots (QDs), furthermore, are also a potential material for charge-trapping sites in floating gate memory devices. Non-volatile OFET memory using a core–shell CdSe–ZnS QD–poly(methyl methacrylate) (PMMA) composite layer as the floating gate has been explored (Figure 10.11).[57] This QD–PMMA hybrid was introduced in a pentacene-based transistor, and the improvement of the memory performance was significant. The memory behavior was mainly attributed to the charge storage/discharge effect in the QD–PMMA composite layer. Under the programming and erasing operations, writable memory devices with a memory window up to 23 V with a long retention time were obtained.

10.2 Functional Moiety Embedded Materials as Charge Storage Layers

In addition to metal nanoparticles, nanostructured non-metallic inorganic materials, organic π-conjugated molecules, and reduced graphene oxide (rGO) sheets have also been introduced into floating gate transistor memory device. Needle-like silicon nanostructures were prepared using a self-assembled honeycomb structure of polystyrene (PS) and poly(methylmethacrylate)

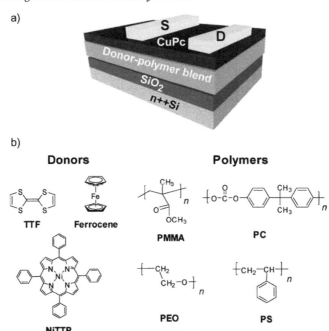

Figure 10.12 (a) Schematic drawing of an organic transistor memory device. (b) Chemical structures of donor molecules and polymers for blend organic memory transistor elements. Reproduced with permission from ref. 59. Copyright 2008 WILEY-VCH Verlag GmbH & Co. KGaA, Weinheim.

(PMMA) block copolymers as a nanomask.[58] π-Conjugated molecules, including electron-donating or -accepting materials, have also been incorporated into gate dielectrics as charge storage sites that are similar to metal NP floating gate devices. Liu et al. achieved a transistor memory device using a donor-polymer blend as a buffer layer between a CuPc semiconductor channel and a SiO_2 gate dielectric.[59] Various organic donor molecules, such as tetrathiafulvalene, ferrocene, or 5,10,15,20-tetraphenyl-21H,23H-porphine nickel(II), were blended with common insulating polymers, including PMMA, poly(ethylene oxide) or PS, as shown in Figure 10.12. The memory characteristics were observed through the charging and discharging operations in the donor-polymer blending layer, and the charging effect was mainly attributed to electric field-induced charge transfer between CuPc and the electron donors. The operation voltage and the memory window were adjustable by tuning the film thickness of the dielectric layer, and could be scaled up to 2 V with a long retention of at least 24 h.

Baeg et al. demonstrated a transistor-type non-volatile memory device based on [6,6]-phenyl-C_{61} butyric acid methyl ester (PCBM) molecules in a poly(4-vinyl phenol) (PVP) matrix.[60] The operating voltage of this memory device was effectively reduced owing to the high capacitance of the PVP/

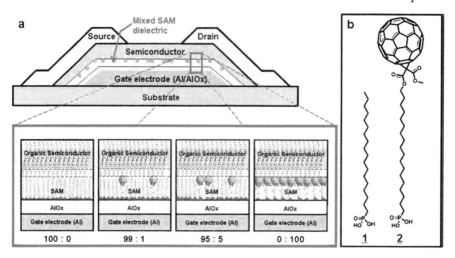

Figure 10.13 (a) Schematic cross-section of an OFET with mixed molecular gate dielectrics; (b) chemical structure of molecules comprising the SAM in the device. Reproduced with permission from ref. 61. Copyright 2010 WILEY-VCH Verlag GmbH & Co. KGaA, Weinheim.

PCBM film and the strong electron withdrawing properties of PCBM. The data storage properties of PVP/PCBM were investigated using pentacene based FETs. This transistor memory device exhibited a field-effect mobility of 0.25–0.3 cm^2 V^{-1} s^{-1} with an on/off ratio of >10^4 s, and a memory window of 20 V could be explored during writing/erasing with an operation voltage of 60 V with a relatively short duration (~1 ms). Moreover, stable data retention ability of more than 40 h was shown. In addition, a single layer of self-assembled organic molecules was explored to improve the uniformity of the charge trapping sites in floating gate devices. Burkhardt *et al.* revealed an electrically programmable self-assembled molecule (functionalized C$_{60}$) which was incorporated into the gate dielectric layer for memory applications. This self-assembled layer could be reversibly charged and discharged, and retain the active state (*i.e.* ON state) even when the power supply was turned off (Figure 10.13).[61] The transistor memory could be operated under very small program and erase voltages of −2 V because of the small thickness of the dielectric stack (~5.7 nm), and its retention time could be kept over 6 h with a continuous −750 mV power source. Comparing the device with and without C$_{60}$ molecules, a larger hysteresis was observed as the content of C$_{60}$ increased. The retention time, however, seemed to be limited by charge leakage back to the channel due to the C$_{60}$ molecules. Generally, the threshold voltage, data storage ability, and on/off ratio could be controlled with different amounts of charge trapping molecules, and optimal device characteristics might be obtained with a suitable molecular doping ratio.

As shown above, transistor memory can be achieved using different approaches. Relatively poor charge-retention time, which is mainly induced

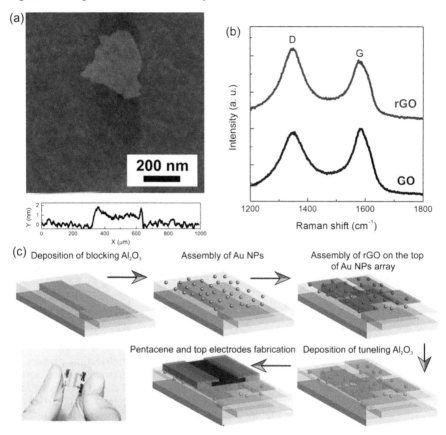

Figure 10.14 (a) Tapping-mode AFM image of rGO sheet with height profile. (b) Raman spectra of rGO and GO. (c) Schematic diagram depicting the basic fabrication process of rGO–Au NP hybrid double-floating gate memory device and an optical image of the flexible memory. Reproduced with permission from ref. 63. Copyright 2012 WILEY-VCH Verlag GmbH & Co. KGaA, Weinheim.

by a thin tunneling dielectric layer, however, is still the most important issue to be solved.[62] The simplest solution is to increase the thickness of the tunneling dielectric layer, but this would degrade the operating speed and increase the power consumption.[15] Instead, a double NP floating gate structure has been proposed to achieve better retention properties. As depicted in Figure 10.14, a new device architecture for low-voltage flash memory based on solution-processed layer-by-layer-assembled rGO sheet–Au NP hybrid double floating gates on flexible PET substrates was presented by Roy and his coworkers.[63] Monolayers of rGO sheets (90% coverage) and Au NP arrays with a density of 1.3×10^{11} cm^{-2} acted as the top and bottom floating gates, respectively. The large area rGO monolayer almost fully covered the Au NP array and overcame the current leakage issue in the vertical direction of the

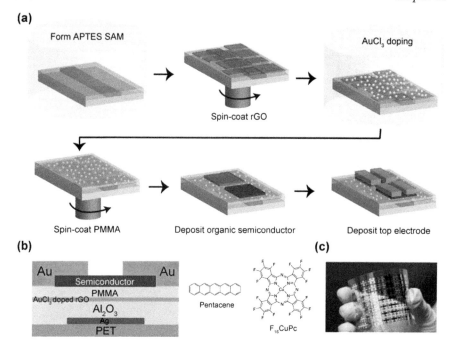

Figure 10.15 (a) Schematic diagram depicting the basic fabrication process of $AuCl_3$ doped rGO floating gate memory device. (b) Fabricated device structure for memory device with $AuCl_3$ doped rGO floating gate and molecular structure of pentacene and $F_{16}CuPc$. (c) The optical image of the flexible memory. Reproduced with permission from ref. 71. Copyright 2014 American Chemical Society.

NP double floating gates. In comparison with floating gate memory devices with only a single layer of Au NPs, the FET characteristics, including mobility, on/off ratio, and sub-threshold swing were significantly improved when the rGO sheets were introduced. In these double-trapping-layer memory devices, furthermore, the memory characteristics, such as the memory window and the retention capability, were also improved.

Two-dimensional (2D) graphene can not only be deposited as large area sheets for floating gate applications, but also possesses the advantages of tunable energy levels (*i.e.* Fermi level).[64] Note that common metallic NPs, such as Au NPs, cannot achieve tunable memory characteristics by energy band engineering due to their fixed Fermi level.[12,65] Approaches, including doping of foreign atoms (boron for p-doping or nitrogen for n-doping), tuning the dipole moment arising from self-assembled monolayers (SAMs), and chemical doping have been proposed to manipulate the work function of graphene.[66–70] A low voltage flash memory device based on $AuCl_3$ doped self-assembled rGO floating gates on flexible substrates was explored by Han *et al.*, as shown in Figure 10.15.[71] Pentacene and $F_{16}CuPc$ served as p-type and n-type semiconductors, respectively. The memory characteristics have been systematically controlled in both p-channel and n-channel memory transistors by embedding rGO monolayers with various doping concentrations of AuCl$_3$ (as a function of doping time of 0 s, 5 s, 25 and 45 s). Compared to a reference floating gate memory device fabricated on a pristine rGO monolayer, the memory window was observed to be significantly enhanced with increasing doping concentration of AuCl$_3$. The charge carrier injection barrier between the Fermi level of the doped rGO layer and the HOMO level of the semiconductor (*i.e.* pentacene and F$_{16}$CuPc) was reduced through the Au doping. The downward shift of the rGO Fermi level and the increased size and density of reduced Au NPs significantly enhanced the memory window. Also, various V_{th}s were achieved in both p- and n-type memory devices due to the control of the AuCl$_3$ doping concentration. The largest memory window was obtained in both p-channel and n-channel memory at a doping time of 25 s. Note that when the doping time was increased up to 45 s, a rough surface morphology was shown with reduced device performance. In addition, such chemically doped rGO flexible floating gate memory devices exhibited fast program/erase speeds (100 ms), long retention time (10^5 s), good operating endurance (>800 cycles), and mechanical stability (>500 bending cycles), which can be applied for the construction of innovative floating gate electronic devices.

10.3 Switching Mechanism

The characteristics of floating gate memory devices are mainly attributed to the channel conductance changed by the stored charges in the floating gates (*i.e.* metallic NPs), leading to reversible shifts in the threshold voltage (V_{th}) of the OFETs. Therefore, such floating gate devices can be utilized not only for non-volatile transistor memory, but also as an effective way to control the V_{th} in OFET devices precisely. The operation mechanism of organic floating gate memory is very similar to that of traditional Si-based flash memory devices. As a strong external electric field is applied to the gate electrode of the floating gate memory, mobile charge carriers are injected and trapped in the floating gates, which are covered by a tunneling or blocking dielectric layer between the gate contact and the semiconducting layer. The threshold voltage (V_{th}) will be increased or decreased by the gate field modulation, and the changed value will depend on the polarity of the floating gate.[13] The basic operation of a p-type non-volatile (for example, pentacene) transistor memory device with a floating gate electret is depicted in Figure 10.16. With application of a positive gate pulse (programming process), a large amount of negative charges are induced through semiconducting layer from the source/drain contacts and trapped in the floating gates. Positive charges (*i.e.* holes) are easily accumulated at the interface between the semiconductor and the floating gate electret layers due to a built-in electric field induced from the retained electrons, even though the external gate voltage is removed. Positive shifts in the transfer curves, hence, are observed and the device reaches a high-conductance state (ON state) at a gate voltage of 0 V. When a negative

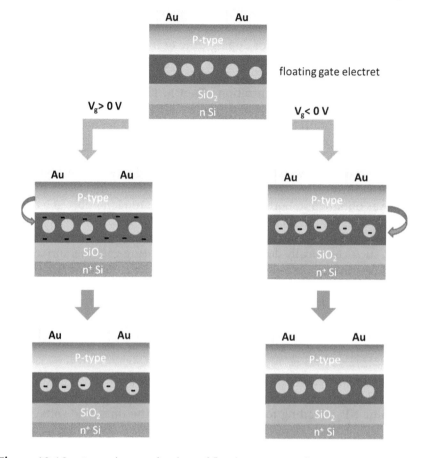

Figure 10.16 Operation mechanism of floating gate transistor memory.

gate pulse (erasing process) is applied, the trapped electrons recombine with the holes from the semiconducting layer. The device then returns to a low-conductance state (OFF state) at V_G = 0 V. Currently, most organic floating gate memory devices exhibit relatively poor device characteristics along with short retention times and slow switching speeds, and require relatively high operating bias because of the inorganic trapping sites.[72,73] Therefore, it is important to select suitable dielectric and metallic trapping sites (*e.g.* nanoparticles) for the development of high performance floating gate memory devices.

10.4 Flexible Electrical Memory Devices

Flexible electronic devices have attracted great attention because of their advantages of good portability, conformal contact with non-flat surfaces, and bio-friendly interfaces compared to conventional silicon technology.[74–76] Many studies have revealed that flexible devices can be integrated based on

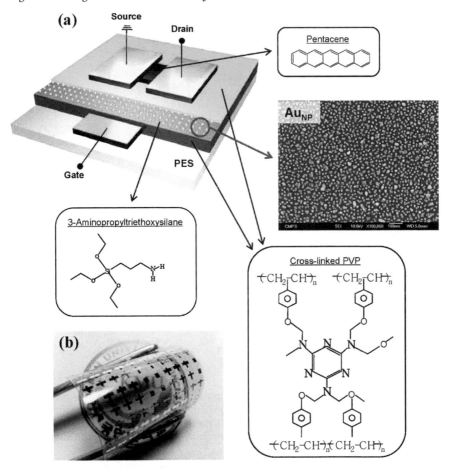

Figure 10.17 (a) 3D schematic diagram of the memory device architecture and (b) photograph of the fabricated flexible organic memory devices. Reproduced with permission from ref. 82. Copyright 2010 American Chemical Society.

organic materials.[9,77–79] Recently, floating gate transistor memory devices have also been investigated for next generation flexible electronic applications using low temperature fabricated metallic NP arrays as charge storage layers.[80,81] Such flexible floating gate memory devices can be integrated on polyether sulfone (PES)[82] or polyethylene naphthalate (PEN) substrates.[63] Kim et al. demonstrated flexible non-volatile organic memory devices with pentacene and Au NPs on a PES substrate (Figure 10.17).[82] The devices showed good memory characteristics, long retention time for more than one year, and outstanding endurance (could be operated for more than 100 programming/erasing cycles). Additionally, the flexibility of the organic memory devices was verified using cyclic tests under bending.

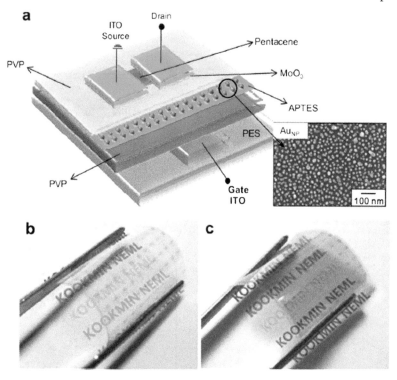

Figure 10.18 (a) Schematic device configuration of transparent organic non-volatile memory devices fabricated on plastic substrates. (b) Transparent flexible organic FETs and (c) the FET memory devices are shown. Reproduced from ref. 85 with permission from The Royal Society of Chemistry.

To further improve the performance and functionality of flexible floating gate devices, transparent electronic devices were then investigated for novel applications such as see-through electronic devices and head-up displays.[83,84] Flexible and transparent organic nano-floating gate memory devices were developed using solution-processed PVP dielectric layers and a pentacene active layer on flexible plastic foil substrates (Figure 10.18).[85] Au NPs, in addition, were synthesized by a citrate reduction method and served as the charge storage elements. Transparent ITO was used as the gate and source/drain contacts in such flexible memory devices to integrate the transparent organic transistor memory devices. The programming/erasing operations exhibited non-volatile memory characteristics, and showed a stable threshold voltage (V_{th}) shift in the programmed/erased states with good retention stability. Furthermore, the flexible organic memory device was characterized in a bent environment to investigate its endurance, and reproducible and reliable memory properties were observed. This example demonstrated that a flexible and transparent data storage device could be easily fabricated using solution processable charge trapping NPs and a polymer dielectric without a high temperature processing step, leading to further high performance electronic applications.

10.5 Conclusion

The recent progress of transistor-type memory devices with floating gate charge storage materials is highlighted in this chapter. Approaches of material design, deposition methods, and device configuration provide a pathway for us to understand the structure–property relationships of organic floating gate transistor memory devices. The current floating gate memory characteristics, such as the charge storage ability, reproducibility of memory cells, data retention time, and mechanical endurance should be further optimized for real applications as discussed above. With technical improvements, organic floating gate transistor memory should possess high potential to be used in next-generation digital data storage electronics with its high charge trapping density and dispersed charge storage sites.

References

1. C.-L. Liu and W.-C. Chen, *Polym. Chem.*, 2011, **2**, 2169.
2. N. G. Kang, B. Cho, B. G. Kang, S. Song, T. Lee and J. S. Lee, *Adv. Mater.*, 2012, **24**, 385.
3. S. G. Hahm, N. G. Kang, W. Kwon, K. Kim, Y. G. Ko, S. Ahn, B. G. Kang, T. Chang, J. S. Lee and M. Ree, *Adv. Mater.*, 2012, **24**, 1062.
4. Q.-D. Ling, D.-J. Liaw, C. Zhu, D. S.-H. Chan, E.-T. Kang and K.-G. Neoh, *Prog. Polym. Sci.*, 2008, **33**, 917.
5. P. Heremans, G. H. Gelinck, R. Müller, K.-J. Baeg, D.-Y. Kim and Y.-Y. Noh, *Chem. Mater.*, 2011, **23**, 341.
6. Y. Guo, G. Yu and Y. Liu, *Adv. Mater.*, 2010, **22**, 4427.
7. D. Kahng and S. M. Sze, *Bell Syst. Tech. J.*, 1967, **46**, 1288.
8. S. M. Sze and K. K. Ng, *Physics of semiconductor devices*, John Wiley & Sons, 2006.
9. T. Sekitani, T. Yokota, U. Zschieschang, H. Klauk, S. Bauer, K. Takeuchi, M. Takamiya, T. Sakurai and T. Someya, *Science*, 2009, **326**, 1516.
10. S. Paydavosi, K. E. Aidala, P. R. Brown, P. Hashemi, G. J. Supran, T. P. Osedach, J. L. Hoyt and V. Bulović, *Nano Lett.*, 2012, **12**, 1260.
11. W. D. Brown and J. E. Brewer, *Nonvolatile semiconductor memory technology: a comprehensive guide to understanding and to using NVSM devices*, Wiley-IEEE Press, 1998.
12. D. V. Talapin, J.-S. Lee, M. V. Kovalenko and E. V. Shevchenko, *Chem. Rev.*, 2009, **110**, 389.
13. K. J. Baeg, Y. Y. Noh, H. Sirringhaus and D. Y. Kim, *Adv. Funct. Mater.*, 2010, **20**, 224.
14. S. Wang, C. Leung and P. K. Chan, *Appl. Phys. Lett.*, 2010, **97**, 023511.
15. J.-S. Lee, J. Cho, C. Lee, I. Kim, J. Park, Y.-M. Kim, H. Shin, J. Lee and F. Caruso, *Nat. Nanotechnol.*, 2007, **2**, 790.
16. S.-J. Kim, Y.-S. Park, S.-H. Lyu and J.-S. Lee, *Appl. Phys. Lett.*, 2010, **96**, 033302.
17. C. Novembre, D. Guerin, K. Lmimouni, C. Gamrat and D. Vuillaume, *Appl. Phys. Lett.*, 2008, **92**, 103314.

18. W. L. Leong, P. S. Lee, A. Lohani, Y. M. Lam, T. Chen, S. Zhang, A. Dodabalapur and S. G. Mhaisalkar, *Adv. Mater.*, 2008, **20**, 2325.
19. W. Leong, N. Mathews, S. Mhaisalkar, T. Chen and P. Lee, *Appl. Phys. Lett.*, 2008, **93**, 222908.
20. M. Kang, K. J. Baeg, D. Khim, Y. Y. Noh and D. Y. Kim, *Adv. Funct. Mater.*, 2013, **23**, 3503.
21. Z. Liu, C. Lee, V. Narayanan, G. Pei and E. C. Kan, *IEEE Trans. Electron Devices*, 2002, **49**, 1606.
22. J.-S. Lee, *Gold Bull.*, 2010, **43**, 189.
23. T.-C. Chang, F.-Y. Jian, S.-C. Chen and Y.-T. Tsai, *Mater. Today*, 2011, **14**, 608.
24. K. Richter, A. Birkner and A.-V. Mudring, *Phys. Chem. Chem. Phys.*, 2011, **13**, 7136.
25. E. Sacher, *Metallization of polymers 2*, Springer, 2002.
26. H. Sirringhaus, T. Kawase, R. Friend, T. Shimoda, M. Inbasekaran, W. Wu and E. Woo, *Science*, 2000, **290**, 2123.
27. Y.-S. Park, S. Chung, S.-J. Kim, S.-H. Lyu, J.-W. Jang, S.-K. Kwon, Y. Hong and J.-S. Lee, *Appl. Phys. Lett.*, 2010, **96**, 213107.
28. V. Santhanam, J. Liu, R. Agarwal and R. P. Andres, *Langmuir*, 2003, **19**, 7881.
29. A. Facchetti, M. H. Yoon and T. J. Marks, *Adv. Mater.*, 2005, **17**, 1705.
30. S. Kolliopoulou, P. Dimitrakis, P. Normand, H.-L. Zhang, N. Cant, S. D. Evans, S. Paul, C. Pearson, A. Molloy and M. Petty, *J. Appl. Phys.*, 2003, **94**, 5234.
31. Z. Liu, F. Xue, Y. Su, Y. M. Lvov and K. Varahramyan, *IEEE Trans. Nanotechnol.*, 2006, **5**, 379.
32. S. T. Han, Y. Zhou, Z. X. Xu, L. B. Huang, X. B. Yang and V. Roy, *Adv. Mater.*, 2012, **24**, 3556.
33. S.-T. Han, Y. Zhou, Z.-X. Xu, V. Roy and T. F. Hung, *J. Mater. Chem.*, 2011, **21**, 14575.
34. M. F. Mabrook, Y. Yun, C. Pearson, D. A. Zeze and M. C. Petty, *Appl. Phys. Lett.*, 2009, **94**, 173302.
35. G. Riess, *Prog. Polym. Sci.*, 2003, **28**, 1107.
36. M. J. Fasolka and A. M. Mayes, *Annu. Rev. Mater. Res.*, 2001, **31**, 323.
37. I. W. Hamley, *Block copolymers in solution: fundamentals and applications*, John Wiley & Sons, 2005.
38. P. Bahadur, *Curr. Sci.*, 2001, **80**, 1002.
39. G. Kästle, H. G. Boyen, F. Weigl, G. Lengl, T. Herzog, P. Ziemann, S. Riethmüller, O. Mayer, C. Hartmann and J. P. Spatz, *Adv. Funct. Mater.*, 2003, **13**, 853.
40. Y.-w. Jun, S.-M. Lee, N.-J. Kang and J. Cheon, *J. Am. Chem. Soc.*, 2001, **123**, 5150.
41. R. Glass, M. Möller and J. P. Spatz, *Nanotechnology*, 2003, **14**, 1153.
42. W. van Zoelen and G. ten Brinke, *Soft Matter*, 2009, **5**, 1568.
43. J. Y. Cheng, C. Ross, V. H. Chan, E. L. Thomas, R. G. Lammertink and G. J. Vancso, *Adv. Mater.*, 2001, **13**, 1174.

44. C.-A. Dai, Y.-L. Wu, Y.-H. Lee, C.-J. Chang and W.-F. Su, *J. Cryst. Growth*, 2006, **288**, 128.
45. W. L. Leong, N. Mathews, S. Mhaisalkar, Y. M. Lam, T. Chen and P. S. Lee, *J. Mater. Chem.*, 2009, **19**, 7354.
46. Y. Lin, V. K. Daga, E. R. Anderson, S. P. Gido and J. J. Watkins, *J. Am. Chem. Soc.*, 2011, **133**, 6513.
47. Q. Wei, Y. Lin, E. R. Anderson, A. L. Briseno, S. P. Gido and J. J. Watkins, *ACS Nano*, 2012, **6**, 1188.
48. Y.-M. Kim, Y.-S. Park, A. O'Reilly and J.-S. Lee, *Electrochem. Solid-State Lett.*, 2010, **13**, H134.
49. Y.-M. Kim, S.-J. Kim and J.-S. Lee, *IEEE Electron Device Lett.*, 2010, **31**, 503.
50. S.-W. Ryu, J.-W. Lee, J.-W. Han, S. Kim and Y.-K. Choi, *IEEE Trans. Electron Devices*, 2009, **56**, 377.
51. C. Collier, E. Wong, M. Belohradský, F. Raymo, J. Stoddart, P. Kuekes, R. Williams and J. Heath, *Science*, 1999, **285**, 391.
52. N. Zhitenev, H. Meng and Z. Bao, *Phys. Rev. Lett.*, 2002, **88**, 226801.
53. E. B. Troughton, C. D. Bain, G. M. Whitesides, R. G. Nuzzo, D. L. Allara and M. D. Porter, *Langmuir*, 1988, **4**, 365.
54. H.-C. Chang, W.-Y. Lee, Y. Tai, K.-W. Wu and W.-C. Chen, *Nanoscale*, 2012, **4**, 6629.
55. S.-T. Han, Y. Zhou, Z.-X. Xu and V. Roy, *Appl. Phys. Lett.*, 2012, **101**, 033306.
56. H. C. Chang, C. L. Liu and W. C. Chen, *Adv. Funct. Mater.*, 2013, **23**, 4960.
57. Y.-C. Chen, C.-Y. Huang, H.-C. Yu and Y.-K. Su, *J. Appl. Phys.*, 2012, **112**, 034518.
58. S. Jung, K. Kim, D.-H. Park, B.-H. Sohn, J. C. Jung, W. C. Zin, S. Hwang, S. Dhungel, J. Yoo and J. Yi, *Mater. Sci. Eng., C*, 2007, **27**, 1452.
59. W. Wu, H. Zhang, Y. Wang, S. Ye, Y. Guo, C. Di, G. Yu, D. Zhu and Y. Liu, *Adv. Funct. Mater.*, 2008, **18**, 2593.
60. K.-J. Baeg, D. Khim, D.-Y. Kim, S.-W. Jung, J. B. Koo and Y.-Y. Noh, *Jpn. J. Appl. Phys.*, 2010, **49**, 05EB01.
61. M. Burkhardt, A. Jedaa, M. Novak, A. Ebel, K. Voïtchovsky, F. Stellacci, A. Hirsch and M. Halik, *Adv. Mater.*, 2010, **22**, 2525.
62. X. Tang, C. Krzeminski, A. Lecavelier des Etangs-Levallois, Z. Chen, E. Dubois, E. Kasper, A. Karmous, N. Reckinger, D. Flandre and L. A. Francis, *Nano Lett.*, 2011, **11**, 4520.
63. S. T. Han, Y. Zhou, C. Wang, L. He, W. Zhang and V. Roy, *Adv. Mater.*, 2013, **25**, 872.
64. X. Huang, X. Qi, F. Boey and H. Zhang, *Chem. Soc. Rev.*, 2012, **41**, 666.
65. J. S. Lee, Y. M. Kim, J. H. Kwon, J. S. Sim, H. Shin, B. H. Sohn and Q. Jia, *Adv. Mater.*, 2011, **23**, 2064.
66. O. Stephan, P. Ajayan, C. Colliex, P. Redlich, J. Lambert, P. Bernier and P. Lefin, *Science*, 1994, **266**, 1683.
67. L. Panchakarla, K. Subrahmanyam, S. Saha, A. Govindaraj, H. Krishnamurthy, U. Waghmare and C. Rao, *Adv. Mater.*, 2009, **21**, 4726.
68. K. C. Kwon, B. J. Kim, J.-L. Lee and S. Y. Kim, *J. Mater. Chem. C*, 2013, **1**, 253.

69. Y. Shi, K. K. Kim, A. Reina, M. Hofmann, L.-J. Li and J. Kong, *ACS Nano*, 2010, **4**, 2689.
70. K. C. Kwon, K. S. Choi and S. Y. Kim, *Adv. Funct. Mater.*, 2012, **22**, 4724.
71. S.-T. Han, Y. Zhou, Q. D. Yang, L. Zhou, L.-B. Huang, Y. Yan, C.-S. Lee and V. A. Roy, *ACS Nano*, 2014, **8**, 1923.
72. B. Cho, J. M. Yun, S. Song, Y. Ji, D. Y. Kim and T. Lee, *Adv. Funct. Mater.*, 2011, **21**, 3976.
73. T.-W. Kim, S.-H. Oh, H. Choi, G. Wang, H. Hwang, D.-Y. Kim and T. Lee, *Appl. Phys. Lett.*, 2008, **92**, 253308.
74. R. H. Reuss, B. R. Chalamala, A. Moussessian, M. G. Kane, A. Kumar, D. C. Zhang, J. A. Rogers, M. Hatalis, D. Temple and G. Moddel, *Proc. IEEE*, 2005, **93**, 1239.
75. T. Tsutsui and K. Fujita, *Adv. Mater.*, 2002, **14**, 949.
76. S. Y. Lee, K.-I. Park, C. Huh, M. Koo, H. G. Yoo, S. Kim, C. S. Ah, G. Y. Sung and K. J. Lee, *Nano Energy*, 2012, **1**, 145.
77. H. Klauk, M. Halik, U. Zschieschang, F. Eder, G. Schmid and C. Dehm, *Appl. Phys. Lett.*, 2003, **82**, 4175.
78. J. Ouyang, C.-W. Chu, C. R. Szmanda, L. Ma and Y. Yang, *Nat. Mater.*, 2004, **3**, 918.
79. W. H. Lee, D. H. Kim, Y. Jang, J. H. Cho, M. Hwang, Y. D. Park, Y. H. Kim, J. I. Han and K. Cho, *Appl. Phys. Lett.*, 2007, **90**, 132106.
80. C. Lee, J.-H. Kwon, J.-S. Lee, Y.-M. Kim, Y. Choi, H. Shin, J. Lee and B.-H. Sohn, *Appl. Phys. Lett.*, 2007, **91**, 153506.
81. J.-S. Lee, Y.-M. Kim, J.-H. Kwon, H. Shin, B.-H. Sohn and J. Lee, *Adv. Mater.*, 2009, **21**, 178.
82. S.-J. Kim and J.-S. Lee, *Nano Lett.*, 2010, **10**, 2884.
83. P. Görrn, M. Sander, J. Meyer, M. Kröger, E. Becker, H. H. Johannes, W. Kowalsky and T. Riedl, *Adv. Mater.*, 2006, **18**, 738.
84. J. W. Seo, J.-W. Park, K. S. Lim, J.-H. Yang and S. J. Kang, *Appl. Phys. Lett.*, 2008, **93**, 223505.
85. S.-J. Kim, J.-M. Song and J.-S. Lee, *J. Mater. Chem.*, 2011, **21**, 14516.

CHAPTER 11

Organic Ferroelectric Memory Devices

HSUAN-CHUN CHANG[a], HUNG-CHIN WU[a], AND WEN-CHANG CHEN*[a]

[a]Department of Chemical Engineering, National Taiwan University, Taipei, 10617, Taiwan
*E-mail: chenwc@ntu.edu.tw

11.1 Introduction

There have been recent developments in portable computing, information management and communication which have grown the demand for technologies enabling the rapid manipulation and storage of large volumes of data. Many of these applications require speed, convenience and reliability. New memory technologies are based on the electrical bistability of materials from changes in certain intrinsic properties. For data storage devices, physical property changes in the memory cells usually lead to current–voltage hysteresis in response to an applied electric field. Such electrical hysteretic characteristics can be defined as stored digital information (*i.e.* "0" or "1") in the circuits. Several kinds of physical phenomena, such as resistance and capacitance changes, can be utilized to create signal hysteresis as well as memory behavior. Recently, ferroelectric materials have received a lot of research attention because this class of material possesses spontaneous electric polarization, which can be reversed by an external electric field. Note that this unique physical property of ferroelectric materials can be defined as ferroelectricity.[1–3]

Since the electric polarization of ferroelectrics can be reversed by inverting the electric field, a critical electric field for reversing such polarization is explored, called the coercive field. In addition, ferroelectric polarization can be described by the mathematical equation:

$$D = \varepsilon_r E + P$$

where P is the polarization, D is the dielectric displacement, E is the electric field, and ε_r is the relative permittivity of the dielectric material. A hysteretic curve (D–E loop) can be drawn consequently by plotting the electric displacement (D) as a function of the electric field strength (E) between opposite polarities, and this electric bistability can be recognized as digital memory information. A standard method for investigating the D–E loop is a Sawyer–Tower circuit (Figure 11.1).[4] A sinusoidal voltage signal is applied to one of the electrodes of the ferroelectric capacitor, and the amount of charge displacement in the other electrode is collected as a voltage is applied. Figure 11.1 also shows the ideal D–E hysteresis curve for a ferroelectric capacitor with two parallel electrodes. In the low-voltage state, only a linear component is measured, which means that the electric field is not large enough to affect P. As the strength of the applied electric field increases, a hysteretic response develops and reaches a saturated state. Based on the above physical characteristics, phase transitions can be found in the solid state of ferroelectric materials, and result in a spontaneous non-zero polarization. The non-linear switching properties can be used to define the digital information storage capacity and applied to memory devices. On the other hand, many studies, including metal–organic semiconductor–metal junctions,[5,6] charge trapping effects in field-effect transistors,[7–9] and electromechanical switches,[10] are ongoing towards high performance memory for practical applications.[11] For memory elements suitable for these applications, the importance lies in

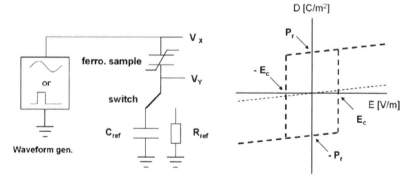

Figure 11.1 Right: hysteretic loop of ferroelectric thin film between two electrodes. Left: measurement of D–E hysteresis and transient switching phenomena using a Sawyer–Tower setup. Reproduced with permission from ref. 79, © 2010 American Chemical Society.

how cost-efficient they are while maintaining the minimum performance requirements. Organic and polymeric materials are promising candidates for future memory devices as they not only satisfy this requirement but can also potentially operate at molecular scale sizes. Most importantly, thin and smooth polymer films can be formed on various substrates by vacuum-free solution processes. The recently emerging interest in ferroelectric polymers for memory applications is obvious and a profound understanding of the microstructures of polymers and the origin of ferroelectricity enables researchers to accelerate the realization of non-volatile memory devices.

In this chapter, we aim to focus on the recent developments in organic memory devices using ferroelectric materials, and the electrical hysteretic and switching properties are presented systematically. This chapter is organized as follows: Section 11.2 presents a short overview about the relevant ferroelectric materials and their given properties. In Section 11.3, the device structures of the organic ferroelectric memory and the electrical parameters will be discussed. Section 11.4 gives a summary and looks at attempts to use organic ferroelectric memory for future applications.

11.2 Materials for Ferroelectricity

Schematic drawings of inorganic (Figure 11.2a) and organic (Figure 11.2b) ferroelectric solids illustrate how conventional materials achieve ferroelectricity on a microscopic level. The inorganic crystals generally can be divided into three different groups. The first group comprises hydrogen bonded systems. In such systems, ferroelectricity is created by preferentially occupied hydrogen sites within the hydrogen bonds. The associated dipole moment creates the polarization. In the second groups, ionic crystals (such as perovskites) are found. Finally, narrow gap semiconductors may also exhibit ferroelectric properties.

The most straightforward mechanism, as exemplified by sodium nitrite ($NaNO_2$) is that the permanent dipoles of the polar molecules or ions generate spontaneous polarization, and their reorientation generates ferroelectricity. The dipole moments are ordered without cancelling each other out in the ferroelectric state, whereas the paraelectric state corresponds to disorder in their orientations.[12] These transformations are mostly classified by order–disorder type phase transitions. The corresponding type of organic solid includes most of the single-component low-molecular-mass compounds as well as polymer ferroelectrics[13] and a vinylidene fluoride oligomer in thin film form.[14] Thiourea was the first example of a genuinely organic ferroelectric of low molecular mass.[15,16] It has been extensively studied since the discovery of its ferroelectricity in 1956. The considerable interest in it arises from its complicated structural transformations, passing through various phases of commensurate and incommensurate superstructures between the paraelectric and

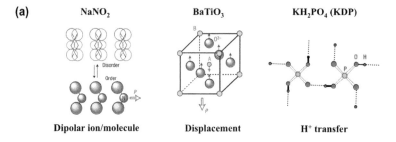
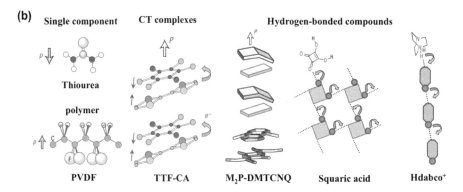

Figure 11.2 Conventional designs of ferroelectric materials and the origin of their dipole moment p or polarization P (open arrows). (a) Typical examples of inorganic ferroelectric materials. (b) Conventional organic ferroelectric or antiferroelectric substances. PVDF = poly(vinylidene difluoride). Reproduced with permission from ref. 80, © 2008, Nature Publishing Group.

ferroelectric phases.[17] Subsequently, the 2,2,6,6-tetramethyl-1-piperidinyloxy (TEMPO, also known as tanane) free radical, well known for its stability, was also found to be an organic ferroelectric of the polar molecular type near room temperature.[18] Another notable organic ferroelectric is 1,6-bis(2,4-dinitrophenoxy)-2,4-hexadiyne,[19–21] a diacetylene monomer that can polymerize while maintaining its single crystal form. The ferroelectric transition at low crystalline temperature (46 K) is found to vanish on polymerization.[19]

In comparison with inorganic compounds, organic substances have been synthesized extensively but ferroelectric properties have been found in them only rarely. They feature a tendency to form highly anisotropic structures with very low lattice symmetry. Despite their occasional crystallization into polar structures, their dielectric properties and possible ferroelectricity have not always been examined. By inspecting the reported organic ferroelectrics, Choudhury and Chitra proposed[22] that transformation between the different degenerate orientations for ferroelectricity would demand a molecular symmetry higher than C_1, which is encountered in as many as 30% of pure organic structures. To seek undisclosed ferroelectric candidates, Zikmund et al.[23] performed a symmetrical analysis of polar crystal structures using the Cambridge Structure Database. The strategy aimed to discover the hidden 'pseudosymmetry', which would survive in paraelectric structures. However, the investigation of polar molecules has not met with much success because ferroelectricity poses some difficult hurdles. The dipole–dipole interactions usually tend to cancel out neighboring molecular dipoles in crystals. Although a large number of non-centrosymmetric molecules have been synthesized, they can occasionally crystallize into polar (and hence acentric) solids without cancelling out their dipoles. A further constraint on the molecular packing is a low enough energy barrier for the molecular reorientations to overcome in the crystalline solid. For the existing polymer ferroelectrics, the steric difficulty in reorienting dipoles gives rise to exceptionally large coercive fields compared with other ferroelectrics.[14]

In spite of dipolar molecules and ion displacement, there is a case in which the dynamic protons on hydrogen bonds trigger the ferroelectric ordering of the lattice. A typical example is the KH_2PO_4 (KDP) family. The collective site-to-site transfer of protons in the O–H···O bonds switches the ferroelectric spontaneous polarization (Figure 11.2a). Such materials are defined as KDP type or proton-transfer type ferroelectrics. The actual ferroelectric crystals may show both order–disorder and displacive characteristics, which are not mutually exclusive. In the case of hydrogen-bonded organic systems, ferroelectricity has been found in tricyclohexylmethanol (TCHM),[24] which had a hydrogen-bonded dimer and showed a minute pyroelectric charge at low temperature. The origin of the ferroelectricity was also a dipole-reorientation process distinct from that of the KDP type: the two O–H···O bonds linking two TCHM molecules, which were simultaneously broken while two OH groups were reoriented around the C–O bonds.[25,26] Correlated proton dynamics, similar to that of KDP, has also been found in other hydrogen-bonded organic molecules. For example, squaric acid[27] (see Figure 11.2b) is a dipolar molecule with two protons, affording four-fold degenerate configurations. Its crystal forms a sheet-like polar network with intermolecular O–H···O hydrogen bonds. When the polarity of the sheet is reversed by correlated proton transfer, the crystal has a very large in-plane dielectric constant value (about 300) above the order–disorder transition temperature (373 K).[28]

Not only organic small molecules but also polymeric materials have been explored for ferroelectric properties. There are in general a number of structural requirements which need to be met to produce useful ferroelectric polymers, including the polar monomer units, conformation, chain packing, the macroscopic polarity of the crystalline grains, etc. Kawai et al. first observed the piezoelectric effect in a polymer, called polyvinylidene fluoride (PVDF), in 1969, as depicted in Figure 11.2(b).[29] PVDF is the most widely used ferroelectric polymer and is manufactured in large quantities for a wide range of applications. Subsequently, many other polymer system containing ferroelectric properties have been investigated, including different

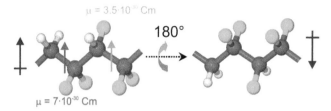

Figure 11.3 Chemical structure of P(VDF-TrFE) and ferroelectric switching mechanism. Reproduced with permission from ref. 79, © 2011 American Chemical Society.

PVDF-based polymer blends and copolymers (*i.e.* PVDF with poly(methylmethacrylate)), polytrifluoroethylene, copolymers of VDF and TrFE, and odd-numbered nylon.[12,30–33] The ferroelectricity of PVDF and P(VDF-TrFE) stems from the dipole moments in the materials that can be aligned with the applied field by a rotation of the polymer chain, as illustrated in Figure 11.3. The dipole moments originate predominantly from the presence of strongly electronegative fluorine atoms. Moreover, the dipole switching involved rotation of the dipoles around the backbone and depended strongly on the temperature and the strength of the applied electric field,[32] implying that the preferred chain configuration was all-*trans*. This phase of bulk material is referred as the β-phase. Due to an optimal alignment of all the C–F dipole moments in the crystal unit cell, such β-phase gives the highest ferroelectric response.

Compared to the β-phase, the α-phase molecular orientation (the polymer chains have alternatively *trans* and *gauche* conformations) can also be explored in PVDF thin films,[30] which would cancel out the overall polarization. To maintain the stable ferroelectric properties of PVDF, it is usually mechanically stretched and/or electrically poled to orient the molecular chains in the PVDF films in the all-*trans* conformation (β-phase). Other methods, involving the addition of hygroscopic salts, epitaxy of KBr, control of cooling and heating or solvent evaporation rate during processing, and blending with poly(methylmethacrylate), can also induce β-phase in the bulk.[35–40]

Figure 11.4 shows D–E hysteresis loops of random copolymers of VDF and TrFE with different TrFE content. The one with less TrFE displays a higher remanent polarization as well as switching current. Note that the coercive field is not affected by the differing TrFE content. The steric difficulty in reorienting the dipoles in PVDF or P(VDF-TrFE) gives rise to a large coercive field (50–60 MV m^{-1}) compared with other ferroelectrics. Thin layers are therefore required in order to achieve low-voltage memory devices. Bune *et al.* prepared P(VDF-TrFE) films as thin as 1 nm by Langmuir–Blodgett deposition and the device could be successfully operated using only 1 V.[31] Such results convince one that P(VDF-TrFE) is a suitable candidate material for low-voltage memory device applications.

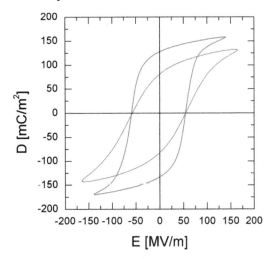

Figure 11.4 *D–E* hysteretic loop of P(VDF-TrFE) films with different VDF to TrFE ratios. In blue, the molar ratio VDF:TrFE was 50:50. In red, it was 80:20. Reproduced with permission from ref. 79, © 2011 American Chemical Society.

11.3 Principles of Organic Ferroelectric Memory Operation

Due to the unique physical polarization behavior as well as the hysteretic properties of the ferroelectric materials discussed above, organic data storage devices, especially non-volatile memory, based on ferroelectricity are a promising approach towards the development of low-cost memory technology. The latest developments in ferroelectric memory, in particular ferroelectric capacitors, field-effect transistors, and diodes, are presented in the following section.

11.3.1 Organic Ferroelectric Capacitors

One of the simplest types of ferroelectric memory devices is the thin film capacitor. The typically implemented ferroelectric data storage element is a capacitor consisting of a thin ferroelectric film in between two conductive electrodes. Therefore, the memory device incorporates a ferroelectric film as a capacitor to hold data. The ferroelectric film has the characteristics of remanent polarization, which can be reversed by an applied electric field and give rise to a *D–E* hysteresis loop. The read-out voltage produced by the charging of the bitline capacitance by this current can be compared with a reference voltage. Digital information can be stored by aligning the internal polarization in either the up or down direction with an applied electric field. As a voltage is applied to the ferroelectric thin film, a high or low charge displacement current response will be observed, depending on whether the internal polarization is aligned or not with the direction of the applied field.

This read-out technique is destructive since the device ends up in the same final state independent of the original data. By using thin film technologies, capacitors with thickness on the submicron scale can be fabricated so that operations are reduced to a level below the standard chip supply voltages. Such ferroelectric capacitors are also called destructive read-out devices since a reading operation can affect the stored information.

Ferroelectric polymers are defined as polymers where the unit cell of the crystal is polar and the direction of polarization can be changed by the application of an electric field. Low-voltage switching using the ferroelectric P(VDF-TrFE) has been explored for electronic device applications. Since the coercive field value of P(VDF-TrFE) (~50 MV m^{-1}) is relatively high, a polymer thin film thinner than 100 nm is required to achieve a switching voltage below 10 V. Also, the sub-100 nm thickness showed an extremely retarded ferroelectric response as compared to the bulk materials,[41,42] implying that P(VDF-TrFE) was unsuitable for low-voltage memory applications. P(VDF-TrFE) thin films prepared using a Langmuir–Blodgett deposition technique showed that a ferroelectric response was retained down to a layer thickness of 15 nm.[31] Unfortunately, these films were probably even less suitable for applications than the aforementioned spin-coated films because the switching times were longer by orders of magnitude.[33,34] Subsequently, some improvement in spin-coated films was achieved by optimizing the annealing conditions,[43,44] but they were relatively limited in polarization and so the prospects for memory applications of P(VDF-TrFE) remained challenging.

However, the situation reached a turnaround due to a serendipitous discovery.[45] Figure 11.5 shows charge displacement measurements that established that the ferroelectric response remains almost the same when the layer thickness is reduced from 210 to 65 nm. These measurements demonstrated that it is possible to have sub-10 V switching while retaining the remanent polarization and switching times of the bulk material. The low-voltage switching behavior was obtained using a bottom electrode stack that includes an interfacial layer of conductive poly(3,4-ethylenedioxythiophene):poly(styrene sulfonic acid) (PEDOT:PSS), and vapor-deposited Au as the top contact. Note that a patent application on a similar idea was also filed at the same time.[46] The above discussion suggests that conductive polymers can lead to memory applications when using P(VDF-TrFE).

Xu et al. later reported a ferroelectric capacitor with the conductive polymer Ppy:PSS [polypyrrole-poly(styrene sulfonic acid)] as a bottom and top interface.[47] This capacitor with a layer thickness of 50 nm was found to have the same ferroelectric properties as the bulk material. The switching cycle endurance properties of the capacitor were also investigated. When a ferroelectric material undergoes a large number of switching cycles, then fatigue may occur, which manifests in the ferroelectric having a lower remanent polarization, higher coercive fields, and longer switching times. The cycle endurance is a critical parameter for ferroelectric capacitors because it directly relates to the normal product lifetime of a ferroelectric capacitor. As presented in Figure 11.6, the polarization degradation is limited to a value of 15% after 10^7 cycles at an elevated temperature of 60 °C.

Organic Ferroelectric Memory Devices

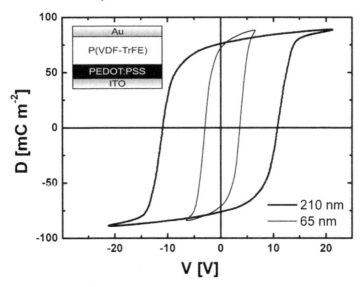

Figure 11.5 Displacement change D versus applied voltage V hysteresis loop measurement on capacitors with a ferroelectric layer thickness of 210 and 65 nm. The capacitors have PEDOT:PSS bottom interfaces, as shown in the inset. Reproduced with permission from ref. 45, © 2004 American Institute of Physics.

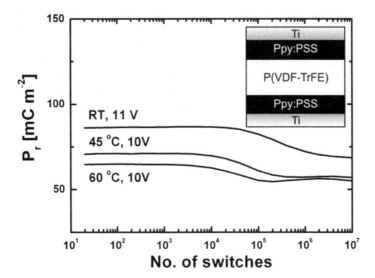

Figure 11.6 Remanent polarization P_r versus number of switching cycles for ferroelectric capacitors with a P(VDF-TrFE) layer thickness of 50 nm, measured at room temperature (RT), 45 and 60 °C. The inset illustrates the device structure with Ppy:PSS top and bottom interfaces. Reproduced with permission from ref. 47, © 2007 American Institute of Physics.

The reason why the observed low-voltage switching behavior as obtained by inserting conductive polymers was not obtained before lies in the fact that all previous investigations used aluminum top and bottom electrodes.[31,33,41–44] X-ray photoelectron spectroscopy (XPS) measurements have indicated that aluminum reacts with P(VDF-TrFE) regardless of whether the aluminum is deposited onto P(VDF-TrFE) or the other way around.[48] The chemical reaction between aluminum and P(VDF-TrFE) leads to the formation of non-ferroelectric or "dead" layers near the electrode interfaces, which can be derived from the retarded switching kinetics that are induced.[49,50] The additional Au layer was inserted to avoid such chemical reactions, and yielded an improvement in the ferroelectric response based on P(VDF-TrFE) thin-film capacitors.[51,52]

11.3.2 Organic Ferroelectric Field-Effect Transistors

Based on the polarization properties of ferroelectric materials, a digital memory device can be also revealed using a field-effect transistor device configuration. Ferroelectric field-effect transistors (FeFETs) typically possess non-volatile memory behavior, particularly flash-type memory characteristics. The straightforward concept of FeFET design is based on a stack comprising a metal gate electrode, a ferroelectric and a semiconductor. This design results in a range of requirements which lead to some extreme challenges for the integration process.

A FeFET is an interesting alternative for non-volatile memory applications. The advantages of FeFETs are non-volatile data storage, non-destructive read-out and their compact cell design with high integration capability. The simplest layout of a FeFET comprises a metal/ferroelectric/semiconductor architecture (Figure 11.7), in which the ferroelectric layer serves as the gate dielectric. The ferroelectric layer, because of its remanent polarization, can adopt either of two stable polarization states. These states persist when no biases are applied, *i.e.* when the power supply is 0. Switching from one polarization state to the other can be observed by applying a sufficiently large gate bias. Depending on the direction of the polarization, positive or negative counter charges are induced in the semiconductor at the semiconductor/

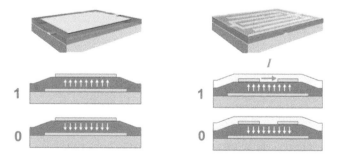

Figure 11.7 FRAM (left) and FeFET (right) basic cells. Reproduced with permission from ref. 79, © 2011 American Chemical Society.

ferroelectric interface, effectively causing a positive or negative onset voltage shift of the transistor. Hence, a gate bias window, defined by the shifted onset voltages, exists wherein the drain current may have either of two levels depending on the actual polarization state of the ferroelectric gate dielectric. The corresponding drain current levels can be used to recognize the "0" and "1" states of a non-volatile memory device with non-destructive read-out (see Figure 11.7 for a schematic illustration).[12] Non-destructive read out provides the advantage that the memory is not subjected to the destructive read/write cycle commonly employed in ferroelectric capacitor memory devices, so the device lifetime is limited only by the number of writing times.

The first FeFET based on a polymeric ferroelectric was demonstrated in 1986. It was shown that ferroelectric polarization within a random copolymer thin film of P(VDF-TrFE)[30] could induce an inversion layer in a bulk silicon semiconductor substrate.[53] Thin film FeFETs using an inorganic ferroelectric and a p-type organic semiconductor were reported by Velu *et al.* and Kodzasa *et al.*[54,55] The inorganic ferroelectric layer in these devices requires processing temperatures above 600 °C, making the process incompatible with the use of plastic substrates. The ratio of remanent channel conductance of the ON and the OFF states was approximately 10, and no switching times or data retention times were explored. Subsequently, all-organic FeFET devices incorporating a ferroelectric-like polymer as the gate insulator and pentacene as the organic semiconductor were reported by Schroeder *et al.*[56] The ferroelectric-like polymer is a nylon (*i.e.* poly(*m*-xylylene adipamide), MXD6).[30] A clear hysteresis in transfer characteristics was observed with an ON/OFF ratio of 200 at gate voltage (V_G) = −2.5 V, and 30 at V_G = 0 V with a retention time of around 3 h. Unni *et al.* fabricated a pentacene-based FeFET memory device with P(VDF-TrFE) (70:30) as the gate insulator.[57] Similar values of the ON/OFF ratio and retention time were exhibited. In a follow-up report, the MXD6 deposition process was improved, and led to much better memory characteristics with an ON/OFF ratio of 10^5 and a data retention time of several days.[58] On the other hand, the programming time can be estimated to exceed 200 ms for nylons typically.[30] On the basis of the switching time alone, one would clearly favor P(VDF-TrFE) over MXD6 as ferroelectric of choice. However, MXD6 is amorphous which may be potentially beneficial as it leads to a smoother interface with the semiconductor, which is particularly relevant for charge transport and low-voltage operation (implying a thin ferroelectric).

In both cases, pentacene was used as active semiconductor. Pentacene was deposited using vacuum processing. High performance solution-processed polymer FeFETs were first reported by Naber *et al.* using P(VDF-TrFE) (65:35) as the gate insulator and MEH-PPV (poly [2-methoxy-5-(2-ethyl-hexyloxy)-*p*-phenylene-vinylene]) as a semiconductor.[59] Identical FETs were prepared with non-ferroelectric PTrFE as the gate insulator. These devices did not show appreciable hysteresis, whereas those made with P(VDF-TrFE) showed hysteretic drain currents with an ON/OFF ratio at zero gate voltage of 10^4 or higher, indicating that the ON/OFF ratio and hysteresis are affected by the ferroelectricity. By directly comparing the transfer curves of the P(VDF-TrFE) and PTrFE transistors, it was estimated that the ferroelectric poling induced a remanent surface charge density of 18 mC m^{-2} in the semiconductor channel. These polymer FeFETs have a retention time longer than 1 week with a programming cycle endurance >1000 cycles, and a short programming time (OFF to ON, 0.3 ms; ON to OFF, 0.5 ms).[59]

In order to avoid a depolarized state in organic FeFETs, the semiconductor needs to supply compensation charges for both ferroelectric states. It has been demonstrated that n-type FeFETs can be achieved with methanofullerene PCBM [(6,6)-phenyl-C_{61}-butyric acid methyl ester].[60] Moreover, both n- and p-type behavior can be obtained in a single FeFET with the use of an ambipolar organic semiconductor comprising a blend of MEH-PPV and PCBM.[61] The transfer measurements of an ambipolar FeFET are depicted in Figure 11.8, which are essentially the same as for the unipolar FeFET based on MEH-PPV,[59] except that the hole accumulation is mirrored on the right side of the figure due to electron accumulation. Evidently, this FeFET can switch between hole and electron accumulation. An obvious disadvantage in this case is that the drain current at zero gate bias is no longer bistable because the hole and electron current response is almost the same. One could address this issue by using a novel way of probing the state of the ferroelectric polarization, such as measuring the sign of the field-effect mobility using a small applied gate voltage. However, it may be simpler to tune the blending ratio of the semiconductor to re-obtain a certain degree of bistability.

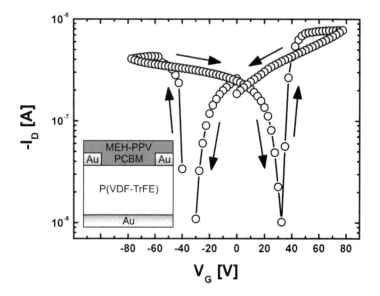

Figure 11.8 Transfer measurements on an ambipolar polymer FeFET with an insulator layer thickness of 0.9 mm. The inset illustrates the device structure with a blend of MEH-PPV and PCBM as the semiconductor. Reproduced with permission from ref. 61, © 2005 WILEY-VCH Verlag GmbH & Co. KGaA, Weinheim.

The operating voltages of the above FeFETs, however, were relatively high (over 60 V). To reduce the programming voltage, the thickness of the gate insulator should be scaled down. In ferroelectric capacitors, using PEDOT:PSS as electrode material results in superior characteristics, particularly at low P(VDF-TrFE) layer thickness. Therefore, Gelinck et al.[60] fabricated all-polymer FeFETs using thin P(VDF-TrFE) films sandwiched between electrodes (gate and source/drain) made of conducting PEDOT:PSS. The electrodes were patterned by photo exposure of the deposited films. Alkoxy-substituted polyphenylene vinylene derivative (OC1C10) and PCBM were used as p-type and n-type semiconductors, respectively. For both unipolar p- or n-type semiconductor channels, the transistors have remanent current modulations of $\sim 1 \times 10^3$ with a retention time of hours. They can be switched within 0.1–1 ms at operating voltages less than 10 V. By measuring the switching current as a function of channel length, the remanent polarization charge was estimated to be 6 mC m^{-2}.

Since then, several groups have reported organic FeFETs based on pentacene and P(VDF-TrFE) as the gate dielectric. Different approaches have been proposed that in one way or another improve the ferroelectric/semiconductor interface (reduced surface roughness leads to higher mobility) and/or leakage current. Nguyen et al. stretched P(VDF-TrFE) thin films to simultaneously enhance the crystallinity and lower the surface roughness.[61] The FeFET had a mobility of 0.1 cm^2 V^{-1} s^{-1} with an ON/OFF ratio of 10^4. Recently, a method was described to fabricate ferroelectric β-type PVDF thin films on a Au substrate by humidity controlled spin-casting combined with rapid thermal treatment. A FeFET with β-PVDF showed a drain current bistability of 100 at zero gate voltage with ±20 V gate voltage sweep, and exhibited a higher thermal stability than P(VDF-TrFE) (i.e. 160 °C).[38] A cross-linkable interlayer inserted between the gate electrode and the PVDF-TrFE layer significantly reduced the gate leakage current, leading to a source–drain OFF current of approximately 1×10^{-11} A. The corresponding FeFET device shows a clockwise I–V hysteresis with a drain current bistability of 10^3 at ±40 V gate voltage.[62] A bottom gate FeFET containing PVDF–PMMA (80 : 20) blend films of 200 nm thickness with low surface roughness exhibited an ON/OFF bistability ratio of 1×10^4 with data retention capability over 15 h at an operation voltage of 15 V.[63] In that work, polycrystalline and single crystalline 6,13-bis(triisopropylsilylethynyl) pentacene (TIPS-PEN) was used as the active semiconductor. Because the low surface roughness was equal to or lower than the width of the accumulation layer, charge transport in the accumulation layer was not inhibited significantly, as evidenced by the high field-effect mobility of 0.65 cm^2 V^{-1} s^{-1}.

Naber et al. also addressed the issue of surface roughness.[64] A top gate configuration was used so that the semiconductor/ferroelectric interface roughness is not determined by the top surface of the P(VDF-TrFE) but by that of P3HT. By further optimizing the P3HT thin film processing, a surface roughness of only 0.7 nm was obtained. This turned out to be a key factor in order to achieve a polymer FeFET with both high charge mobility

(0.1 cm^2 V^{-1} s^{-1}) and high charge density (28 mC m^{-2}). MIS diode measurements with P3HT as active layer demonstrated that the bistability in these devices originates from switching between two states in which the ferroelectric gate dielectric is either polarized or depolarized. Pulsed charge displacement measurements of these diodes in the polarized state gave accumulated charge values of 40 ± 3 mC m^{-2}.[65]

In order to further simplify the device manufacturing process and to make integration easier, reconfigurable circuits are required alongside the development of new high performance organic ambipolar materials. Fabiano et al. demonstrated selective remanent ambipolar charge transport in polymeric FETs operable under ambient conditions.[66] The channel polarity of FETs based on an ambipolar dithienocoronenediimide-thiophene copolymer can be permanently switched to a p- or n-type mode using the remanent polarization of the ferroelectric gate insulator. The extremely high charge carrier density (15–20 mC m^{-2}) induced by the ferroelectric polarization is the origin of the high drain currents (10^{-6} A) observed at low source–drain voltages (5 V). This result is in contrast to FETs where conventional dielectric gate insulators are used. Such ferroelectric polarization enhanced FETs can be exploited to enhance the performance of complementary devices like inverters, and non-destructive read-out operation, CMOS-like compatible one transistor–one transistor (1T1T) memory cells.

P(VDF-TrFE) based FeFETs with semiconductors such as pentacene, poly(3-hexylthiophene), and poly[2-methoxy-5-(2′-ethyl-hexyloxy)-p-phenylenevinylene] exhibited promising memory functionalities including high resolution of programming/erasing conductance states, long-term retention time, and excellent endurance ability. Based on the above discussion, Naber et al. further improved the functionalities of P(VDF-TrFE)-based FeFETs by optimizing the thickness and deposition method for the P(VDF-TrFE) dielectric to operate the device at low operating voltage and with long-term data storage capabilities. However, heat treatment is often required to enhance the extent of β-phase crystallinity to improve the ferroelectricity of the P(VDF-TrFE) layer in memory devices. On the other hand, one-dimensional (1D) nanostructures such as nanotubes, nanowires, or nanofibers have received much attention because of their promising advantages such as the perfect order of molecules, the absence of grain boundaries, and good contact interfaces. Park et al. investigated a bottom-gate FeFET memory device containing a P(VDF-TrFE) gate dielectric in which a 1D ribbon-type TIPS-PEN single crystal was grown through a solvent exchange method. The data storage ability was improved to at least 5 × 10^4 s under ambient conditions with excellent environment stability for more than 40 days by incorporating an interlayer between the gate electrode and the P(VDF-TrFE) thin film.[67] Nanodot P(VDF-TrFE) non-volatile FeFET memory devices with single-walled carbon nanotubes (SWNTs) was precisely controlled and deposited at the center of an SWNT channel by utilizing a dip-pen nanolithography (DPN) technique.[68] It was indicated that the modulation of the channel conductance by the bipolar properties of the P(VDF-TrFE) nanodots enhanced the resolution of the bistability to perform superior switching for ~10^{10} endurance cycles. The electrospinning process was also used to induce improved crystalline packing of polymers, such as the ferroelectric β-phase.[69] Chen et al. reported the characteristics of FeFET non-volatile flash memory devices using aligned P(VDF-TrFE) electrospun nanofibers as the dielectric layer.[70] These FeFET devices showed reliable memory behavior and a memory window proportional to the quantity of aligned nanofibers containing the ferroelectric β-phase crystalline domain. Moreover, the FeFET devices using nanofibers exhibited long-term stability in data retention longer than 10^4 s with an ON/OFF ratio of ~10^3, and multiple switching operation stability for up to 100 cycles.

Many interface and materials aspects of FeFET devices have been addressed in the past. Ferroelectric capacitors also tend to show significant problems that include fatigue, imprint and retention loss. The device performance still suffers due to interface trapping states and some fabrication issues. Newly developed ferroelectric materials and advanced layer sequences may help to overcome the existing issues.

11.3.3 Organic Ferroelectric Diodes

Diodes alleviate the destructive read-out problem of ferroelectric capacitors. By switching the direction of the polarization the resistance of the diode is switched between a high and low resistance state. The first ferroelectric diodes were produced by blending the organic semiconductor P3HT with P(VDF-TrFE).[71] Atomic force microscopy (AFM) and scanning transmission X-ray microscopy (STXM) images (Figure 11.9) show that the blend phase separates into amorphous P3HT domains embedded in a crystalline P(VDF-TrFE) matrix.[72,73] The microstructure is independent of the type of substrate and does not change upon annealing. The P3HT domains are continuous throughout the film and protrude from the surface of the film with a hemispherical shape. The lateral size of the domains is mono-disperse and increases linearly with P3HT content. Concomitantly, the number of domains decreases, indicating that the morphology coarsens with increasing P3HT content. These observations exclude standard nucleation and growth solidification mechanisms, and point to phase separation by spinodal decomposition.[71]

Formation of this phase separated microstructure is crucial for the operation of the diodes. The ferroelectric phase provides polarization and bistability, and current transport takes place through the semiconductor phase which is modulated by the polarization field of the P(VDF-TrFE) phase.

To fabricate bistable diodes, the phase separated blend is sandwiched between two electrodes. The current can only flow through the semiconductor phase, and an ohmic contact is required to inject charges efficiently. At the same time, the Fermi-level of the contacts should align with the valence or conduction band of the semiconductor. When the Fermi-level is not aligned,

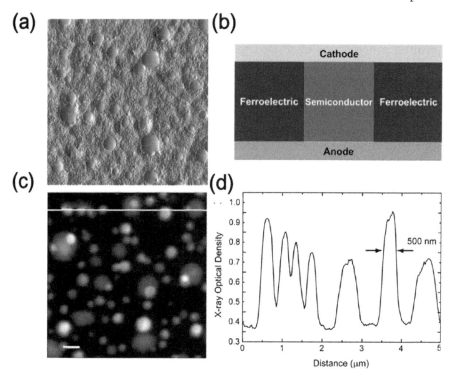

Figure 11.9 Blend microstructure. (a) AFM phase image of a blend with 10 wt% rir-P3HT after annealing at 140 °C. The image shows randomly scattered, rather mono-disperse rir-P3HT domains in a P(VDF-TrFE) matrix. (b) Schematic representation of the blend microstructure. (c) Scanning transmission X-ray microscopy (STXM) image of the blend. The image and (d) the corresponding cross-section trace of the 287.2 eV X-ray optical density show that rir-P3HT forms columnar phases through the film that are continuous from the substrate/blend interface to the blend/air interface. Reproduced with permission from ref. 74, © 2011 Elsevier Ltd.

the injection of charge carriers is limited, then the current in the device is low. The operation principle of the diode is based on deliberately using a contact that poorly injects charges into the semiconductor.

Figure 11.10 demonstrates the lowering of the injection barrier by the ferroelectric polarization, resulting in high injection current, then the device reaches the ON-state. The top electrode is the injecting contact, characterized by a certain hole injection barrier, and the current collecting bottom contact is grounded. The electrical transport is calculated by numerically solving the coupled drift-diffusion, Poisson, and current continuity equations on a rectangular grid. The 3D phase separated morphology is therefore mapped onto a simplified 2D structure of alternating ferroelectric and semiconducting slabs, under periodic boundary conditions (Figure 11.10). The electric current only runs through the

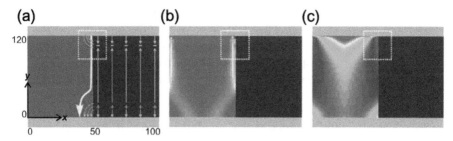

Figure 11.10 Operation principle of a bistable ferroelectric diode. (a) Schematic mechanism of injection barrier lowering and current injection. The ferroelectric, semiconductor and electrodes are indicated by blue, red, and gray planes, respectively. The top electrode at $y = 120$ nm is the injecting contact characterized by an injection barrier of 0.7 eV. The collecting bottom electrode at $y = 0$ nm is grounded. Blue and yellow arrows indicate electric fields and current flow, respectively. +/− indicate polarization charge. (b) Current density and (c) hole density (10 log-scale) of a ferroelectric diode in the ON-state. Reproduced with permission from ref. 74, © 2011 Elsevier Ltd.

semiconducting phase since the ferroelectric P(VDF-TrFE) is an insulator. Figure 11.10a shows the polarization charges in the ferroelectric phase. The direction of the electric field lines (blue arrows) is from positive to negative polarization. Importantly, near the top contact, field lines also run from the positive image charges in the electrode to the negative polarization charges in the ferroelectric. Similar field line movement is shown near the bottom contact. At the injecting top contact, it is this stray field of the positive image charges and the negative polarization charges, shown by the curved arrows, that causes a strong lowering of the hole injection barrier by the image force effect.[74] As a result, the contact becomes ohmic and charges can be injected into the semiconductor phase. Since the lateral x-component of the stray field is directed towards the ferroelectric phase, the injected holes (Figure 11.10c) accumulate at the phase boundary and consequently the current, shown by the white rectangle, will be confined to a narrow region at the phase boundary (Figure 11.10b). This spatial confinement causes space charge effects to limit the diode current in the ON-state. In the lower half of the semiconductor phase the lateral x-component of the stray field becomes smaller and the current spreads over the whole semiconductor phase before it reaches the collecting contact at $y = 0$.[74]

From the above discussion, we find that charge injection can be switched between the ON- and OFF-states at the injecting contact of the diode, depending on the direction of the ferroelectric polarization. The current in the ON-state is a space-charge limited current (SCLC), which is the maximum electrostatic current that the semiconductor can supply. In the OFF-state of the diode, the current is an injection limited current (ILC). The ON/OFF ratio of the switchable diode depends exponentially on the injection barrier at the injecting contact. Asadi *et al.* have demonstrated that injection barriers up to 1.6 eV can be surmounted by the stray field of the ferroelectric polarization, yielding ON/OFF current ratios of more than five orders of magnitude.[75] The current modulation depends exponentially on the magnitude of the injection barrier with a slope of 0.25 eV decade^{-1}.

On the other hand, both cathode and anode (*i.e.* an injection limited contact with the semiconductor) are needed in order to make a bistable rectifying diode. An example is given in Figure 11.11, where the current density of a discrete blend diode is presented as a function of bias. LiF/Al and Au are used to form injection limited contacts for both electron and holes on the polymeric semiconductor PFO.[76] The barriers for electron and hole injection under forward bias are 0.8 eV and 1.3 eV, respectively. In the forward direction, holes are injected from Au and electrons from LiF/Al, although in the case of an unpoled ferroelectric these currents are severely injection limited by high injection barriers. In the reverse direction, holes and electrons cannot be injected from LiF/Al and Au because there is no current flow. The result is a rectifying diode that can be poled using an electric field exceeding the coercive field. In the ON-state, the ferroelectric polarization points towards the anode, and the injection barriers for holes and electrons are simultaneously lowered by the ferroelectric stray fields, which can effectively be disregarded. In contrast, the ferroelectric polarization points towards the cathode in the OFF-state. Effectively, both injection

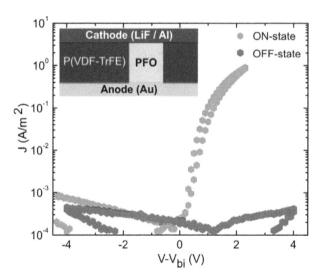

Figure 11.11 Bistable rectifying diode. Current density–voltage characteristics of a bistable rectifying diode based on a phase separated 10:90 wt% blend of a ferroelectric polymer, P(VDF-TrFE) and a semiconducting polymer, PFO. The inset shows the device layout and the schematic microstructure. Reproduced with permission from ref. 76, © 2010 American Institute of Physics.

barriers then increase and no current flows, irrespective of the bias. The resistance of the diode can be read out nondestructively at any bias lower than the coercive field.

11.4 Application and Summary of Organic Ferroelectric Memory Devices

Here, we mainly focus on various materials and fabrication issues of ferroelectric thin films for organic ferroelectric memory devices. Followed by an overview of ferroelectric memory architectures, the basic characteristics and recent development of three main memory elements are elaborated. For high performance ferroelectric memory devices, a good quality film should be guaranteed which can not only form proper interfaces with other device components but also at the same time maximize its ferroelectric properties. Most of the recent work on ferroelectric memory devices seems to address embedded memory applications building on the strengths of memory devices that facilitate simple processability and non-destructive read-out functionality, and possess long retention endurance, short switching time and low-voltage operation, as presented in this chapter. The current challenge for future ferroelectric memory design is reduction of the lateral dimensions of the devices, especially for FeFETs. The difficulty in scaling to ultrasmall cell size is still a weakness. The minimum channel size will most likely be limited by short channel effects, which are known to occur at significantly larger lateral dimensions.[77] The information density of integrated memory circuits based on FeFETs, thus, may also be restricted. However, such FeFETs are still suitable for low-cost and large-area applications based on integrated circuits of organic transistors. The first integrated memory circuits based on organic FeFETs were made with printing techniques recently.[78,79]

Not only device processing parameters but also materials have been innovated to explore high performance ferroelectric devices. Driven by cost reduction combined with fast access and storage times and an infinite number of endurance cycles, new polymeric ferroelectric materials with dipole moments should be suggested in thin film devices for non-volatile memory applications. Along this line, polymer memory could effectively provide a very low cost per bit with a high chip capacity. New ferroelectric compounds are design to reduce the surface roughness between the dielectric and semiconductor layers, leading to a higher charge carrier mobility of the device. Moreover, ambipolar FeFETs are investigated because ambipolar semiconductors can effectively supply compensation charges to the ferroelectric in both polarization states. Extension to flexible memory arrays fabricated on plastic substrates or specially functionalized circuit elements embedded into large area electronics systems are the ultimate goals for polymer FeFETs. By combining the native advantages of ferroelectric devices (*i.e.* outstanding stability, low power consumption) with new organic material innovations, high performance digital information storage devices could be demonstrated in the near future.

References

1. M. Lines and A. Glass, *Principles and Applications of Ferroelectrics and Related Materials*, Clarendon Press, Oxford, 1979.
2. J. Valasek, *Phys. Rev.*, 1920, **15**, 537.
3. J. Valasek, *Phys. Rev.*, 1921, **17**, 475.
4. C. B. Sawyer and C. W. Tower, *Phys. Rev.*, 1930, **35**, 269.
5. Y. Yang, J. Ouyang, L. Ma, R. J.-H. Tseng and C.-W. Chu, *Adv. Funct. Mater.*, 2006, **16**, 1001.
6. J. C. Scott and L. D. Bozano, *Adv. Mater.*, 2007, **19**, 1452.
7. T. B. Singh, N. Marjanovic, G. J. Matt, N. S. Sariciftci, R. Schwodlauer and S. Bauer, *Appl. Phys. Lett.*, 2004, **85**, 5409.
8. K.-J. Baeg, Y.-Y. Noh, J. Ghim, S.-J. Kang, H. Lee and D.-Y. Kim, *Adv. Mater.*, 2006, **18**, 3179.
9. W. Wu, H. Zhang, Y. Wang, S. Ye, Y. Guo, C. Di, G. Yu, D. Zhu and Y. Liuj, *Adv. Funct. Mater.*, 2008, **18**, 1.
10. Y. Li, A. Sinitskii and J. M. Tour, *Nat. Mater.*, 2008, **7**, 966.
11. Q.-D. Ling, D.-J. Liaw, C. Zhu, D. S.-H. Chan, E.-T. Kang and K.-G. Neoh, *Prog. Polym. Sci.*, 2008, **33**, 917.
12. J. F. Scott, Ferroelectric Memories, in *Advanced Microelectronics*, Springer-Verlag, Berlin, 2000.
13. T. Furukawa, M. Date and E. Fukada, *J. Appl. Phys.*, 1980, **51**, 1135.
14. K. Noda, K. Ishida, A. Kubono, T. Horiuchi, H. Yamada and K. Matsushige, *J. Appl. Phys.*, 2003, **93**, 2866.
15. A. L. Solomon, *Phys. Rev.*, 1956, **104**, 1191.
16. G. J. Goldsmith and J. G. White, *J. Chem. Phys.*, 1959, **31**, 1175.
17. F. Dénoyer and R. Currat, in *Incommensurate Phases in Dielectrics*, North-Holland, Amsterdam, 1986, pp. 129–160.
18. D. Bordeaux, J. Bornarel, A. Capiomont and J. Lajzerowicz-Bonneteau, *Phys. Rev. Lett.*, 1973, **31**, 314.
19. G. F. Lipscomb, A. F. Garito and T. S. Wei, *Ferroelectrics*, 1980, **23**, 161.
20. H. Schultes, P. Strohriegl and E. Dormann, *Ferroelectrics*, 1986, **70**, 161.
21. P. Gruner-Bauer and E. Dormann, *J. Phys.: Condens. Matter*, 1992, **4**, 5599.
22. R. R. Choudhury and R. Chitra, *Cryst. Res. Technol.*, 2006, **41**, 1045.
23. Z. Zikmund, P. Vaněk, M. Havránková, B. Březina, M. člermák and M. Vássa, *Ferroelectrics*, 1994, **158**, 223.
24. P. Szklarz and G. Bator, *J. Phys. Chem. Solids*, 2005, **66**, 121.
25. Y. Yamamura, H. Saitoh, M. Sumita and K. Saito, *J. Phys.: Condens. Matter*, 2007, **19**, 176219.
26. G. Bator, R. Jakubas and Z. Malarski, *J. Phys. C*, 1986, **19**, 2799.
27. D. Semmingsen and J. Feder, *Solid State Commun.*, 1974, **15**, 1369.
28. J. Feder, *Ferroelectrics*, 1976, **12**, 71.
29. H. Kawai, *Jpn. J. Appl. Phys.*, 1969, **8**, 975.
30. H. S. Nalwa, *Ferroelectric Polymers: Chemistry, Physics and Applications*, Marcel Dekker, New York, 1995.
31. A. V. Bune, V. M. Fridkin, S. Ducharme, L. M. Blinov, S. P. Palto, A. V. Sorokin, S. G. Yudin and A. Zlatkin, *Nature*, 1998, **391**, 874.

32. M. Dawber, K. M. Rabe and J. F. Scott, *Rev. Mod. Phys.*, 2005, **77**, 1083.
33. G. Vizdrik, S. Ducharme, V. M. Fridkin and S. G. Yudin, *Phys. Rev. B*, 2003, **68**, 094113.
34. S. Ducharme, T. J. Reece, C. M. Othon and R. K. Rannow, *IEEE Trans. Device Mater. Reliab.*, 2005, **5**, 720.
35. A. Salimi and A. A. Yousefi, *Polym. Test.*, 2003, **22**, 699.
36. J. Scheinbeim, C. Nakafuku, B. A. Newman and K. D. Pae, *J. Appl. Phys.*, 1979, **50**, 4399.
37. A. J. Lovinger, *Polymer*, 1981, **22**, 412.
38. S. J. Kang, Y. J. Park, J. Sung, P. S. Jo, C. Park, K. J. Kim and B. O. Cho, *Appl. Phys. Lett.*, 2008, **92**, 012921.
39. D. C. Yang and E. L. Thomas, *J. Mater. Sci. Lett.*, 1987, **6**, 593.
40. C. Leonard, J. L. Halary and L. Monnerie, *Macromolecules*, 1988, **21**, 2988.
41. K. Kimura and H. Ohigashi, *Jpn. J. Appl. Phys.*, 1986, **25**, 383.
42. Y. Tajitsu, *Jpn. J. Appl. Phys.*, 1995, **34**, 5418.
43. Q. M. Zhang, H. Xu, F. Fang, Z.-Y. Cheng and F. Xia, *J. Appl. Phys.*, 2001, **89**, 2613.
44. F. Xia, B. Razavi, H. Xu, Z.-Y. Cheng and Q. M. Zhang, *J. Appl. Phys.*, 2002, **92**, 3111.
45. R. C. G. Naber, P. W. M. Blom, A. W. Marsman and D. M. deLeeuw, *Appl. Phys. Lett.*, 2004, **85**, 2032.
46. N. Johansson and L. Chen, WO 02/43071 A1, 2002.
47. H. Xu, J. Zhong, X. Liu, J. Chen and D. Shen, *Appl. Phys. Lett.*, 2007, **90**, 092903.
48. K. Muller, Y. Burkov, D. Mandal, K. Henkel, I. Paloumpa, A. Goryachko and D. Schmeisser, *Phys. Status Solidi A*, 2008, **205**, 600.
49. T. Furukawa, K. Nakajima and Y. Takahashi, *IEEE Trans. Dielectr. Electr. Insul.*, 2006, **13**, 1120.
50. R. Gysel, I. Stolitchnov, A. K. Taganstev, N. Setter and P. Mokry, *J. Appl. Phys.*, 2008, **103**, 084120.
51. T. Nakajima, R. Abe, K. Takahashi and T. Furukawa, *Jpn. J. Appl. Phys.*, 2005, **44**, L1385.
52. R. C. G. Naber, P. W. M. Blom and D. M. de Leeuw, *J. Phys. D: Appl. Phys.*, 2006, **39**, 1984.
53. N. Yamauchi, *Jpn. J. Appl. Phys.*, 1986, **25**, 590.
54. G. Velu, C. Legrand, O. Tharaud, A. Chapoton, D. Remiens and G. Horowitz, *Appl. Phys. Lett.*, 2001, **79**, 659.
55. T. Kodzasa, M. Yoshida, S. Uemura and T. Kamata, *Synth. Met.*, 2003, **137**, 943.
56. R. Schroeder, L. A. Majewski and M. Grell, *Adv. Mater.*, 2004, **16**, 633.
57. K. N. NarayananUnni, R. deBettignies, S. Dabos-Seignon and J.-M. Nunzi, *Appl. Phys. Lett.*, 2004, **85**, 1823.
58. R. Schroeder, L. A. Majewski, M. Voigt and M. Grell, *IEEE Electron Device Lett.*, 2005, **26**, 69.
59. R. C. G. Naber, C. Tanase, P. W. M. Blom, G. H. Gelinck, A. W. Marsman, F. J. Touwslager, S. Setayesh and D. M. de Leeuw, *Nat. Mater.*, 2005, **4**, 243.

60. G. H. Gelinck, A. W. Marsman, F. J. Touwslager, S. Setayesh, D. M. de Leeuw, R. C. G. Naber and P. W. M. Blom, *Appl. Phys. Lett.*, 2005, **87**, 092903.
61. C. A. Nguyen, S. G. Mhaisalkar, J. Ma and P. S. Lee, *Org. Electronics.*, 2008, **9**, 1087.
62. J. Chang, C. H. Shin, Y. J. Park, S. J. Kang, H. J. Jeong, K. J. Kim, C. J. Hawker, T. P. Russell, D. Y. Ryu and C. Park, *Org. Electron.*, 2009, **10**, 849.
63. S. J. Kang, Y. J. Park, I. Bae, K. J. Kim, H.-C. Kim, S. Bauer, E. L. Thomas and C. Park, *Adv. Funct. Mater.*, 2009, **19**, 2812.
64. R. C. G. Naber, M. Mulder, B. deBoer, P. W. M. Blom and D. M. deLeeuw, *Org. Electron.*, 2006, **7**, 132.
65. R. C. G. Naber, J. Massolt, M. Spijkman, K. Asadi, P. W. M. Blom and D. M. de Leeuw, *Appl. Phys. Lett.*, 2007, **90**, 113509.
66. S. Fabiano, H. Usta, R. Forchheimer, X. Crispin, A. Facchetti and M. Berggren, *Adv. Mater.*, 2014, **26**, 7438.
67. S. J. Kang, I. Bae, Y. J. Park, T. H. Park, J. Sung, S. C. Yoon, K. H. Kim, D. H. Choi and C. Park, *Adv. Funct. Mater.*, 2009, **19**, 1609.
68. J. Y. Son, S. Ryu, Y. C. Park, Y. T. Lim, Y. S. Shin, Y. H. Shin and H. M. Jang, *ACS Nano*, 2010, **4**, 7315.
69. T. Lei, X. Cai, X. Wang, L. Yu, X. Hu, G. Zheng, W. Lv, L. Wang, D. Wu, D. Sun and L. Lin, *RSC Adv.*, 2013, **3**, 24952.
70. M.-S. Lu, C. Lu, M.-H. Li, C.-L. Liu and W.-C. Chen, *Proc. SPIE*, 2014, **9185**, 91850N.
71. K. Asadi, D. M. de Leeuw, B. de Boer and P. W. M. Blom, *Nat. Mater.*, 2008, **7**, 547.
72. K. Asadi, H. J. Wondergem, R. S. Moghaddam, C. R. McNeill, N. Stingelin, B. Noheda, P. W. M. Blom and D. M. de Leeuw, *Adv. Funct. Mater.*, 2011, **21**, 1887.
73. C. R. McNeill, K. Asadi, B. Watts, P. W. M. Blom and D. M. de Leeuw, *Small*, 2010, **6**, 508.
74. K. Asadi, M. Li, P. W. M. Blom, M. Kemerink and D. M. de Leeuw, *Mater. Today*, 2011, **14**, 592.
75. K. Asadi, T. G. de Boer, P. W. M. Blom and D. M. de Leeuw, *Adv. Funct. Mater.*, 2009, **19**, 3173.
76. K. Asadi, M. Li, N. Stingelin, P. W. M. Blom and D. M. de Leeuw, *Appl. Phys. Lett.*, 2010, **97**, 193308.
77. H. Sirringhaus, *Adv. Mater.*, 2005, **17**, 2411.
78. T. Sekitani, K. Zaitsu, Y. Noguchi, K. Ishibe, M. Takamiya, T. Sakurai and T. Someya, *IEEE Trans. Electron Devices*, 2009, **56**, 1027.
79. P. Heremans, G. H. Gelinck, R. Müller, K.-J. Baeg, D.-Y. Kim and Y.-Y. Noh, *Chem. Mater.*, 2010, **23**, 341.
80. S. Horiuchi and Y. Tokura, *Nat. Mater.*, 2008, **7**, 357.

CHAPTER 12

Summary and Outlook

WEN-CHANG CHEN[*,a]

[a]Department of Chemical Engineering, National Taiwan University, Taipei 10617, Taiwan
*E-mail: chenwc@ntu.edu.tw

In this book, the recent developments of organic electrical memory materials and devices are summarized systematically. Types of new charge storage materials, such as organic small molecules, functional polyimides, nonconjugated polymers with pendent chromophores, conjugated polymers, and polymer composites, have been innovated to create high capacity and stable digital information storage environments. Different device configurations, including resistor- and transistor-type memory, were also discussed in detail along with defined charge storage mechanisms. The charge transfer strength, charge trapping/detrapping ability, surface polarity, interfacial energy barrier, and morphology of the memory active layer could significantly affect the electrical switching behavior of such electrical data storage devices.

High density data storage, and stable and maintainable electrical memory devices could be realized through molecular design. In resistive memory devices, electron acceptor moieties are usually incorporated into donor-type materials to create charge trapping sites or induce a charge transfer conducting channel, which can significantly affect the bistable (or multistable) resistance switching characteristics and lead to various memory behaviors. From the summarized structure–property relationships of the resistive switching properties (Chapter 1 and 2), three main approaches to the emerging

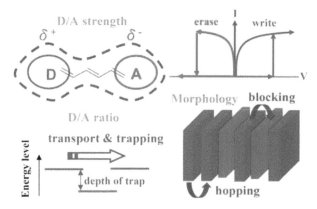

Figure 12.1 Contributing effects of donor–acceptor materials to memory switching behavior.

chemical structures of novel donor–acceptor materials for memory devices are explored (Figure 12.1):[1] (1) materials with suitable donor/acceptor strength to exhibit charge transfer as well as switching behavior, such as the donor–acceptor small molecules described in Chapter 3, the functional polyimides in Chapter 4, and the nonconjugated or conjugated polymers in Chapters 5 and 7, respectively; (2) materials with an optimized relative ratio of donor–acceptor motifs to define the volatile/non-volatile nature, including the polyimide–inorganic hybrid materials and polymer composites described in Chapters 4 and 6; and (3) systems with a well-defined morphology of the active layer to control the charge transporting/blocking ability for designing memory devices, for example, the block copolymers with specific pendent groups outlined in Chapter 5, the donor–acceptor polymer composites in Chapter 6, and the donor–acceptor block copolymers in Chapter 8.

For transistor-type memory, on the other hand, several factors contributing to materials innovation are also discussed (Chapter 9) to vary the performance of memory devices (Figure 12.2),[2] including (1) surface hydrophobicity, (2) structural architecture (*i.e.* linear-, branch-, or star-shape), (3) conjugation length, (4) strength of donor–acceptor interaction, and (5) interfacial energy barrier between the semiconducting and charge storage layer. In most cases, such as the nanoparticle embedded materials and functional polymer electrets discussed in Chapter 10 or the ferroelectric materials in Chapter 11, charge storage layer materials with a longer side-chain conjugation length, a more hydrophobic surface, and a smaller energy barrier at the interface can achieve efficient charge trapping/blocking to obtain a larger memory window of the devices.

The development of future high-performance organic memory devices with low working voltage, low power consumption, long-term stability, and high reliability relies on appropriate chemical structural design, high density storage and a large memory window without a sneak current path, and solution-processing on a flexible/stretchable substrate.[3–8] In particular,

Summary and Outlook

stretchable electrical memory devices could be important for the development of wearable electronics. Such devices possess extensive potential to be integrated with other organic electronic or optoelectronic devices (*i.e.* radio frequency identification tags or electronic skins) toward practical applications, as proposed in Figure 12.3.

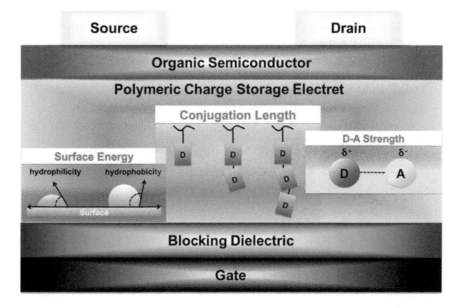

Figure 12.2 Structural factors of charge storage materials for the performance of organic transistor memory devices.

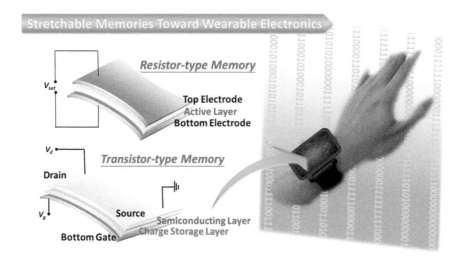

Figure 12.3 Schematic illustration of flexible/stretchable memory devices for future wearable electronics applications.

References

1. C.-L. Liu and W.-C. Chen, *Polym. Chem.*, 2011, **2**, 2169.
2. Y.-H. Chou, H.-C. Chang, C.-L. Liu and W.-C. Chen, *Polym. Chem.*, 2015, **6**, 341.
3. H.-C. Wu, A.-D. Yu, W.-Y. Lee, C.-L. Liu and W.-C. Chen, *Chem. Commun.*, 2012, **48**, 9135.
4. H.-C. Wu, C.-L. Liu and W.-C. Chen, *Polym. Chem.*, 2013, **4**, 5261.
5. S.-J. Kim and J.-S. Lee, *Nano Lett.*, 2010, **10**, 2884.
6. H.-C. Chang, C.-L. Liu and W.-C. Chen, *Adv. Funct. Mater.*, 2013, **23**, 4960.
7. Y.-C. Lai, F.-C. Hsu, J.-Y. Chen, J.-H. He, T.-C. Chang, Y.-P. Hsieh, T.-Y. Lin, Y.-J. Yang and Y.-F. Chen, *Adv. Mater.*, 2013, **25**, 2733.
8. Y.-C. Lai, Y.-C. Huang, T.-Y. Lin, Y.-X. Wang, C.-Y. Chang, Y. Li, T.-Y. Lin, B.-W. Ye, Y.-P. Hsieh, W.-F. Su, Y.-J. Yang and Y.-F. Chen, *NPG Asia Mater.*, 2014, **6**, e87.

Subject Index

References to figures are given in *italic* type. References to tables are given in **bold** type.

4N4OPZ, 122
6F-HAB, 170–4
6F-HAB-AM, 188–9
6F-HAB-CBZ, 25, 74
6F-HAB-DPC, 176
6F-HAB-TPAIE, 174, *176*, 194–6, *199*
6F-HABCZ, 169, 170, 194
6F-HTPA, *73*
6F-TPA, 175–7, *175*, **178**, 179, 181, 194, **198**
6FPI, *138*, 139, 156
8HQ, 65

AIDCN, *16*, 17, 210–11
aluminium, *281*, 283
2-amino-4,5-imidazoledicarbonitrile *see* AIDCN
APTA-3FA, 66, *67*
APTA-6FDA, 66, *67*
APTT, 150
APTT-6FDA, 59, 61, *73*, 74
atomic force microscopy (AFM), 369–70
Azo1, *20*, *63*, *106*, 113–14, *114*
Azo2, *20*, *63*, *106*
Azo3, *17*, *104*, *106*
Azo4, *17*
azobenzenes, 103–5
AZTA-Pia, 61, 74, 85, *86*

BDOYM, 17, *17*, 102–3
BDT, 27, *29*

benzenetetracarboxydiimide (BTI), 157–8
benzodithiophene (BDT), 27, *29*
biocompatible devices 21–2
biphenyls, 14, 151–3, *152*
3,3′-bis[9-carbazole(ethyloxy) biphenyl]-4,4′-diamine *see* HABCZ
2,2′-bis(3, 4-dicarboxyphenyl)hexafluoropropane dianhydride *see* 6F
9,10-bis-(9,9-diphenyl-9H-fluoren-2-yl)-anthracene (DPFA), 15
block copolymers, 26, 214, 218, 222–4, 239, 260–1, 269–70, 317
  coil–coil, 275–90
  rod–coil, 261–7
  triblock, 285–7
BTI, 157–8
BTVCz-NO$_2$, *63*
buckyballs *see* fullerenes

cadmium selenide, 320
cadmium sulfide, *77*
CANACA, 126, *129*
capacitor-type memory
  device structure, 9–10
  ferroelectric, 361–4
CAPyNA, 107
2-(9H-carbazol-9-yl)ethyl methacrylate *see* PCzMA
carbazole polymers, 14, 19, 168–70
carbon nanotubes (CNT), 5, 35–7, 220–2, 271, 277

382

catanane, 63
CDPzN, *20*, 122–3
charge transfer effects, 13–14, 17–18, *17*, *28*, 85–7, 200–2, 208–9
    6FDA group, 108–9
    azo group, 103–5
    block copolymers, 289–90
    carbazole group, 109–10
    naphthalimide group, 105–8
    polyimides, 137–40, 152–61
    see also charge transfer effects
charge trapping, 303–16
    molecule–polymer hybrids as trapping sites, 306–8
chemical vapor deposition (CVD), 94, 207
cobalt(III) complexes, 6, 33–4
complementary metal-oxide semiconductor (CMOS), 11–12
conformational changes, 14–15, 87–8, 201–2, 236, 301
    ferroelectric materials, 359–60
    polyimides, 138–9, 143
    Rose Bengal, 64
conjugated polymers, 27–30, 70–2, 245–6
    metal complexes, 246–51
    polyazomethines, 243–5
    polycarbazoles, 239–42
    polyelectrolytes, 251–2
    polyfluorenes, 234–8
    polythiophenes, 238–40
copolymers, 5, 218–19
copper phthalocyanine (CuPc), 15, 16, 211, 310–12, 343, 346
copper tetracyanoquinodimethane *see* CuTCNQ
CP, 69
CuPc, 15, 16, 211, 310, 310–12, 343, 346
current-sensing atomic force microscopy (CS-AFM), 200–1
CuTCNQ, 64–5, 110
cyclobis(paraquat-*p*-phenylene) *see* CBPQT

DAT, 159
DBC, 15, *16*, *63*
DDQ, *63*, 64
density functional theory (DFT), 83–4, *84*, 143, 152, 159, 189
1,4-dibenzyl C60 (DBC), 15, *16*, *63*
2,3-dichloro-5,6-dicyano-1,4-benzoquinone (DDQ), 64
dimethylamino group, 115, 182
9,10-dimethylanthracene, *77*
diodes, 369–73
    see also organic light-emitting diodes
1,5-dioxynapthalene (NP), 64
N-(2,4-diaminophenyl)-*N*,*N*-diphenylamine (DAT), 159
disperse red 1-functionalized pVK copolymer (PVD), 26
3,5-di(trifluoromethyl)phenyl group, 182
1-dodecanethiol, 76, *77*, 213
donor–acceptor complexes *see* charge transfer complexes
donor–bridge–acceptor (DBA), 20–1, *22*
DPAPIT, *20*, *63*
DPKAZO, *20*, 116, *117*
DRAM, 3, 5, 24, 54–5, 56–7, 131, 257–8
    charger transfer effects, 85
    polyfluorene, 72
    polyimide, 73–6, 142, 158, 160, 178
    triphenylamine, 178, 179
DSDA, 155
DSDA-TPA, 177–8, *180*, **181**
DSPI, *138*, 156
dynamic random access memory *see* DRAM

electron acceptors
    6FDA group, 108–9
    azo group, 103–5
    naphthalimide group, 105–8

Subject Index

electron donor-acceptor complexes *see* charge transfer complexes
electron donors, TPA group, 102–3
eosin Y, 64

F8T2, 303, 320
ferrocenes, 250–1, 343
  see also metal complexes
ferroelectric random access memory (FeRAM), 2, 9
ferroelectrics, 300 3, 355–6
  principles of operation, 357–60
  applications, 373
  diodes, 369–73
  field-effect transistors, 364–9
field effect transistors *see* organic field effect transistors
filament conduction, 12–13, 78–82, *81*, 200
  polyfluorenes, 234–8
  polyimides, 140, 149, 150
  see also hopping conduction
FKAZO, 20–1, *20*, 63, 116, *117*
flash memory, 3, 59, 74–5, 257–8, 299
  hybrid materials, 76, 212, 340
  nonconjugated polymers with chromophore pendants, 183, 184, *188*, 192
  polyfluorene, 32–3, 85, **235**
  polyimide, 148–51, 153, 156, 158, 160
  PVK, 68
  triphenylamine, 118, 125, 174, 175, 178
flexible memory devices, 36–7, 163
fluoroscein sodium, *63*
Fowler–Nordheim tunnelling, 79
Frenkel–Poole emission, 79
fullerenes, 30–1, 183–4, 192–4, 217–20, 269–70

GO-DDAT, *273*
gold nanoparticles, 5, 42, 76, 213–14, 277, 316–17, **332–3**, *333*

graphene, 222–4, 263–4, 277, 342–3, 346–7
graphene macromolecular memory label (GMML), 93

HABCZ, 169–74, *171*, 180, 192
HATT, *130*, 131
HDTT, *130*, 131
HETT, *130*, 131
4,4′-hexafluoroisopropylidenebis-[4-(N,N-diphenylamino)phenyl phthalimide] (APTA-6FDA), 66, *67*
hexafluoroisopropylidene (6F) group, 23
highest occupied molecular orbital *see* HOMO/LUMO
HOMO/LUMO, 170–3, 177–8, 201, 213–14
hopping conduction, 79
  see also filamentary conduction
HRTT, *130*, 131
hybrid materials, 34–42
  carbon allotropes, 34–7, 217–23
  flash memory, 76, 212, 340
  graphene, 222–4, 263–4, 277, 342–3, 346–7
  organic/inorganic memory devices, 76–8, *162*, 215–18
  polyimide, 161–3
  polymer/nanoparticle, 340–1
  see also nanoparticles
hybrid memory, 3
8-hydroxyquinoline (8HQ), 5, *63*, 65
hysteresis loops, 10

iamP1, *33*, 69, 85
iamP2, *33*
iamP7, *69*
iamP8, *69*
insulators, conduction processes, **79**
International Technology Roadmap for Semiconductors (ITRS), 6–7

ionic liquids, 223–4
iridium complexes, 249–50
    block copolymers, 265–6
isoindigo, 237

Kumada catalyst-transfer polymerization, 72

Langmuir–Blodgett (LB) process, 5
Langmuir–Schaefer deposition, 335
layer-by-layer assembly, 337–8
light-charge organic molecules (LCOM), 311
linkage effect, polyimides, 158–60

MEH-PPV, 6, 303, 365–6
memory devices
    basic concepts, 2–4
    capacitor-type, 9–11
    resistor-type, 11
    transistor-type, 7–9
memristors, 6
    biocompatible, 21–2
metal complexes, 31–4, 110–11, 110–13, 343
    conjugated polymers, 246–51
methanofullerene 6,6-phenyl C61-butyric acid methyl ester (PCBM), 15
molecular design, polyimides, 151–61
MXD6, 301, 365

NACA, 106
NACANA, *129*
nanocomposites, 39–42
nanoparticles (NP)
    as charge storage layers, 330–42
        chemical assembly, 335–9
        thermal evaporation, 331–4
        gold, 5, 42, 76, 277, 316–17, **332–3**, *333*
        silver, *77*, *83*, 161

nanowire field-effect transistors (NW-FET), 308–9
naphthalenetetracarboxydiimide, 157–8
napthalimides, 19, 70, 105–7
m-nitrobenzal malononitrile
    see m-NBMN
nonconjugated polymers, 25–30, 183–9, 189–90
NPPI, *138*
N(PTPMA), *307*
NTI, 157–8
NTPI, 156

ODPI, *138*
OFET *see* organic field-effect transistor
OLED *see* organic light-emitting diodes
(OMe)2TPPA-6FPI, *73*
ON/OFF ratio, 125–31
organic field-effect transistors (OFET), 5, 7, *8*
    basic principles, 298–9
        polymer electrets as, 303–6
    charge trapping, 303–16
        self-assembled monolayers, 312–16
    embedded nanoparticles
        chemical assembly, 335–9
        thermal evaporation, 331–4
    ferroelectric, 364–9
    nanoparticles, 316–22
        at semiconductor/dielectric interface, 321–3
        gate dielectric, 316–18
organic light-emitting diodes (OLED), 110–11
organic memory devices, 4–7
    hybrid materials, 76–8
oxadiazoles, 143–4

Subject Index

P1-C60, *183*
P2-C60, 183–4, *183*
P2VP, 262
P3-DSPI, 139, 156
P3-PMPI, 139, 156
P3HT, 5, 34–6, 39–42, 72, 77, 316–17, 337, 369–70
P3HT-*b*-P3PT, 56–7, 218
P3HT-*b*-POXD, *71*, 72, 262–3
P3PT, 56–7
P6FBEu, 69
P6OMe, *71*
PAA, 136–7
PAM, 243–5
PANI-PARA, *71*
PARA, *71*
PBC, 69
PBCz, 69
PbPc, 16
P(BPPO)-PI, *58*, 59, *73*, *86*
PCBM, *16*, 35–6, *63*, 65, 217, 262, *264*, 343
PCFO, 27, *28*
PCPCz-*b*-P2VP, 279–80
PCz, 69
PCzMA, 192, 265
PCzOxEu, *33*, 69
PDA, 63, 310
PDMS, 317
PEDOT:PSS, 28, 110
perylene bisimide, 74
perylenetetracarboxydiimide *see* PTI
PET, 163, 335
PF-*b*-P2VP, 263–4, *264*
PF6Eu, 69
PF8Eu, 69
PFOxPy, *28*, *71*, 84
PFT-PI, *28*
PFT2-Fc, 69
PFTPACN, 27, *28*
phenanthro[9, 10-d]-imidazole (PFT-PI), 27
6,6-phenyl C$_{61}$-butyric acid methyl ester *see* PCBM
p-phenylenediamine *see* pDA
photoelectric effects, 122–3

phthalocyanines (Pc), 4, 16
PI-PBIX, *73*
PI(AAPT-6FDA), *73*
PI(AMTPA), 74–5, *75*, 208
PI(AMTPA-APAP), *73*
PI(APAP), 74–5, 158
PKEu, *33*, 69
platinum complexes, 249
PMCz, *89*
PMMA, 301, 342–3
PMPI, *138*
poly[4-(diphenylamino)benzyl methacrylate], 217, 306–7
poly(allylamine hydrochloride) (PAH), 64
poly(amic acid) *see* PAA
poly(4,4′-aminotriphenylene3,3′,4,4′-diphenylsulfonyltetracarboximide) *see* DSDA-TPA
polyaniline (PANI), 39
polyazomethines, 243–5
poly(2-(*N*-carbazolyl)ethyl methacrylate) *see* PCzMA
poly(4-(9,9-dihexylfloren-2-yl)styrene) *see* P(St-Fl)
poly(4-di(9,9-dihexylfluoren-2-yl) styrene)-*b*-poly(2-vinylpyridine), 276
poly[2,7-(9,9-dihexylfluorene)]-*b*-poly(2-vinylpyridine), 263
poly(dimethylsiloxane) (PDMS), 317
poly(9,9-dioctylfluorene-*co*-bithiophene) *see* F8T2
polyethylene terephthalate *see* PET
poly(3,4-ethylenedioxythiophene):poly(styrenesulfonate) (PEDOT:PSS), 28, 110
poly(3,3′-bis(*N*-ethylenyloxycarbazole)-4,4′-biphenylene hexafluoroisopropylidenediphthalimide) *see* 6F-HABCZ
poly(ethyleneterephthalate) *see* PET
poly[10-(2′-ethylhexyl)phenothiazine-3,7-diyl] *see* PPTZ
polyfluorenes, 85, 233–8

poly-3-hexylthiophene (P3HT) see P3HT
polyimides, 2, 23–5, *23*, *73*, 136–7
  chromophore pendants, 168–70
  filament conduction, 140
  flexible devices, 163
  hybrid materials, 161–3
  inorganic hybrids, 161–3
  molecular design, 151–61
  phthalimide, 156–7
  space charges and traps, 140
  volatile memory, 140–5
poly(*m*-xylylene) adipamide, 301
polymer composites, 206–7
  fullerenes, 218–20
  graphene, 222–3
  inorganic nanomaterials, 210–17
  organic molecules, 207–10
polymer memory devices
  history, 4–7
  memory mechanisms, 197–202
poly(2-methoxy-5-(2′-ethylhexyloxy)-p-phenylenevinylene) see MEH-PPV
poly(methyl methacrylate) see PMMA
poly(*N*-vinylcarbazole) see PVK
poly(10,12-pentacosadiynoic acid) (PDA), 63, 310
poly(5-phenyl-1,3,4-oxadiazol-2-yl-[1,1′-biphenyl]carboxyloxy-*n*-noyl acrylate) see PPOXBPA
poly(3-phenoxymethylthiophene) (P3PT), 56–7
poly(2-phenyl-5-(4-vinylphenyl)-1,3,4-oxazole) see POXD
polypyridyl complexes, *112*
polypyrroles, 207
polystyrene (PS), 4, 35, 42, 77, *77*, 208, 262, 268–9
poly(styrene-*block-para*phenylene) see PS-*b*-PPP
poly(*tert*-butyl acrylate) PtBA, 223
polythiophenes, 238–40

poly(vinyl alcohol) (PVA), 207, 304
poly(9-(2-(4-vinyl(benzyloxy)ethyl)-9*H*-carbazole)) see PVBEC
polyvinylidene fluoride (PVDF), 301, 359
poly(4-vinyl phenol) (PVP), 35
poly(9-(4-vinylphenyl)carbazole)-*b*-poly(2-vinylpyridine) see PCPCz-*b*-P2VP
poly(2-vinylpyridine) see P2VP
polyvinylpyrrolidone (PVP), 42, 78
porphyrins (Por), 16, 111–12, 308, *309*
POXD, 72, 262–3
  P3HT-*b*-POXD, *71*, 72, 240, 262–3
PPOXBPA, 184, 184–6, 271–2, *274*
PPTZ, 208
PPV, *71*
programmable read-only-memory (PROM), 3
protonic-acid-doped polyazomethine (PA-TsOH), 30
PS, 4, 35, 42, 77, *77*, 208, 262
PS-*b*-P4VP, 218, 317, 337
PS-*b*-PPP, 265
P(St-Fl), 275
P(St-Fl)-*b*-P2VP, 277, *278*
PStFl2-*b*-P2VP, 275
PtBA, 223
PTI, 157–8
PTPMA, 217, 306, *307*
push–pull organic molecules, 311
PVBEC, 269, *270*
PVDF, 301–3, 359–60
PVDF-TrFE, 207, 217, 360, 367
PVF, 301
PVK, 4, 35, 40, 59, 68, *69*, 88, *89*, 110, 223
  graphene composites, 220, 222, 223, 271
  nanoparticle composites, 212, 213, 215
PVK-AZO-2CN, 26
PVK-AZO-NO2, 26
PVK-C60, 30, 59, *60*, 68, 85, 220

PVK-PBC, 69
PVK-PF, 69
PVP (poly(vinyl phenol)), 35, 190, *191*, 279
PVP (polyvinylpyrrolidone), 42, 78, 216
PVPCz-*b*-P2VP, 279–80, *282*
PVPEP-*b*-PVBPTPA-*b*-PVPEP, 287

quantum dots, 42, 277, 320, **332**, 342

RB, *16*
reduced graphene oxide, 342–3
resistor random access memory (RRAM), 257
resistor-type memory, 11–15
    applications, 88–90
        scale down, 91–2
        three-dimensional devices, 90
    conformation effect, 87–8
    device structure, 12–13, 54–5
    filamentary conduction, 78–82
    materials design, 62
    metal complexes, 110–13
    multilevel, 113–21
        conjugated, 70–2
        with pendant groups, 68–70
        polyimides, 72–6
    operation mechanism, 12–13
    organic *see* organic resistor memory
    polymeric, 66–76
    switching characteristics, 56–62
        voltage, 125–31
reversible addition fragmentation chain transfer (RAFT), 271
Rose Bengal, *63*, 64, 87
rotaxane, *63*, 64, 91
ruthenium complexes, 110–11

Saran wrap, 4
Sawyer–Tower circuit, 300
scale down, 91–2

scanning probe microscopy (SPM), 64, 91
scanning transmission X-ray microscopy (STXM), 369–70
scanning tunnelling microscopy (STM), 17
Schottky emission, 79
SDA, 150
self-assembled monolayers (SAM), 308, 312–16, *320*, 321–2, 339
self-assembled nanostructures, 260–1, 308–9
self-propagating molecular-based assembly (SPMA), 112–13
sexithiophenes, 5
silver, 77, *83*, 161
SNACA, 106, *106*
sodium nitrite, 357
space charges and traps, 13, 82–5, 140
space-charge limited conduction (SCLC), 170, 269, 371–2
SRAM, 58–9, 143
    polyimide, 143–4, *144*
stability, 194–7
static RAM *see* SRAM

TABM, 102
TAPA, *63*, 64
TCHM, 359
TCNQ, 5, 15, *16*
TCz, 208
tetracenes, 4
7,7,8,8-tetracyanoquinodimethane (TCNQ), 5, 15, 237
tetracyanoethylene (TCNE), 237
(+)-2-(2,4,5,7-tetranitro-9-fluorenylideneaminoxy)propionic acid *see* TAPA
tetrathiafulvalene *see* TTF
thermal stability, 195–6
thermionic emission, 79
three-dimensional devices, 90
titania nanoparticles, 40
TP6F-PI, 73, *73*, *86*
TPA, 155

TPA-AC, *20*
TPA-BAP, *20*
TPA-NAP, *20*
TPAPAM, 223
TPDBCN, *18*, *63*, 66
TPDYCN1, *18*, *63*, 66
TPDYCN2, *18*, *63*, 66
transistor-type memory
  devices, 7–9
    operation mechanism, 8–9
    structure, 7–8
transmission electron microscopy (TEM), 88
tricyclohexylmethanol (TCHM), 359
triphenylamine unit, 102–3, 174–82
triphenylamine-based polyazomethine *see* TPAPAM

TTF, 15, *16*, *63*, 64, 65, 208, 217, 306, 308, 310
TTF-TCNQ, 101–2
Tunnel of field emission, 79

WORM memory, 3, 15, 145–8
  organic resistive, 59–62, 131
  polyimide, 74, 145–8
WPF-BT-FEO, *71*
write-once-read-many *see* WORM

yolk–shell nanospheres, 42

zinc oxide, *77*
zinc phthalocyanine (ZnPc), 15–16, *16*, *63*, 65
zinc sulfide, 258–9
ZnPC, 15–16, *16*, *63*, 65
ZnPor, *16*, 308